Seismic Resistant Structures

WIT*PRESS*
WIT Press publishes leading books in Science and Technology.
Visit our website for the current list of titles.
www.witpress.com

WIT*eLibrary*
Home of the Transactions of the Wessex Institute.
The WIT electronic-library provides the international scientific community with immediate and permanent access to individual papers presented at WIT conferences.
http://library.witpress.com

Seismic Resistant Structures

Editors

S. Ivorra
University of Alicante, Spain

C.A. Brebbia
Wessex Institute, UK

WIT*PRESS* Southampton, Boston

Editors:

S. Ivorra
University of Alicante, Spain

C.A. Brebbia
Wessex Institute, UK

Published by

WIT Press
Ashurst Lodge, Ashurst, Southampton, SO40 7AA, UK
Tel: 44 (0) 238 029 3223; Fax: 44 (0) 238 029 2853
E-Mail: witpress@witpress.com
http://www.witpress.com

For USA, Canada and Mexico

Computational Mechanics International Inc
25 Bridge Street, Billerica, MA 01821, USA
Tel: 978 667 5841; Fax: 978 667 7582
E-Mail: infousa@witpress.com
http://www.witpress.com

British Library Cataloguing-in-Publication Data

A Catalogue record for this book is available
from the British Library

Library of Congress Catalog Card Number: 2018932132

ISBN: 978-1-78466-315-5
eISBN: 978-1-78466-316-2

The texts of the papers in this volume were set individually by the authors or under their supervision. Only minor corrections to the text may have been carried out by the publisher.

Printed in Great Britain by Lightning Source, UK.

The material contained herein is reprinted from special editions of the International Journal of Computational Methods and Experimental Measurements, Volume 6, Number 5 and the International Journal of Safety and Security Engineering, Volume 7, Number 4 published by WIT Press.

Preface

This volume contains a series of selected papers on the topic of Seismic Resistant Structures.

The papers deal with basic and applied research in the various fields of earthquake engineering relevant to the design of structures. The problem of protecting the built environment in earthquake-prone regions involves not only the optimal design and construction of new facilities, but also the upgrading and rehabilitation of existing structures including heritage buildings. The type of highly specialized retrofitting employed to protect the built heritage is an important area of research and appropriate to the conference objective of designing better earthquake resistant buildings.

The Editors of the volume are grateful to the authors for their valuable contribution, as well as to the colleagues who helped during the review process.

The Editors
Alicante, Spain

Contents

TRM REINFORCEMENT OF MASONRY SPECIMENS FOR SEISMIC AREAS

S. IVORRA[1], D. BRU[1], A. GALVAÑ[1], STEFANO SILVESTRI[2], CRISTINA APERA[2] & DORA FOTI[3]
[1]Civil Engineering Department, University of Alicante, Spain.
[2]Dipartimento di Ingegneria Civile, Chimica, Ambientale e dei Materiali, Unversità di Bologna, Italy.
[3]DICAR, Politecnico di Bari, Italy.

ABSTRACT

This document analyses the resistant behaviour to diagonal compression and direct compression of brick masonry specimens, reinforced with textile reinforced mortars (TRM) and without them. The numerical models have been calibrated with experimental results in order to have a suitable technique to reinforce masonry historic constructions where extraordinary tension stresses could occur in seismic situations. For the numerical modelling, FEM models have been developed using non-linear layered bi-dimensional shell elements. Moreover, a comparative analysis has been developed between the numerical models and the experimental ones. There are two different types of reinforced specimens: two layers to reinforce both surfaces (i), and one layer for only one surface (ii). The point of this (ii) is to respect the Italian legislation indications for the protection of historic constructions.
Keywords: reinforced masonry TRM walls FEM

1 INTRODUCTION

Many residential areas in the historical centres of European Seismic areas have a very high vulnerability due to of their structural system based in masonry or brick masonry walls. These structures have a very low capacity to resist tension or shear stresses. The use of textile reinforced mortar (TRM) as a reinforcing element can provide an additional capacity to solve these structural problems. Mortar cementitious materials as matrix of these reinforcement systems are more compatible with masonry than other polymeric matrices where its use in historical constructions is not permitted by the authorities in charge of protecting these constructions.

Brick masonry walls are building systems made up of brick elements joined by mortar with a specific disposition. Due to its configuration, with elements of different properties, their simulation by finite element method becomes a complicated task. Brick and mortar have different strength capacities, and the very different strength behaviour in tension and in compression generates an anisotropic behaviour. Another important factor is the influence of the elastic modulus of both materials. By having different elastic modulus, their strain, for a given stress, will be different and the compatibility equations should solve this interaction. For these reasons, the ultimate load behaviour of brick masonry specimens are very dependent on the direction of the applied load and the plane on which the loads act.

Some studies, such as those carried out at the Polytechnic University of Catalonia [1], analyse the out-of-plane behaviour, although the present paper will only focus on the loads acting on the plane of the wall. Masonry structures are especially weak against lateral cyclic loads such as those caused by earthquakes and this work analyses them. The difference of elastic modulus and the differences between the maximum stresses in tension and compression for brick and mortar are the cause of the specimen fault in the weakest areas of the wall. It is necessary to understand that the weakest areas will be the joints between the bricks: the mortar. These joints are in charge of the masonry specimens' cohesion. Many authors have tried to define the properties of the masonry walls against vertical, horizontal, in plane or out

DOI: 10.2495/SAFE-V7-N4-463-474

of plane loads, in order to be able to characterize the existing ones, since this type of construction has fallen into disuse for new constructions. At present, some research lines focus on the reinforcement of these constructions, generally of historical character, for different types of loads, being able to find a variety of reinforcements from bars or steels meshes to laminated materials of diverse nature Ref. [2–6].

The modelling of masonry walls with finite element models such as those performed by several authors Ref. [7–13] has allowed a qualitative leap in the field. In this case, simple FEM models will be developed in order to create a model to be use it in large structural models of complete constructions. In this paper, a numerical study will be carried out using the SAP2000 program [14] as a preliminary phase to develop an experimental campaign in the Civil Engineering Laboratory at the University of Alicante. For these models, the masonry wall specimens have been modelled under different load cases (diagonal compression and simple compression), characterizing them properly, to provide a relatively simple and sufficiently approximated to reality calculation model. In addition, models of masonry walls with fiberglass reinforced mortar (TRM) have been made in order to obtain adequate approximation values and to estimate the improvements provided by this type of reinforcement, applying to the walls in both surfaces as in one.

TRM reinforcement may be implemented according to two different procedures: (i) the double layer one, and (ii) the single layer one. The use of a single layer is frequently imposed by the Italian Regional Boards of the Ministry of Cultural Heritage to preserve the memory of the facades of historic buildings, by imposing the reinforcement only in the inner face. These models have been carried out by taking into account the values of real experimental tests in order to have an approximate idea of whether the numerical calculations are sufficiently approximate or not. Other results of experimental articles have also been contrasted [15, 16].

2 METHODOLOGY AND MATERIAL CHARACTERIZATION

2.1 Geometric definition of masonry walls and reinforcement

The selected geometry for the numerical compression model is developed according to the proposal of DIN EN 1052-1:1998-12 [17]. With this geometry, the results of numerical and the future experimental tests can be compared. Figure 1 shows the geometry.

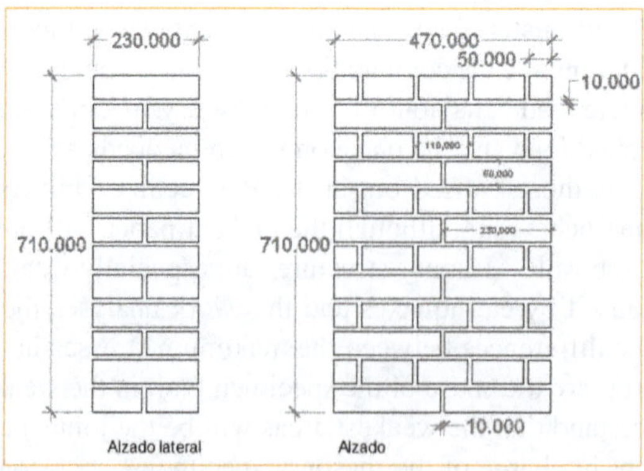

Figure 1: Geometry of the compression specimen (mm).

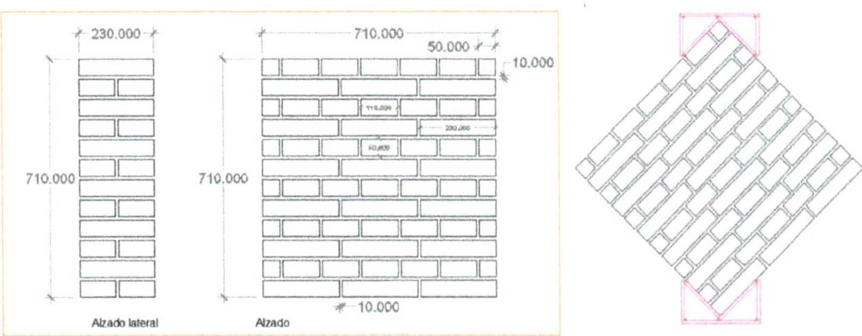

Figure 2: Diagonal Compression specimen's geometry (left) and test configuration (right).

The selected geometry for the numerical diagonal compression model (Fig. 2) has been developed according to the indications of ASTM E519/E519M. 2010 [18].

Two TRM layers – one at each side of each masonry specimen – have been considered for the reinforcement. The general procedure for each TRM layer is:

- Mortar with a 10-mm thick layer in the whole surface (in the simulation, the mechanical characteristics of Mapei Planitop HDM Restoration have been considered).
- Fiberglass in mesh form at 0 and 90 degrees established in the middle thickness of the reinforcing mortar (Mapegrid G220).

2.2 Failure criterion

The failure criterion used by SAP2000 [14] is the Mohr-Coulomb failure criterion (Ec. 1), in which two terms are considered:

- A first term, which is a function of the normal stress applied perpendicular to the shear strength and a function of the internal friction angle of the masonry material.
- A second term, which, in this case, is the cohesion of the masonry material.

$$\tau = \sigma^* \tan\theta + c \tag{1}$$

2.3 Model properties

To model unreinforced masonry specimens, non-linear layered shell elements have been considered. To simulate the masonry specimens two non-lineal layers have been used in these shells:

- A first layer, called the friction layer, to simulate the behaviour to normal stresses (S11 and S22 -horizontally and vertically within the plane, respectively -), as well as the first failure criterion term of Mohr-Coulomb failure criterion. S11 and S22 stresses will have the non-linear behaviour showed in Fig. 3. S12 is inactive.
- A second layer, called cohesion layer, responsible for providing the shear strength due to the cohesion of the wall, which would be the second addend to the criterion discussed above. S11 and S22 stresses will be inactive and in S12 will have the non-linear characteristics showed in Fig. 4.

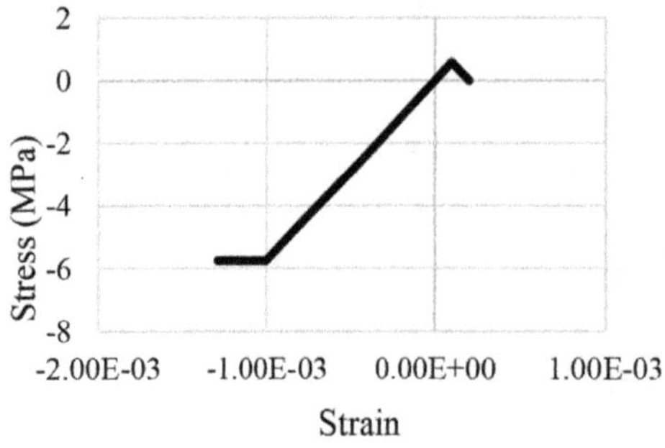

Figure 3: Masonry behaviour for the friction layer S11–S22 (Units: MPa).

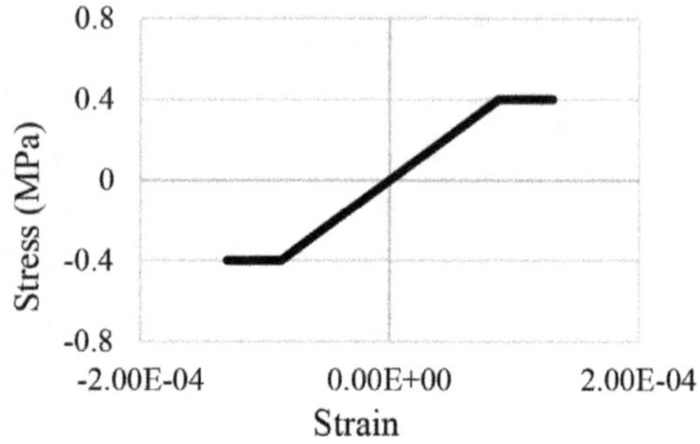

Figure 4: Masonry behaviour for the cohesion layer S12 (Units: MPa).

2.3.1 Compression test

For reinforced masonry walls, additional layers have been added to the layered shell elements. Two different models have been developed to compare their capability to simulate the real tests:

i. Fiber mesh reinforcing has been considered as continuous layer with a thickness of 0.035 mm.
ii. Fiber mesh reinforcing has been considered like a rebar mesh using the effective dimension of the Mapegrid G220.

2.3.2 Diagonal compression

For reinforced models, two/four layers have been used depending on whether the model was reinforced, on one/two sides:

- A layer where mortar with non-linear behaviour will be introduced in all directions (S11–S22–S12). This layer will be situated at a distance from the axis of half the wall thickness without reinforcing plus half the mortar layer thickness (Fig. 5).
- Another layer for the fiberglass mesh with non-linear characteristics in S11 and S22 and inactive in S12 placed at a distance from the axis of half of the thickness of the wall without reinforcing plus half the thickness of the layer of mortar and half of the fiber thickness (Fig. 6).

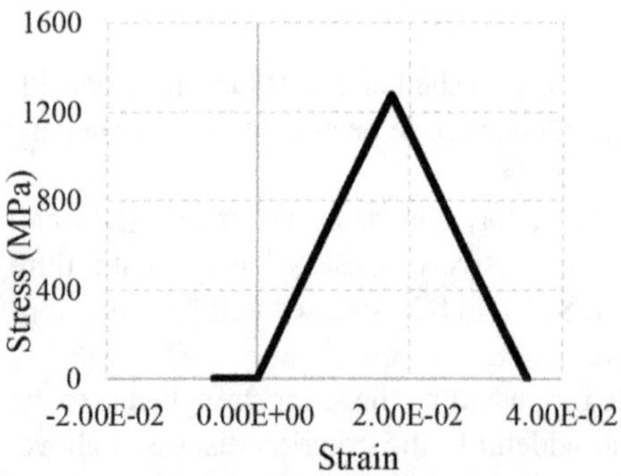

Figure 5:Fiberglass reinforcing mesh, stress-strain behaviour (Units: MPa).

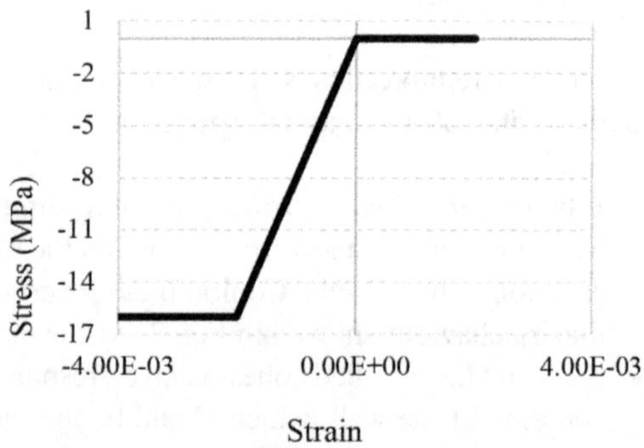

Figure 6: Reinforcing mortar, stress-strain behaviour (Units: MPa).

Table 1: Material properties

Materials	Properties				
	f_{ck} (MPa)	E (MPa)	f_t (MPa)	G (MPa)	τ (MPa)
Brick	15.00	-	-	-	-
Mortar	4.50	-	-	-	-
Masonry	5.75	5750	0.57	2300	0.20

2.4 Material definition

2.4.1 Masonry material

2.4.2 Reinforcing materials

The characteristics for the mortar used in the reinforcement have been provided by the manufacturer Mapei for their mortar Planitop HDM Restoration, and the characteristics for the mesh used in the reinforcement have been provided by the manufacturer Mapei for their product Mapegrid G220:

3 SPECIMENS MODELLED FOR THE SIMULATED TEST

3.1 Compression test

The models for simple compression are based on the use of linear and non-linear shell elements, alternating themselves vertically, with the purpose of having a symmetrical section. The non-linear shell elements are located where failure can occur. All models use layered shell elements of the maximum size of 24 mm. The boundary conditions for all the nodes in the base are pinned connections. For the top of the specimen, free displacements are considered for all nodes (Fig. 7).

For the linear layer, fiber and mortar have been considered with linear behavior in S11 and S22, inactive in S12. Instead in the non-linear section they have non-linear behavior in S11 and S12, inactive in S12.

Table 2: Reinforcing material properties

Materials	Properties						
	E (MPa)	f_{ck} (MPa)	ρ (kg/m^3)	ε (%)	Mesh size (mm)	Resistant area per unit length (mm^2/m)	Tensile strength per unit length (kN/m)
Reinforcing mortar	8000	16	1900	-	-	-	-
Fiberglass mesh	72000	-	2500	1.80	25x25	35.27	45

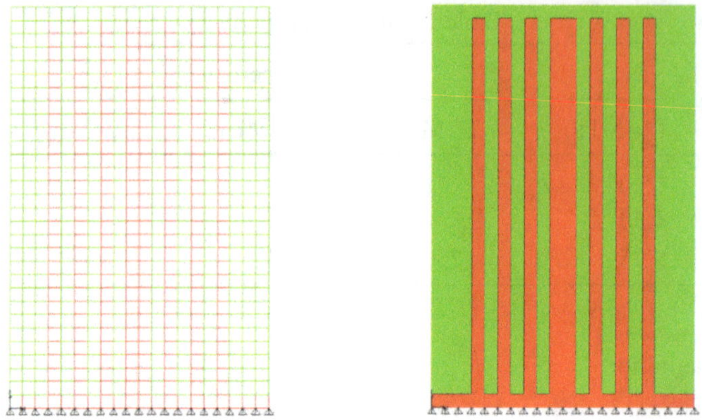

Figure 7: Numerical model of brick masonry wall for compression test. Left: Mesh. Right: layers, red nonlinear, green linear.

3.2 Diagonal compression

To develop a numerical model for the diagonal compression test, some restrictions are imposed both at the top and at the bottom, as well as the 45 degree rotated speciment. Restrictions have been placed distributedly at the top and at the bottom to avoid local stresses on the shell elements and do that an undesirable type of breakage does not occur (Fig. 8).

In the lower part, recesses have been established and in the upper one only the displacement in the vertical direction or Z axis has been allowed, applying a target displacement in the upper point. In addition, the local axes of the shells have been rotated 45 degrees.

4 RESULTS

A nonlinear quasi static load test has been developed controlling the vertical displacement of a top symmetric node in each specimen, for the compression model and for the diagonal compression model.

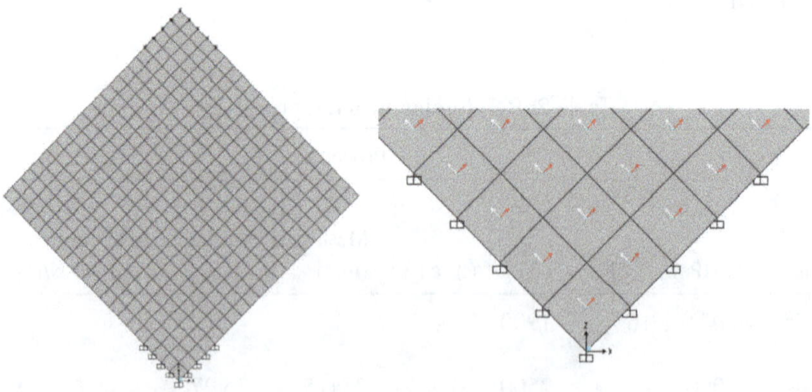

Figure 8: Numerical model of brick masonry wall for diagonal compression test and detail of the shell element local axes.

4.1 Compression results

The model shows the elastic behaviour and the Poisson effect during all the loading process (Fig. 9).

Analysing the results of the reinforced masonry specimens some stages can be detected. Initially, glass fiber mesh does not start to work in tension until the breakage of the masonry wall occurs (Fig. 10).

In a second stage, when the masonry specimen is broken, the glass fiber mess starts to work in tension reinforcing the specimen (Fig. 10).

Figure 11 shows the increment of the ultimate compression load and the ductility of the reinforced specimen. The glass fiber mesh is the source of that small increment on the ultimate load.

Figure 12 shows the deformed model, with and without reinforcing fiber glass mesh. The showed (Fig. 12, right) specimen has two layers reinforcing the masonry specimen, one at each side, of the masonry unit.

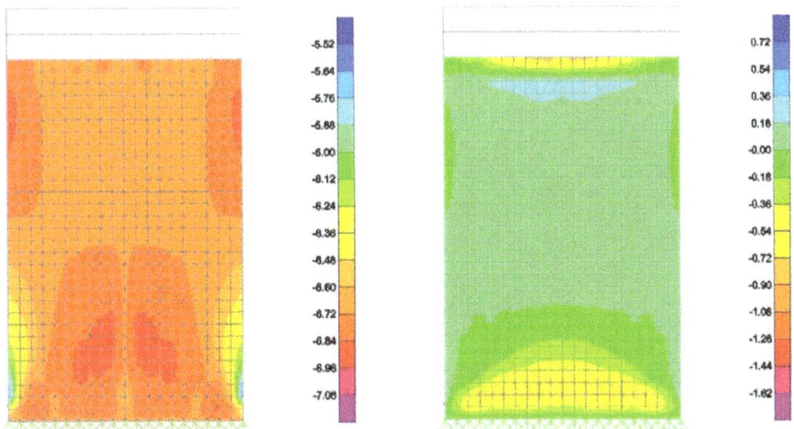

Figure 9: Brick masonry specimens. Numerical results for the compression test. Left: Vertical Compression stresses (S22). Right: Tension stress (S11). (Units: MPa).

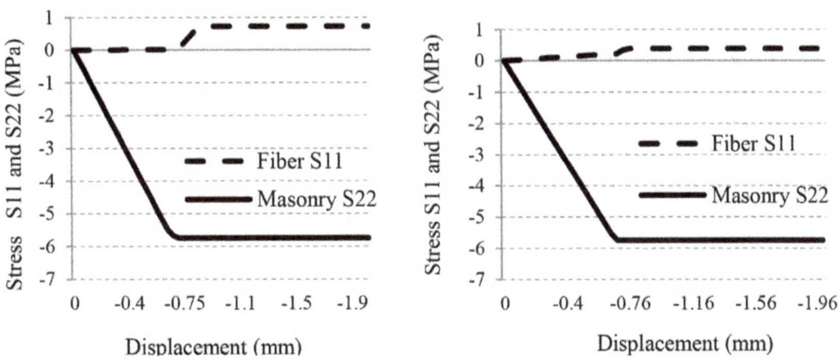

Figure 10: Stress vs displacement curve. Reinforced masonry specimen under compression loads. Left: Reinforcement modelled as continuous shell. Right: Reinforcement modelled as rebar mesh. (Units MPa-mm).

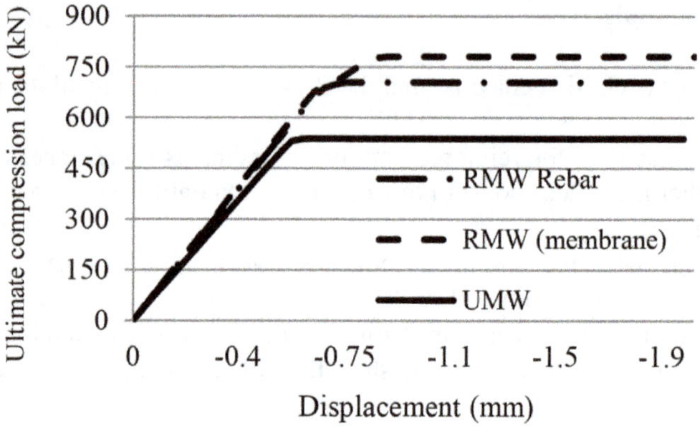

Figure 11: Base shear and displacement in the middle and in the upper point of the wall respectively (Units: kN-mm).

Figure 12: Deformation of the masonry specimen at the moment of the breakage. Left: unreinforcement. Right: 2 sides reinforced (glass fiber as a layer).

4.2 Diagonal compression results

In the development of this test it was possible to observe how the wall was compressed until it reached a sudden breakage diagonally, and with this, some sudden tensile efforts in that area. In the following image, representative of the S22 stresses in the brick wall, it is shown the wall compressed state and the breakage beginning (Figs. 13–16).

In the following image, for the reinforced two-surface model, we can observe how the layer of the fiber starts having no stress in S22 and then starts to work in traction. This way of working of the reinforcement, as well as its breakage, is similar to that studied by other authors from the University of Nizwa [5].

All this can be observed with more precision in the Fig.15, perceiving the breakage of the brick factory and the loading of the TRM reinforcement.

Moreover, Fig. 17 shows a comparison between the horizontal displacements of two specimens when a vertical load is applied. The use of one-surface to reinforce the model does not show the expected precision because it is working with a model of shell elements which does

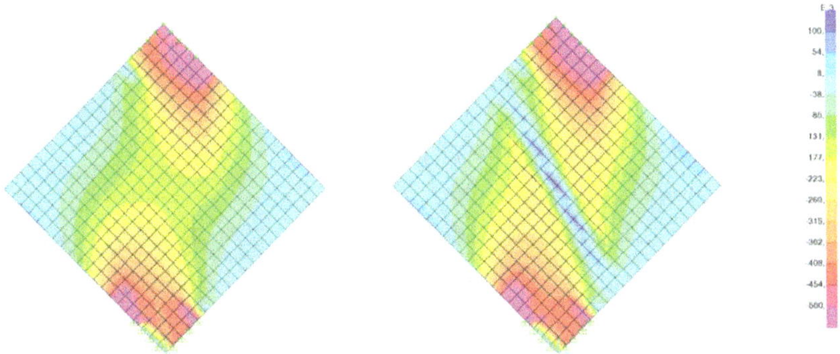

Figure 13: S22 stresses in the wall layer. On the left, compressions before breakage; to the right, breakage of the wall and tensile crack. (Units: MPa).

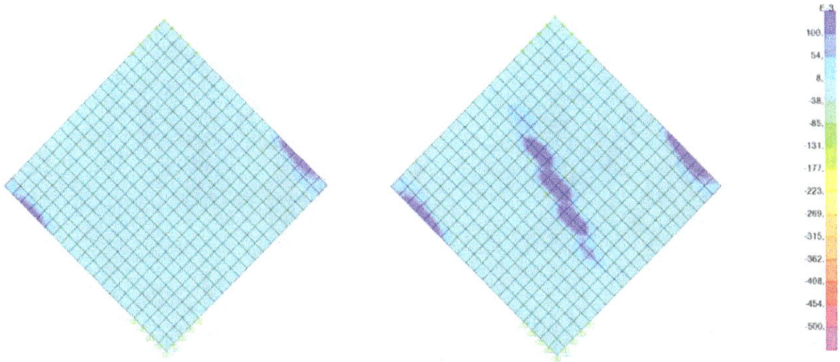

Figure 14: S22 stresses in the fiber layer. Left, fiberglass mesh without entering in charge; right, fiberglass mesh starts to work once the masonry wall cracks. (Units: MPa).

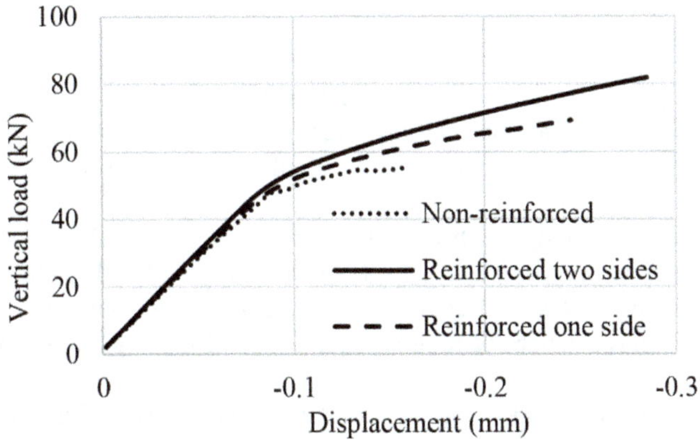

Figure 15: Base loads vs vertical top displacement. (Units: kN-mm).

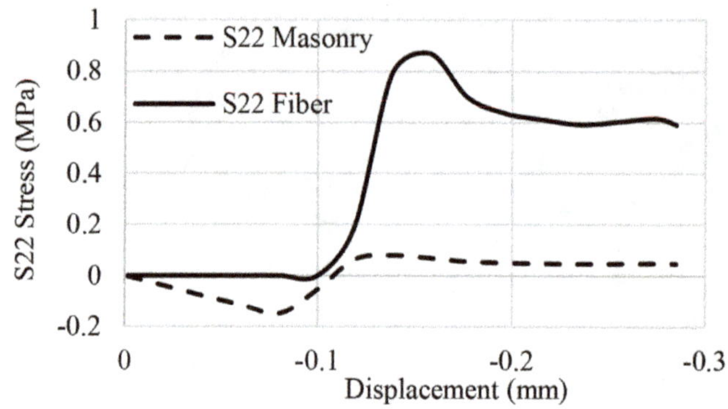

Figure 16: S22 stresses vs vertical top displacement. (Units: MPa-mm)

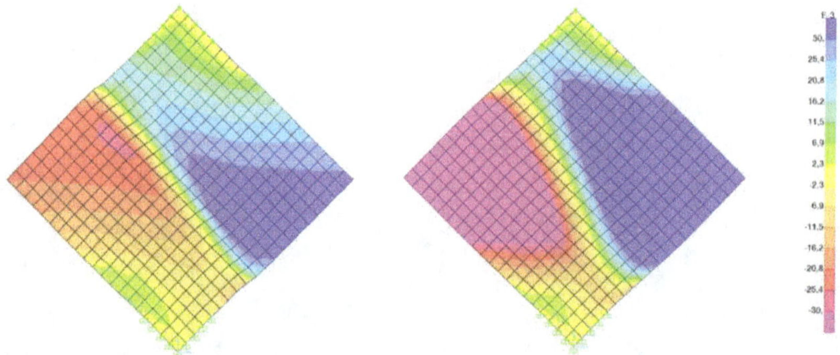

Figure 17: Deformed shape. Left: Specimen without reinforcing; right: Double-sided reinforced masonry specimen. (mm)

not consider the asymmetry but the perfect contact surfaces between wall and reinforcement.

Observing the unreinforced wall deformed shape as well as the reinforced on two sides deformed shape, we can see how the reinforcement allows to reach bigger displacements in the specimens.

5 CONCLUSIONS

The model created with the finite element software SAP2000 [14] using non-linear shell-layered elements has proved to be a valid technique to simulate masonry specimens reinforced with TRM.

With this preliminary analysis, through the appropriate calibrations with experimental results, an accurate maximum capacity of specimens reinforced with TRM can be obtained. This model allows to professional engineers to have a simple calculation tool for modelling structures that have generally required more powerful software to obtain consistent results.

This technique allows to model historical structures under seismic loads in order to know the reinforcement necessary to be able to resist these loads.

REFERENCES

[1] Bernat-Maso, E., Gil, L. & Roca, P., Numerical analysis of the load-bearing capacity of brick masonry walls strengthened with textile reinforced mortar and subjected to eccentric compressive loading. *Engineering Structures*, **91**, pp. 96–111, 2015. https://doi.org/10.1016/j.engstruct.2015.02.032

[2] Corradi, M., Borri, A., Castori, G. & Sisti, R., Shear strengthening of wall panels through jacketing with cement mortar reinforced by GFRP grids. *Composites Part B: Engineering*, **64**, pp. 33–42, 2014. https://doi.org/10.1016/j.compositesb.2014.03.022

[3] Ismail, N. & Ingham, J.M., Polymer textiles as a retrofit material for masonry walls. *Proceedings of the Institution of Civil Engineers - Structures and Buildings*, **167**(1), pp. 15–25, 2014. https://doi.org/10.1680/stbu.11.00084

[4] Papanicolaou, C.G., Triantafillou, T.C., Karlos, K. & Papathanasiou, M., Textile-reinforced mortar (TRM) versus FRP as strengthening material of URM walls: in-plane cyclic loading. *Materials and Structures*, **40**(10), pp. 1081–1097, 2007. https://doi.org/10.1617/s11527-006-9207-8

[5] Yardim, Y. & Lalaj, O., Shear strengthening of unreinforced masonry wall with different fiber reinforced mortar jacketing. *Construction and Building Materials*, **102**, pp. 149–154, 2016. https://doi.org/10.1016/j.conbuildmat.2015.10.095

[6] Foti, D. & Romanazzi, A., Experimental analysis of fiber-reinforced mortar for walls in rectified brick blocks [Analisi sperimentale di malte fibrorinforzate per pareti in blocchi di laterizio rettificati]. *C e Ca*, **41**(2), pp. 109–118, 2011.

[7] Bilgin, H. & Korini, O., *A new modeling approach in the pushover analysis of masonry structures*. International Students' Conference of Civil Engineering, ISCCE, EpokaUniversity, Tirana, Albania, May 2012.

[8] Chaimoon, K. & Attard, M.M., Modeling of unreinforced masonry walls under shear and compression. *Engineering Structures*, **29**(9), pp. 2056–2068, 2007. https://doi.org/10.1016/j.engstruct.2006.10.019

[9] Gabor, A., Ferrier, E., Jacquelin, E. & Hamelin, P., Analysis and modelling of the in-plane shear behaviour of hollow brick masonry panels. *Construction and Building Materials*, **20**(5), pp. 308–321, 2006. https://doi.org/10.1016/j.conbuildmat.2005.01.032

[10] Akhaveissy, A.H., The DSC model for the nonlinear analysis of in-plane loaded masonry structures. *The Open Civil Engineering Journal*, **6**(1), pp. 200–214, 2012. https://doi.org/10.2174/1874149501206010200

[11] Koutas, L., Triantafillou, T. & Bousias, S., Analytical modeling of masonry-infilled RC frames retrofitted with textile-reinforced mortar. *Journal of Composites for Construction*, pp. 1–14, 2014.

[12] Lee, J.S., Pande, G.N., Middleton, J. & Kralj, B., Numerical modelling of brick masonry panels subject to lateral loadings. *Computers & Structures*, **61**(4), pp. 735–745, 1996. https://doi.org/10.1016/0045-7949(95)00361-4

[13] Foti, D., On the numerical and experimental strengthening assessment of tufa masonry with FRP. *Mechanics of Advanced Materials and Structures*, **20**(2), pp. 163–175, 2013. https://doi.org/10.1080/15376494.2012.743634

[14] SAP2000, v. 14, *Computers and Structures, Inc*. Structural Sotware for analysis and Design, Berkeley, CA, USA, 2012.
[15] Corradi, M., Borri, A. & Vignoli, A., Experimental study on the determination of strength of masonry walls. *Construction and Building Materials*, **17**(5), pp. 325–337, 2003.
https://doi.org/10.1016/s0950-0618(03)00007-2
[16] Corradi, M., Tedeschi, C., Binda, L. & Borri, A., Experimental evaluation of shear and compression strength of masonry wall before and after reinforcement: deep repointing. *Construction and Building Materials*, **22**(4), pp. 463–472, 2008.
https://doi.org/10.1016/j.conbuildmat.2006.11.021
[17] EN 1052-1, Methods of test for masonry - Part 1: Determination of compressive strength. *European Committee for standardization*, p. 11, 1999.
[18] ASTM E519/E519M. *Standard Test Method for Diagonal Tension (Shear) in masonry assemblages*. ASTM International, West Conshohocken, PA, 2010.

SEISMIC REHABILITATION OF MASONRY HERITAGE STRUCTURES WITH BASE-ISOLATION AND WITH SELECTED CONTEMPORARY STRENGTHENING MEASURES

VOJKO KILAR & SIMON PETROVČIČ
University of Ljubljana, Faculty of Architecture, Slovenia.

ABSTRACT
In the seismically active zones of southern and central Europe, unreinforced masonry (URM) structures are one of the most common types of buildings. Some of them possess a high historical value and could therefore be classified as part of the architectural heritage, requiring special attention with regard to their preservation and retrofitting measures. In practice, however, the accurate prediction of the seismic response of such structures often proves to be difficult, not only due to the complex geometrical features of their individual architectural parts, but also due to the composite and non-homogeneous nature of URM. Due to the great variety of structural configurations and materials used, rational approaches for the assessment of the seismic safety levels of such buildings are needed. The paper analyses the applicability of two contemporary seismic strengthening measures, namely the use of fibre-reinforced polymer composites and the implementation of base-isolation to achieve the desired, code-based seismic protection levels. A three-storey masonry building was considered in the study. Non-linear static analyses for different levels of seismic intensities were conducted on a mathematical model of the fixed-base, FRP-strengthened and base-isolated variants of the structure.
Keywords: base isolation, fibre-reinforced polymers, historic masonry structures, seismic rehabilitation

1 INTRODUCTION

In general, architectural conservation guidelines (e.g. ICOMOS Charters) and guidelines for structural restoration (e.g. Eurocodes) usually lead to contradictory approaches to heritage conservation. On the one hand, architectural conservation guidelines follow the minimum intervention concept by which the applied interventions should be limited, reversible and should strive towards minizing the impact into the heritage substance. On the other hand, the aim of structural restoration measures is to maximise safety and resilience of the structure by using strengthening techniques which are often very invasive and irreversible.

In this paper, two contemporary seismic strengthening measures with varying degrees of invasiveness and strengthening efficiency are presented and employed in a case study numerical simulation on a typical neo-renaissance masonry heritage building. Firstly, the initial building is strengthened with fibre-reinforced polymer (FRP) composites. Compared to traditional techniques, FRP composites are lightweight, noncorrosive and exhibit high tensile strength and stiffness [1]. FRP materials usually are applied in different geometric layouts on masonry panels. In the context of architectural conservation, this strengthening method is moderately invasive.

Secondly, seismic base isolation is implemented in the initial structure. A base isolation system consists of a decoupling isolation layer between the building and the ground. The isolation layer is made of devices that support the building against static actions and, in case of seismic actions, implement a decoupling effect of the dynamic response of the building compared to the ground. A base-isolated building is characterised by a smaller acceleration response, leading to a smaller amount of forces of inertia, a smaller amount of interstorey drifts

DOI: 10.2495/SAFE-V7-N4-475-485

and internal forces in structural elements [2]. The implementation of base isolation in existing buildings is usually less invasive then FRP strengthening, since all major interventions are concentrated on the base storey, where isolation devices are usually installed. The installation of base isolation is therefore very limited: installation issues and accessibility to the equipment for inspection, maintenance, or replacement purposes are the most challenging factors to be met in the implementation of the base-isolation of existing buildings. The results for BI variant of the building are taken from our previous papers and are not presented here in detail.

The paper aims to compare the two strengthening techniques in terms of the achieved level seismic resistance. In the final part of the paper, comparison of results based on static pushover analyses for various ranges of seismic intensity are presented and discussed. Each selected strengthening measure is assessed in terms of its strengthening efficiency.

What is more, the authors of the paper have in their previous research [3] proposed a refined technique for the modelling and analysis of regular unreinforced masonry (URM) structures, which is based on the equivalent frame approach, and incorporates linear beam elements and the plastic hinge concept. Therefore, this study also aims in applying this technique also to URM structures that have been strengthened with FRP composites. An application of the technique to base-isolated structures has already been successfully undertaken [4].

2 ANALYSED MASONRY BUILDING

2.1 Building characteristics

In the conducted case study, a base-isolated masonry building typical for the neo-renaissance era in Europe was analysed. The building is a three-storey URM structure with floor plan dimensions 24.6m × 12.9m (directions X and Y, respectively) and a total height of 15.8m. The examined masonry building is schematically presented in Fig. 1.

Detailed information regarding its geometry, mechanical parameters of masonry and modelling are presented in a previous paper by the authors]. The structure can be divided into four different types of planar wall assemblies, i.e. Wx-1, Wx-2, Wy-1 and Wy-2. Each planar wall of the structure has been modelled based on the equivalent frame approach, using piers, spandrels and rigid zones as indicated in Fig. 1. The storey masses amount to m_{base} = 878 tons, m_1 = 748 tons and m_2 = 499 tons, whereas the fundamental periods of the structure are T_X = 0.39s (in the X-direction) and T_Y = 0.29s (in the Y-direction).

The building is located in a moderately active seismic area, with a design ground acceleration equal to a_g = 0.25 g and a soft soil site that corresponds to sub-soil of class C in accordance with Eurocode 8-1. Through a preliminary pushover analysis of the fixed-base structure, it has been determined that the behaviour factor of the building equals to q = 2.5 for the X-direction of seismic loading and q = 2.2 for the Y-direction [3].

2.2 Equivalent frame modelling

In general, masonry walls often incorporate irregularities in the form of an irregular layout of openings, so that special modelling considerations have to be taken into account [5]. In the cases of irregular geometry, a macro-element discretisation of masonry elements is usually adopted, which is able to predict the seismic response with sufficient accuracy and with relatively low computational costs compared to those involved in the case of conventional nonlinear finite element modelling approaches [6].

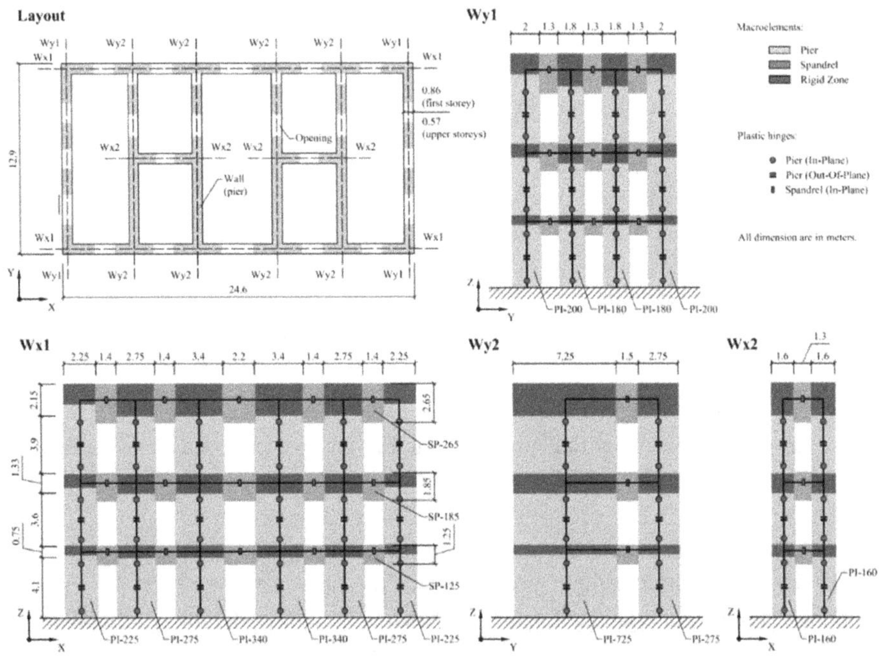

Figure 1: Examined masonry building. *(Source: Petrovčič & Kilar, 2013.)*

The in-plane lateral resistance of URM buildings is provided by the piers and spandrels. In addition to the vertical force in the piers due to the dead loads, these resisting elements are under horizontal seismic actions at the base of the structure subjected to shear and bending [7]. Depending on the width to height ratio of a masonry element and on the respective values of the normal force, bending moment and shear force, three failure mechanisms might be observed (Table 1).

The symbol θ_{cr} in Table 1 denotes the pier's chord rotation at the formation of the first crack (elastic limit). In the case of the analysed building values of θ_{cr} lie between 0.003 and 0.01. The symbol α_V is the pier's shear ratio, considering the zero moment contra-flexure point and the height-to-width ratio of the pier (for details see [3]). The ratio l_{sp}/h_{sp} is to length-to-height ratio of the spandrel.

The equivalent frame discretisation of URM and seismic failure modes are schematically presented in Fig. 2. In order to accommodate different failure modes in as simple as possible and simultaneously computationally effective way, the authors proposed a new technique for the modelling and analysis of regular URM structures [3]. This technique is based on the

Table 1: Seismic failure modes of URM.

	Limit rotations for piers			Limit rotations for spandrels		
Failure mode	θ_{DL}	θ_{SD}	θ_{NC}	θ_{DL}	θ_{SD}	θ_{NC}
Rocking	θ_{cr}	$0.008\alpha_V$	$0.011\alpha_V$	0.002	$0.008\ l_{sp}/hsp$	0.015
Diagonal cracking	θ_{cr}	0.004	0.005	0.001	0.004	0.020
Shear sliding	θ_{cr}	0.004	0.005			

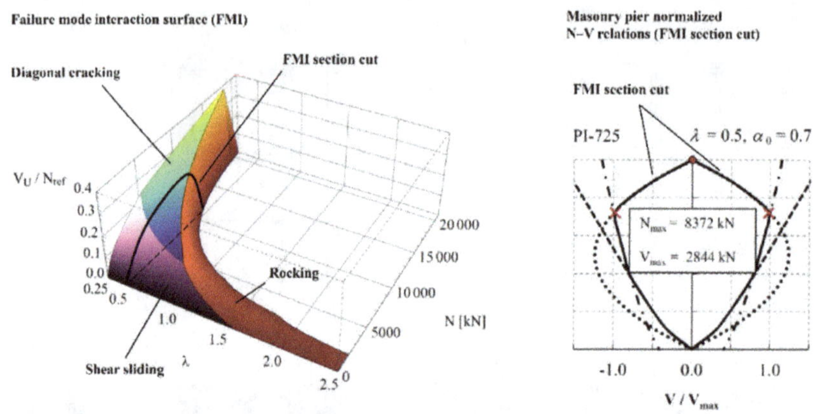

Figure 2: An example of a failure mode interaction surface of a masonry pier (left) and normalised pier N-V relations (right).

equivalent frame approach and incorporates linear beam elements and the plastic hinge concept. The complex seismic failure mechanism of masonry piers is expressed by a single failure mode interaction surface (an 'FMI surface'), taking into account the influence of variation in the pier's vertical loading, and its bending moment distribution. The ultimate lateral strength of a masonry element is expressed as a section which cuts through the FMI surface. A single failure mode interaction plastic hinge (an 'FMI hinge') for each masonry frame element is introduced by combining the specific failure modes, taking into account their minimum envelope. In the present paper, a nonlinear equivalent frame 3D mathematical model of the initial (fixed-base) structure was created in the structural analysis program SAP2000 [6] based on this technique. The model was extended to include the FRP-strengthened variant and the base-isolated variant of the building.

3 SEISMIC STRENGTHENING

3.1 Fiber- reinforced polymers

The primary objective of FRP strengthening is to increase the capacity of each member as well as the overall capacity of the masonry structure. For the design of the FRP strengthening of URM piers, the guidelines CNR-DT 200/2004 [9] were considered.

The shear capacity of masonry pier strengthened with FRP applied on both sides of the panel can be seen as the combination of two resisting mechanisms [9]: (i) shear forces due to friction in presence of compression loads and (ii) for elements capable of resisting tensile stress a truss mechanism becomes active, and shear forces are carried out by equilibrium.

The CNR-DT 200/2004 defines the design shear capacity, V_{Rd}, of the FRP strengthened masonry panel in the following manner:

$$V_{Rd} = \min(V_{Rd} + V_{Rd,f}, V_{Rd,\max}) \tag{1}$$

In the above expression, the $V_{Rd,m}$ denotes the masonry contribution to the shear strength of the panel, defined as:

$$V_{Rd,m} = \frac{1}{\gamma_{Rd}} \cdot f_{vd} \cdot d \cdot t \tag{2}$$

while the $V_{Rd,f}$ denotes the FRP contribution, which may be evaluated as follows:

$$V_{Rd,f} = \frac{1}{\gamma_{Rd}} \cdot 0.6 \cdot \frac{A_{fw}}{p_f} \cdot f_{fd} \cdot d \tag{3}$$

The symbols in eqns (13) have the following definitions: g_{Rd} is the partial factor to be assumed equal to 1.20, *d* is the distance between the compression side of the masonry and the centroid of FRP flexural strengthening, t is the masonry panel thickness, f_{vd} is the design shear strength of the masonry, equal to f_{vk} / γ_M, A_{fw} is the area of FRP shear strengthening in the direction parallel to the shear force, pf is the center-to-center spacing of FRP reinforcement measured orthogonally to the direction of the shear force, f_{fd} is the design strength of FRP reinforcement, defined as the lesser between FRP tensile failure strength and debonding strength and *t* is the thickness of the masonry panel.

The maximum value $V_{Rd,max}$ inducing failure of the compressed strut of the truss mechanism corresponds to:

$$V_{Rd,\max} = 0.3 \cdot f_{md,h} \cdot d \cdot t \tag{4}$$

where $f_{md,h}$ is the design compressive strength of the masonry parallel to the mortar joints. It should be noted that the equations above are only valid for FRP strengthening placed parallel to the mortar joints, i.e. grid layout of FRP strips.

In the conducted numerical simulation, the strengthening of masonry piers with carbon FRP (CFRP) strips has been considered. It has been considered in the study that the FRP strips have been installed on both sides of the panels and have been arranged in a grid pattern consisting of two vertical FRP strips at the edges of the pier and three horizontal FRP strips. It has been assumed that the vertical FRP strips run continuously from the bottom to the top storey (Fig. 3a).

The selection of FRP stripes was based on the effective axial stiffness, defined as the product Ef. ρ_H, which represents a measure of the effective axial stiffness of the external FRP reinforcement of a masonry panel, where E_f is the Young's modulus of FRP and ρ_H is the horizontal FRP reinforcement ratio, defined as the area of horizontal FRP strips on both sides of the panel divided by the product of the height and the thickness of the panel. Based on experimental results by Marcari and co-autohrs [10], the effective axial stiffness equal to 60 MPa has been considered. The authors have concluded that effective axial stiffness of the FRP strips larger than 15 MPa changes the original failure mode of the panels clearly changed from shear to a shear/flexural mode; consequently, larger gains in lateral strength were

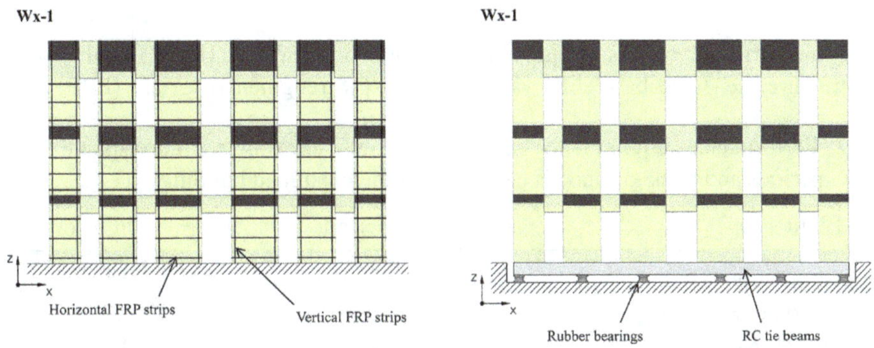

Figure 3: FRP strengthened variant (a) and BI variant (b).

achieved. Based on this assumption, CFRP strips with the width/thickness ratio 120 mm/1.4 mm have been selected (Ef = 21 000 kN/cm2, $e_f = 0.014$).

The selected FRP characteristics have been included in the mathematical model of the initial, unstrengthened variant by using the same methodology as described in Section 2.2.

3.2 Base isolation

The results of base isolation (BI) variants have been added in the paper only for the sake of comparison. The analysis methods and models for BI variants of this building have already been presented by the authors extensively elsewhere [2, 3] and are therefore not repeated in this paper.

4 ASSESSMENT OF SEISMIC PERFORMANCE

4.1 Eurocode 8 requirements

Part 3 of Eurocode 8, i.e. EC8-3 [11], which deals with the assessment and retrofitting of existing buildings, is based on recent trends regarding performance requirements and checks of compliance in terms of displacements, providing also a degree of flexibility to cover the large variety of situations arising in practice. The fundamental requirements of EC8-3 refer to the state of damage in the structure that is defined by means of three limit states presented in Table 2.

In order to achieve an appropriate level of protection, EC8-3 takes into account a different seismic return period for each of the limit states. Longer return periods are associated with greater seismic intensities. Although it should be pointed out, that many heritage buildings have resisted earthquakes with limited damage only, even when their calculated resistance has not completely met code requirements.

This allows for the code-based ground acceleration values that are in use in new buildings to be reduced, meaning that a higher probability of exceedance is admitted. In turn, the considered return periods vary from country to country and are given in the corresponding National Annexes to EC8-3. Table presents the return periods and corresponding peak ground acceleration (PGA) values, referring to the site of the analysed building, which is located on sub-soil class C (soft site, soil factor S = 1.15). The PGAs indicated in the table refer to a hard site (foundation on rock).

4.2 Nonlinear static analyses and comparison of results

The seismic performance was examined for the following variants: (i) the initial unstrengthened structure, i.e. fixed-base (FB) variant, (ii) FRP-Strengthened variant (FRP) and (iii)

Table 2: Performance requirements based on EC8-3, indicating the corresponding return periods and peak ground accelerations of the examined building.

EC8-3 Limit state	Return period	PGA on rock
Damage limitation (DL)	95 years	0.15g
Significant damage (SD)	475 years	0.25g
Near collapse (NC)	2475 years	0.43g

Base-Isolated variant (BI). Since it was concluded in the previous study [3] that the X-direction seismic response is more critical than that in the Y-direction, in this paper only the seismic performance for the loading in X-direction of the building was analysed.

4.2.1 Capacity curve comparison

A nonlinear static (pushover) analysis was conducted. The demand parameter that was compared to the capacity was the top (roof) displacement. The N2 method [12] was used to determine the demand top displacement by comparing the intersection between the capacity obtained by pushover analysis and by an inelastic response spectrum. A lateral load pattern proportional to the first mode of vibration was considered. Lateral loads (weighted, taking into account the nodal masses) were applied to the nodes corresponding to the individual storey levels.

Pushover curves of analysed variants, indicating the base shear vs. the lateral roof (top) displacement relationship, are presented in Fig. 4. The protection levels which are defined in EC8-3 as the limit states DL, SD and NC are also indicated in the figure. It can be seen that the capacities of the FB and BI variants are directly comparable. It should be noted that in order to assess the effects of BI on the initial structure, the relative displacements need to be observed and compared. The relative displacements indicate the lateral top displacement with respect to the BI level, i.e. subtracting the lateral displacements due to the base isolation. In the

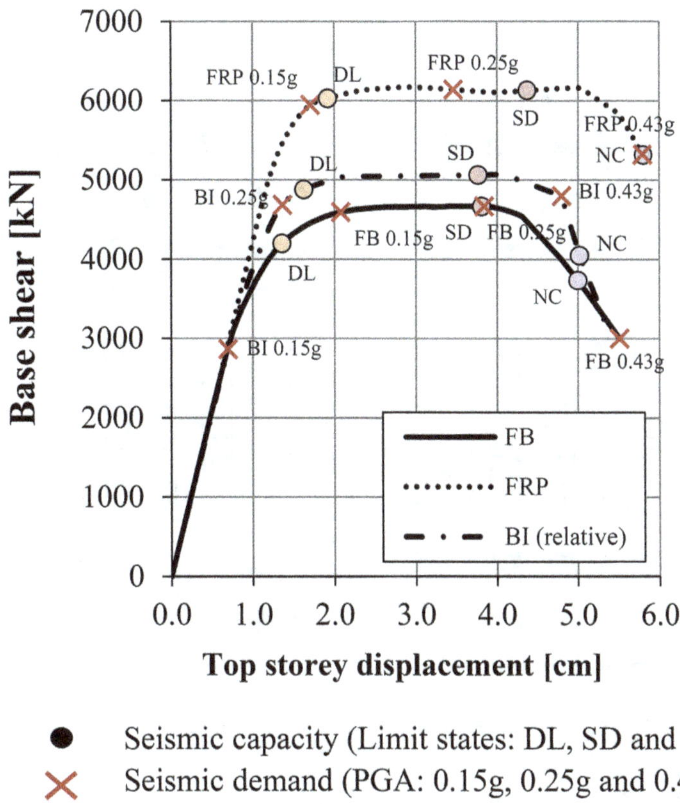

Figure 4: Comparison of pushover curves.

BI case, the maximum base shear is slightly larger due to the different lateral force distribution, and due the difference in the total seismic mass of the FB and BI variants. The FRP variant, on the other hand, can withstand an approximately 30% larger maximum base shear force when compared to the initial FB variant.

4.2.2 Seismic demand displacements and damage patterns

The seismic demand displacements were calculated by employing the N2 method for PGA levels corresponding to the observed limit states DL, SD and NC (Table 2). It can be seen from Fig. 5 that the FB structure exhibits inadequate seismic performance for all three observed limit states, i.e. the corresponding demand exceeds the capacity requirements.

In the case of the FRP and BI variants for the PGA = 0.15g and 0.25g (the DL and SD limit states, respectively), the structure's top displacement is smaller than the displacement of the observed limit state. This means that in both cases, the structure has been adequately strengthened for the analysed limit state. Moreover, the displacement of the BI variant is for both PGAs smaller than the DL limit state, indicating limited and repairable damage to the structure also in the case of PGA = 0.25 g.

When observing the demand displacement for PGA = 0.43 g, the structure's performance is inadequate for all three examined variants. In the case of the FB and also FRP strengthened variants, the structure would for this PGA value reach its maximum displacement, indicating its collapse. The BI variant would, based on the results, be heavily damaged and would be very close to collapsing.

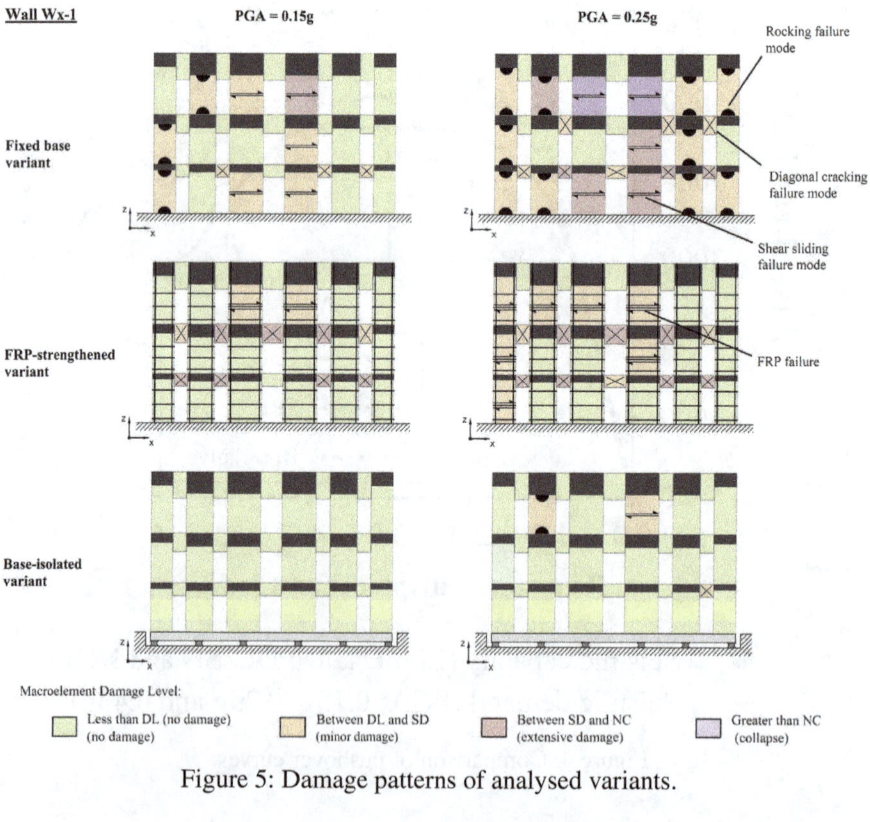

Figure 5: Damage patterns of analysed variants.

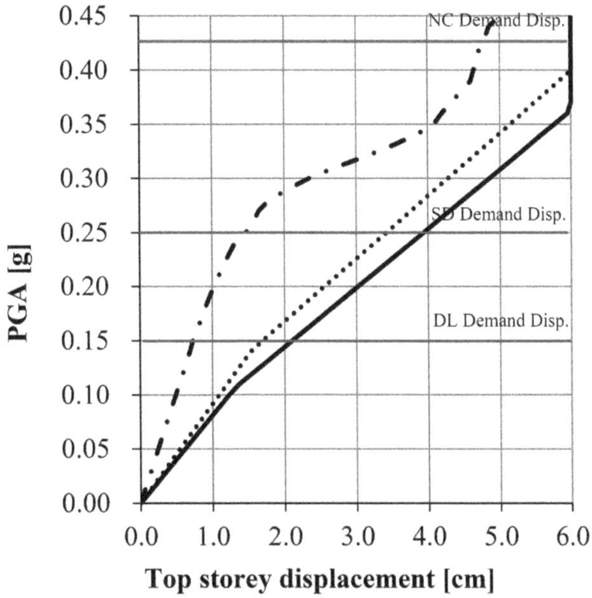

Figure 6: Comparison of IN2 curves.

For PGA = 0.15g and 0.25g, the damage patterns of the structure for all three variants are presented in Fig. 6 for the wall Wx-1. The type of failure mechanism and the level of damage are indicated for each macroelement. The specific macroelement limit chord rotations are given in Table 1. It can be seen from Fig. 6 that the FB variant is heavily damaged in the case of PGA = 0.25g. The piers of the top storey are all damaged, some of them considerably. A similar observation can be made for the piers and spandrels in the base storey. All the damage grades which are higher than the SD macroelement limit state indicate damage that is very difficult and uneconomic to repair.

The FRP strengthened variant receives significantly less damage in the case of PGA = 0.25g. The damage mostly concentrated in the top storey, where the axials forces in the pier are smaller. It should also be noted that since the strengthening of spandrels was not considered, most of the spandrels in the model are heavily damaged.

The least amount of damage was recorded in the case of the BI variant. For the target PGA = 0.25g, it can be seen that the structure undergoes minor damage, which is likely to be repairable. Damage is localised in two piers in the top storey. This is consistent with the EC8-3 requirements for the DL limit state, which allows the occurrence of local damage, where repair measures are usually not needed or are very limited.

4.2.3 Top storey displacements for varying PGA levels

In the final part, the relationship between the top storey displacements and the PGA was analysed by progressively increasing the PGA in a stepwise fashion. In each step, the seismic demand displacement has been calculated by using the N2 method. Such an analysis, which has been given the name 'incremental N2 analysis' (IN2), was first introduced and applied to infilled RC frames [13]. In the conducted IN2 analyses, forty-five consecutive intensity levels were considered, ranging from PGA = 0.01g to PGA = 0.45g, with a 0.01g step size.

The results of the analyses are presented in Fig. 6. It can be seen from the figure that the IN2 curves for the FB and FRP variants follow the same trend and that in the PGA range from approximately 0.15g to 0.37g the FRP variant yields about 15% smaller seismic demand displacements, leading to reduced damage in the piers and different damage pattern distributions. In the case of the BI variant, this difference is more profound. It is largest at PGA = 0.20g where the seismic demand displacement of the BI variant equals to 34% of that of the FB variant.

5 CONCLUSIONS

Two seismic strengthening measures of an existing unreinforced masonry structure have been considered. In the first variant, FRP composites have been used, whereas in the second variant seismic BI has been implemented. It has been shown that the BI variant of the building obtains significantly less seismically induced damage than the FRP variant. It was concluded that both strengthening measures provide sufficient lateral strength increases to meet basic code requirements.

It is, however, not easy to implement the base isolated system under the existing building. Such endeavour is possible, but seldom financially feasible and reserved only to buildings that meet specific requirements of value and importance. The presented results for base isolated building are to be seen more as a presentation of maximum earthquake resistance capacity, which could be achieved for the structure without additional invasive retrofitting measures. It usually depends on the conservation plan set up by conservator, architect and structural engineer that determines the measures suitable for particular building. The strengthening with FRP is usually appropriate for most buildings, since it can be applied on inner/outer or on bout sided of existing walls.

The presented microelement modelling technique for URM structures developed by the authors enables also a comparison of effectiveness of different seismic retrofitting techniques and selection of the most suitable for particular building and location, which was demonstrated on two selected retrofitting techniques applied in this paper.

ACKNOWLEDGEMENTS

The results presented in this article are based on work that was supported by the Slovenian Research Agency (grant number P5-0068). The authors greatly acknowledge the financial support for this research.

REFERENCES

[1] Prota, A., Manfredi, G. & Nardone, F., Assessment of design formulas for in-plane FRP strengthening of masonry walls. *Journal of Composites for Construction,* **12**(6), pp. 643–649, 2008.
https://doi.org/10.1061/(asce)1090-0268(2008)12:6(643)

[2] Kelly, J.M., Robinson, B.W.H. & Skinner, R.I., *Seismic Isolation for Designers and Structural Engineers*, Robinson Seismic Limited, New Zealand, 2007.

[3] Petrovčič, S. & Kilar, V., Seismic failure mode interaction for the equivalent frame modeling of unreinforced masonry structures. *Engineering Structures,* **30**(54), pp. 9–22, 2013.
https://doi.org/10.1016/j.engstruct.2013.03.050

[4] Petrovčič, S. & Kilar, V., Seismic retrofitting of historic masonry structures with the use of base isolation—modeling and analysis aspects. *International Journal of Architectural Heritage*, **11**(2), pp. 229–246, 2016.
https://doi.org/10.1080/15583058.2016.1190881

[5] Parisi, F. & Augenti, N., Seismic capacity of irregular unreinforced masonry walls with openings. *Earthquake Engineering & Structural Dynamics*, **42**(1), pp. 101–121, 2013. https://doi.org/10.1002/eqe.2195

[6] Kappos, A.J. & Papanikolaou, V.K., Nonlinear dynamic analysis of masonry buildings and definition of seismic damage states. *The Open Construction and Building Technology Journal*, **10**(1), pp. 192–209, 2016. https://doi.org/10.2174/1874836801610010192

[7] Magenes, G. & Calvi, G.M., In-plane seismic response of brick masonry walls. *Earthquake Engineering & Structural Dynamics*, **26**(11), pp. 1091–1112, 1997. https://doi.org/10.1002/(sici)1096-9845(199711)26:11<1091::aid-eqe693>3.3.co;2-y

[8] CSI – Computers and Structures, *SAP2000 Ultimate (v18.2.0) – Structural Analysis Program*. Berkeley, USA, 2016.

[9] CNR – Advisory Committee on Technical Recommendations for Construction, *Guide for the Design and Construction of Externally Bonded FRP Systems for Strengthening Existing Structures*. CNR, Rome, 2016.

[10] Marcari, G., Manfredi, G., Prota, A. & Pecce, M., In-plane shear performance of masonry panels strengthened with FRP. *Composites Part B: Engineering*, **38**(7), pp. 887–901, 2007. https://doi.org/10.1016/j.compositesb.2006.11.004

[11] CEN – European Committee for Standardisation, *Eurocode 8: Design of structures for earthquake resistance – Part 3: General rules, seismic actions and rules for buildings, Design Code EN 1998-3*. CEN, Brussels, 2005.

[12] Fajfar, P., A nonlinear analysis method for performance-based seismic design. *Earthquake Spectra*, **16**(3), pp. 573–592, 2000. https://doi.org/10.1193/1.1586128

[13] Dolšek, M. & Fajfar, P., Simplified non-linear seismic analysis of infilled reinforced concrete frames. *Earthquake Engineering & Structural Dynamics*, **34**, pp. 49–66, 2005. https://doi.org/10.1002/eqe.411

SEISMIC BEHAVIOR OF A MASONRY CHIMNEY RETROFITTED WITH COMPOSITE MATERIALS: A PRELIMINARY APPROACH

D. BRU, S. IVORRA & F. J. BAEZA
Civil Engineering Department, University of Alicante, Spain.

ABSTRACT

This paper presents a structural analysis of a masonry chimney, which is being catalogued as local interest heritage, according Eurocode 8. The chimney, located in Alicante (Spain), is severely damaged, and it shows several longitudinal cracks, and mortar loss between bricks. In order to guarantee the structural safety under seismic forces, the chimney was retrofitted with composite materials. This reinforcement comprised an internal textile reinforced mortar (TRM) layer, i.e. glass fiber mesh and cement matrix, and local reinforcement with longitudinal carbon fiber bands. This study consisted of two stages: first, numerical and experimental analyses of the original chimney were done. Second, design of an internal reinforcement scheme was done. The experimental text includes acceleration measures under ambient vibration for an operational modal analysis. And laboratory test for bricks and mortar to study the mineralogical composition and mechanical properties. The numerical analysis includes, preliminary pushover analysis before and after the reinforcement was done, and second, linear response spectrum analysis to evaluate the structural stability under the seismic demand.

Keywords: FEM, Reinforced masonry TRM walls, seismic loads, seismic retrofitting, slender masonry structures

1 INTRODUCTION

The recent Lorca earthquake increased the interest in the seismic behavior of masonry structures in Spain. In particular, the focus of the present paper is on the seismic behavior improvement of slender historical masonry constructions [1]. Nowadays, the existing literature related to the numerical analysis of chimneys is rather scarce. Some recent and insightful 3D nonlinear analyses of a collapsed reinforced concrete chimney were presented in Huang *et al.* [2]. Regarding to masonry chimneys, numerical results were presented by Pallarés, Ivorra, Lourenço and Foti ([3–5]). Moreover, a critical comparison between nonlinear static and dynamic analysis methods is presented with reference to a stone masonry minaret [6]. About the structural characterization of masonry chimneys and experimental test, different examples were presented in Aoki *et al.* (2006) [7] and Pallarés *et al.* (2009) [8]. This researches show the experimental vibration mode shapes and the relationship between the natural frequencies and the stiffness of the structure.

The present paper is focused on the analysis of a 25 m tall masonry chimney located in Agost (Alicante, Spain), Fig. 1. The chimney was built in the beginning of the 20th century in the service of a brick factory. The main objective of this research is to analyze the seismic vulnerability of the structure before and after the reinforcing process, according to Eurocode 8 [9] and NCSE Spanish [10] standard. Both linear and nonlinear pushover analyses were developed [11], for the unreinforced and reinforced structure. This reinforcement comprised an internal textile reinforced mortar (TRM) layer, i.e. glass fiber mesh and cement matrix, and local reinforcement with longitudinal carbon fiber composites.

DOI: 10.2495/SAFE-V7-N4-486-497

2 GEOMETRY AND EXPERIMENTAL TESTS

2.1 Geometrical properties

The chimney was built in masonry bricks (24x11.5x5 cm), with mortar joints (1.50 cm), for a total height of 24.70 m. It can be divided in three different parts: 2.84x2.84 m squared section base with 0.80 m thickness; and the stack and crown, both with a regular octagonal cross section with variable dimensions and thickness (from 0.6 m at the stack-base connection to 0.25 m at the crown). The main characteristics of this structure are the drift of the top of the shaft and the longitudinal crack on the central part of the chimney, Fig. 1. Nowadays, the top of the shaft has a slope of 3%, besides the 10 m of longitudinal crack in the shaft of the chimney that shows the lack of cohesion between masonry bricks.

2.2 Laboratory tests

A survey campaign was carried out to estimate the chemical and physical properties of the masonry materials (bricks and mortars). Twelve samples were taken from the structure at

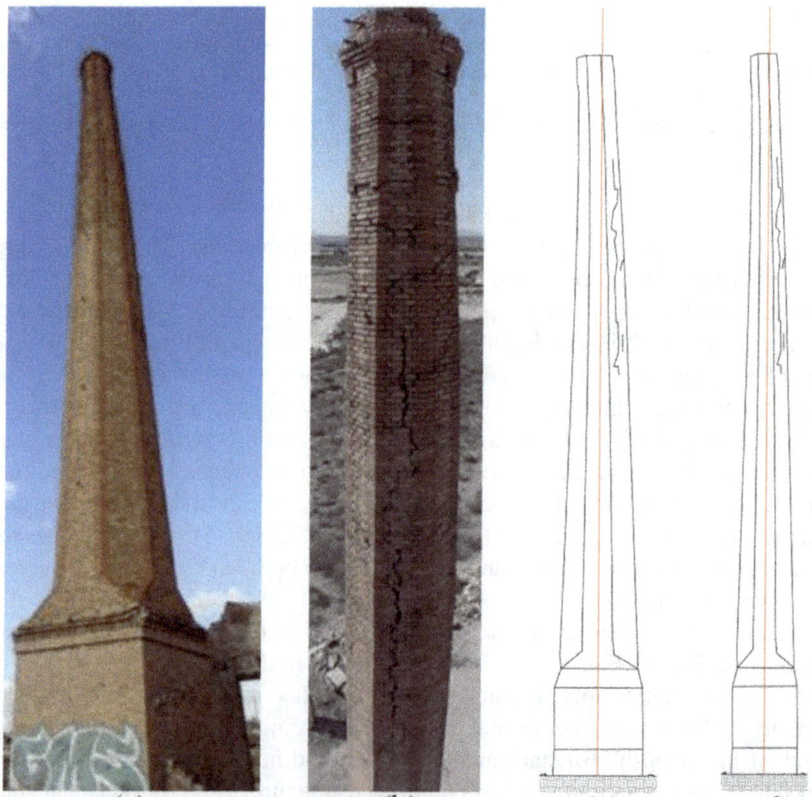

(a) (b) (c) (d)

Figure 1: a) General view of the chimney, b) vertical cracking pattern along the stack, c) and d) NE and SE elevations.

different positions. Samples number M1, M2 and M6 were taken from the inner part of the chimney. All samples were analyzed using scanning optical microscopy technique and X-ray diffraction. The results showed a high concentration of sulphur in the interior of the chimney, Fig. 2, due to combustion processes during the industrial activity of the facility. This fact caused significant mortar and brick deterioration due to the acid action of the sulphur, if water was present because vapor condensation. The mortar was a lime mortar with silica sand, as showed the X-ray diffraction results included in Table 1.

Finally, mechanical and physical properties were analyzed. In particular, density and compression strength for mortars and bricks were measured, according to Magalhães

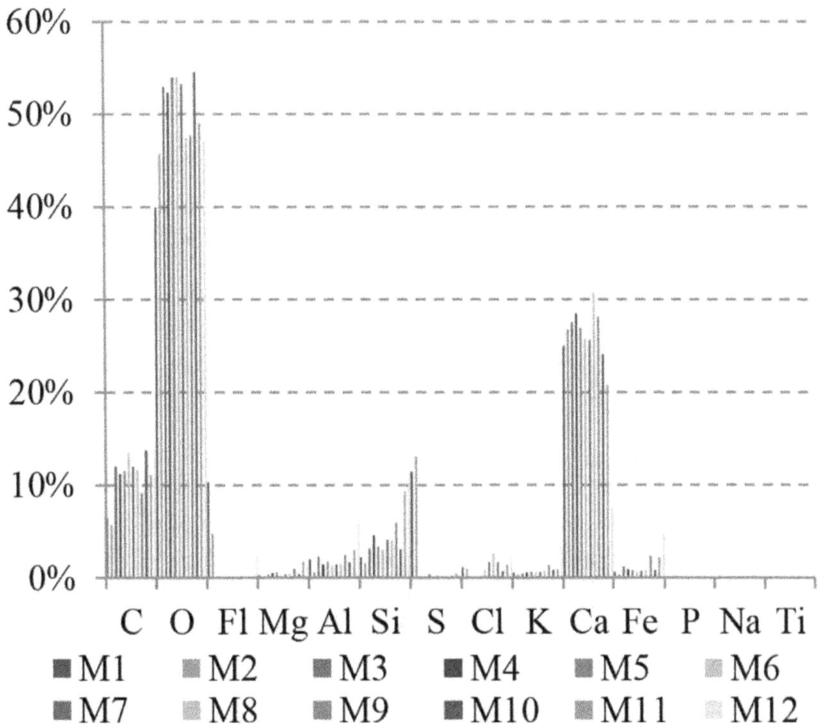

Figure 2: Results of SEM analysis.

Table 1: Results of DRX analysis.

Samples	Crystalline phases				
	Gympsum ($CaSO_4{\cdot}2H_2O$)	Quartz (SiO_2)	Anhydrite ($CaSO_4$)	Calcite ($CaCO_3$)	Dolomite ($CaMgCO_3$)
M1	8.05%	14.10%	75.94%		
M3		28.97%		59.76%	6.85%
M6	14.22%	25.55%		60.23%	
M8		28.67%		67.58%	
M12	34.43%	19.53%		46.04%	

research [12]. The average density was equal to 1893 kg/m³ for the mortar and 2018 kg/m³ for the bricks. Moreover, the average compression strength for mortar was equal to 7.63 MPa and 37.79 MPa for brick. According to Eurocode 6 [13], for masonry walls, the compression strength was equal to 11.16 MPa, the tensile strength was equal to 0.2 MPa and the shear initial strength was equal to 0.2 MPa. Finally, the described survey ensures a KL1 knowledge level, according to Eurocode 8, leading to a confidence factor CF_{KL1}, 1.35.

2.3 General properties of the repair process

The reinforcement of the structure has been divided into several phases. First, the most damaged areas of mortar were repaired, i.e. all joints were cleaned with compressed air and damaged joints were completely removed. In the second phase, these damaged joints were repaired by means of new lime mortars, dosage 1/3 for slaked lime and 0–3 mm sand. In the third phase, the area where the cracks affected the whole wall thickness was repaired. In particular, all bricks near the crack were exchanged for new similar bricks and mortar.

After this preliminary repairing process of the masonry wall, the structural reinforcement was placed. The reinforcement comprised an internal textile reinforced mortar (TRM) and local reinforcement with longitudinal carbon fiber bands. In particular, one layer of fiberglass, Mapegrid G220, and Planitop Restauro mortar were placed together. Regarding the carbon fiber reinforcement, eight internal and longitudinal FRP bands – Carboplate E170/100/1.4 and E170/80/1.2 – were placed along the whole chimney. A special adhesive, specifically Adhesilex PG-1, guaranteed the bonding between the carbon fiber laminates and the repair mortar. MAPEI provided all referred materials.

2.4 Dynamic properties of the repair process

Ambient vibration tests were conducted on the chimney in June 2014 to assess the dynamic response at 12 different points only under ambient vibrations (Fig. 4). The structure was instrumented with eight 393B12 PCB Piezotronics uniaxial accelerometers, placed at 2, 10, 16 and 23.24 m from the structure's base, and connected to a centralized data acquisition system by means of long transducer cables. The accelerometers position and orientation are referred in Fig. 4, together with the experimental model used for OMA. Two PCB 482A22 signal

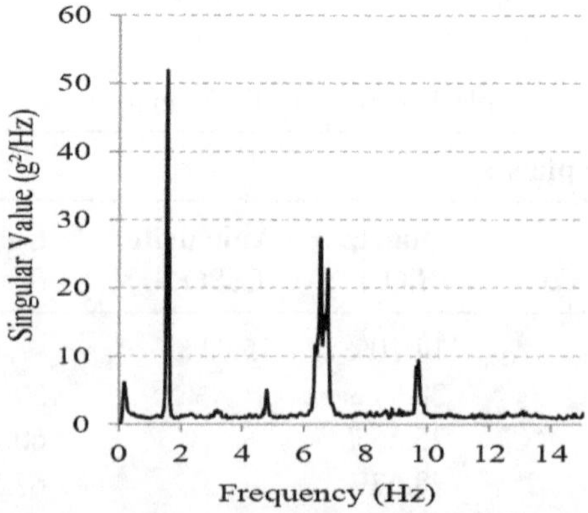

Figure 3: Natural frequencies of the mode shapes.

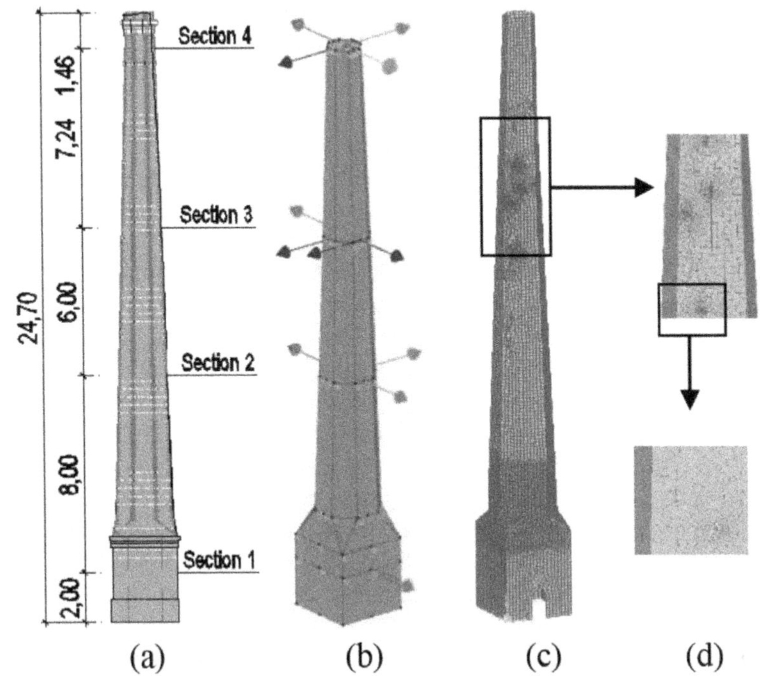

Figure 4: a) Position of the sensors, b) experimental model for OMA, c) numerical model for FEM, d) numerical model: Details of cracks.

conditioner and two Kyowa PCD 320 data logger registered the response during ambient vibration tests.

The measurement of the dynamic behavior of the structure was performed in two phases. In the first phase, eight accelerometers were used, and in the second phase, four of these were fixed as reference signals and the remaining four were moved between predefined positions and directions. In each setup, a continuous signal acquisition was carried out for approximately 10 min, with 100 Hz sampling frequency. Figure 5 shows the four vibration mode

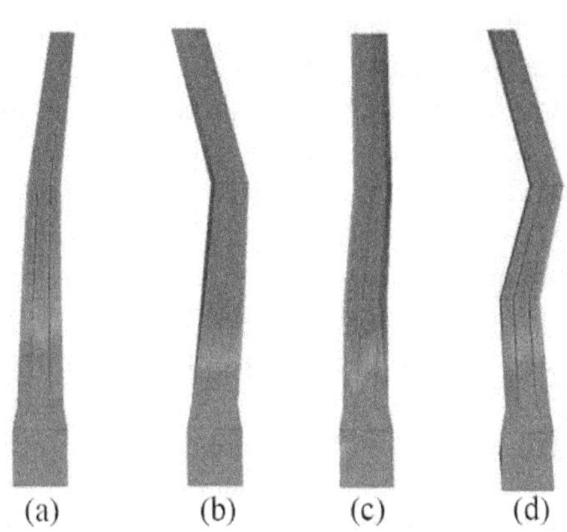

Figure 5: Modal shapes estimated from experimental tests: a) mode 1, b) mode 2, c) mode 3, and d) mode 4.

shapes estimated from the recorded measurements. The natural frequencies for these modes were 1.55Hz, 4.81Hz, 6.83Hz and 9.98Hz. All modes were detected in two orthogonal directions.

3 NUMERICAL MODEL

Two different numerical models were considered to evaluate the structural behavior of the masonry chimney. Both models use the Finite Element Method by means of the SAP2000 software [14], and they were modeled using four nodes shell elements. The main difference between the two models is that the first one considered the main crack in the chimney and the second model presented the reinforced structure, Fig. 4. The first model had a total 2543 nodes and 2538 shell elements. The second model had a total 15249 nodes and 14936 shell elements. In the model definition, the structural footing was considered fixed, the masonry weight per unit volume equal to 14.49 kN/m³, a constant Poisson's ratio equal to 0.25, and isotropic behavior was assumed for the whole chimney.

The material nonlinearities for masonry were reproduced by means of a multi-directional elastic and plastic curve. The compressive strength was estimated using the mean value 4.59 MPa obtained from experimental tests on masonry units and considering a safety factor. The behavior for compression curve was elastic with an initial stiffness of 3654 MPa up to a 4.59 MPa stress, and perfect plasticity afterwards. The reinforcement material is considered linear for tension and zero for compression.

After the analyses, the natural frequencies of the unreinforced model were 1.46, 5.10, 7.47 and 10.11 Hz and for reinforced model were 1.66, 5.93, 10.72, and 11.88 Hz.

4 RESPONSE SPECTRUM ANALYSIS

4.1 Elastic response spectrum

Seismic evaluation of the masonry chimney was done according to Eurocode 8 by means of a response spectrum analysis. Elastic response spectrum, for a 5% damping factor and special relevance region, has been calculated with the Spanish standard NCSE 02, Fig. 6. The constant acceleration part of the spectrum was located between 0.13 s and 0.54 s periods. A type II ground was selected according to geotechnical parameters and preliminary excavation near the foundation of the chimney. The results of the excavation showed that the first 1.5 m of the excavation was natural loose ground, and from 1.5 m to 2.25 m of the excavation, were sandy loams with good consistency. According to the ground properties, basic acceleration and maximum acceleration values are equal to 0.11 g and 0.15 g.

Figure 6 shows the natural periods for unreinforced and reinforced structures. In particular, seismic acceleration, for the first mode of vibration, increases by 11.33% for reinforced structure. However, seismic acceleration decreases by 10% for the fourth vibration mode and remains constant for the second mode of vibration. These results show that a high structural reinforcement may reduce the seismic demand, although the actions for the first mode were increased.

Finally, ultimate limit state verification for bending and shear has been done by means of the Eurocode 8. In particular, ultimate bending moment before the structural collapse has been evaluated as the overturning stability. This moment was calculated by equilibrium with the compressive stress resultant of the masonry section, with maximum compressive strain equal to 0.35%, and the tensile stress resultant of the reinforced area, with maximum tensile

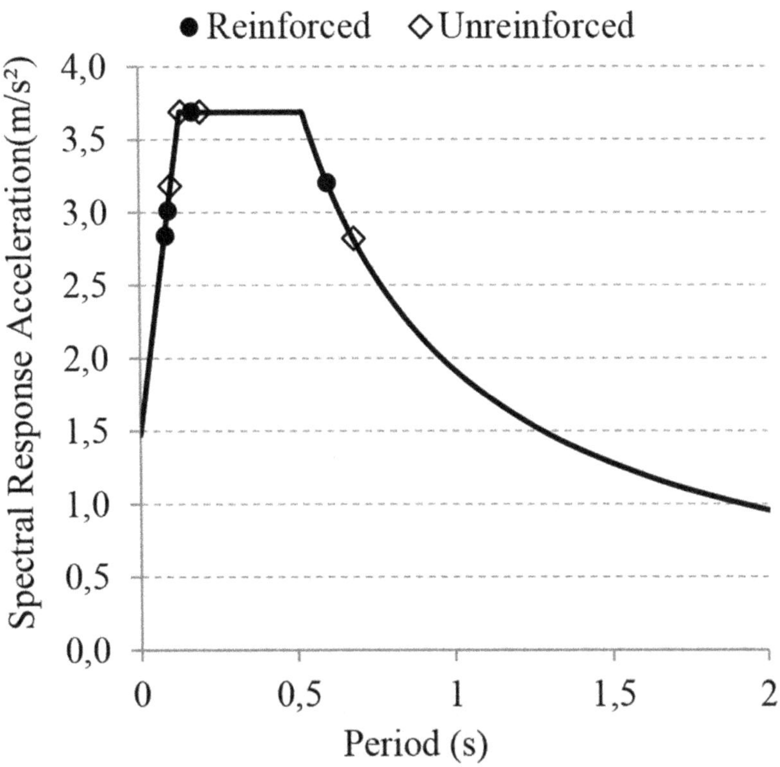

Figure 6: Elastic response spectrum.

strain equal to 1%. Furthermore, shear capacity of the masonry section was assessed by means of the Mohr-Coulomb criteria, with cohesion equal to 0.2 MPa and friction coefficient equal to 0.4.

Figure 7 shows the bending results for linear analysis and bending resistance moment for reinforced and unreinforced structure. Before reinforcement of the structure, the masonry chimney shows high risk of bending collapse. However, for the reinforced chimney, all bending moments of the seismic actions are lower than the bending capacity of the reinforced sections. Furthermore, in case of shear seismic actions, Fig. 8 shows similar conclusions than bending analysis. In particular, the critical region is the upper 10 m of the masonry chimney.

5 PUSHOVER ANALYSIS

In order to analyze the nonlinear response of the system versus the seismic actions, a modal pushover analysis (MPA) was done. In this case, nonlinear response was considered for mode 1 and linear response for higher modes, according to Chopra *et al.* 2002 [11]. The basic idea of MPA is to combine the results of N pushover analyses, the nth of which is based on the invariant forces distribution proportional to the effective mass and modal shape, Equation 1.

$$s_n = M \cdot \Phi_n \tag{1}$$

where M = modal mass; Φ_n = n^{th} elastic mode shape.; and s_n = invariant force distribution for mode n.

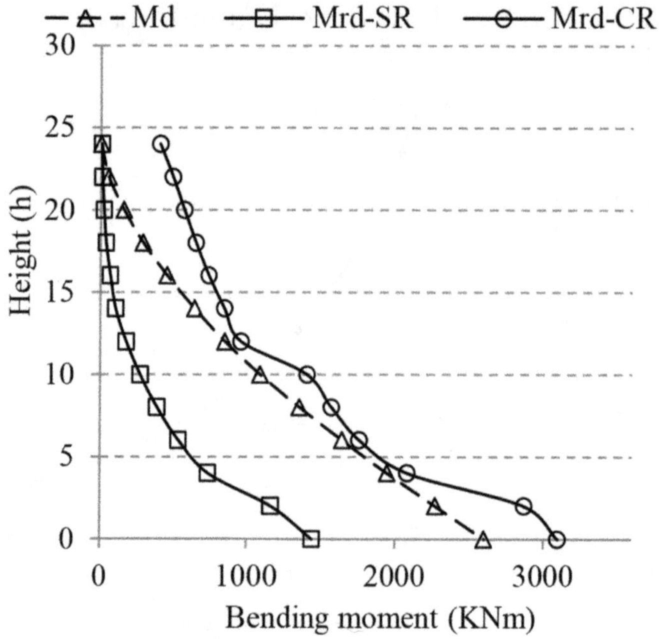

Figure 7: Linear analysis results: Computed bending moment diagrams (Md); Moment resistance without reinforcement (Mrd-Sr); Moment resistance with reinforcement (Mrd-CR).

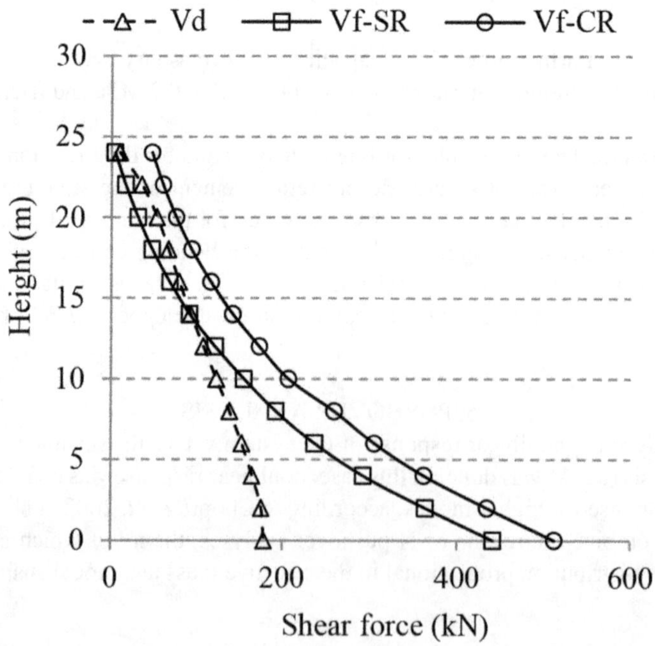

Figure 8: Linear analysis results: Computed shear force diagram (Vd); Shear resistance without reinforcement (Vf-SR); Shear resistance with reinforcement (Vf-CR).

Figure 9 shows the capacity curve for reinforced and unreinforced first mode of vibration. The collapse point (CP), shows the maximum displacement, before the structure collapse due to the second-order effects. This failure mode is characterized by the cracking of the upper zone of the stack, Fig. 10, reaching the minimum value of the compression area, which prevents the collapse of the structure. Moreover, this failure mode shows that the structure has a small resistance capacity as cracked structure. The maximum displacement is seven times greater than the elastic displacement.

Furthermore, Fig. 9 shows that the stiffness of the cracked chimney and the maximum displacement for the reinforced structure are bigger than the unreinforced structure. In particular, the cracked stiffness of the reinforced structure is 1.45 times higher than unreinforced chimney. Moreover, reinforced structure shows a new crack pattern compared with unreinforced structure, Fig. 10. This new pattern has much more cracks at the bottom of the stack.

Figure 11, shows the relationship between capacity and demand curve, for the reinforced structure in the ADSR system. In this case, the performance point (PP) is 0.15 m and 63.28 kN, Fig. 9, in the real capacity curve. The first result of the Fig. 11 is that the period of the cracked structure, $T_{1NL,}$ is three times bigger than the elastic period for linear analysis, T_{1L}. Moreover, the base shear demand for nonlinear analysis is equal to 30% of the base shear for linear analysis.

Finally, Equation 2 allows to analyze the real effect of the highest modes of vibration. In particular, for lineal analysis, the maximum displacement for modes 2 and 4 are 3.59 mm and 0.74 mm. In other words, the higher mode contribution to maximum displacement of the structure is less than 10% of the full displacement. For this reason, if the ultimate limit state, according to Eurocode 8, depends on the maximum displacement of the structure, the effect

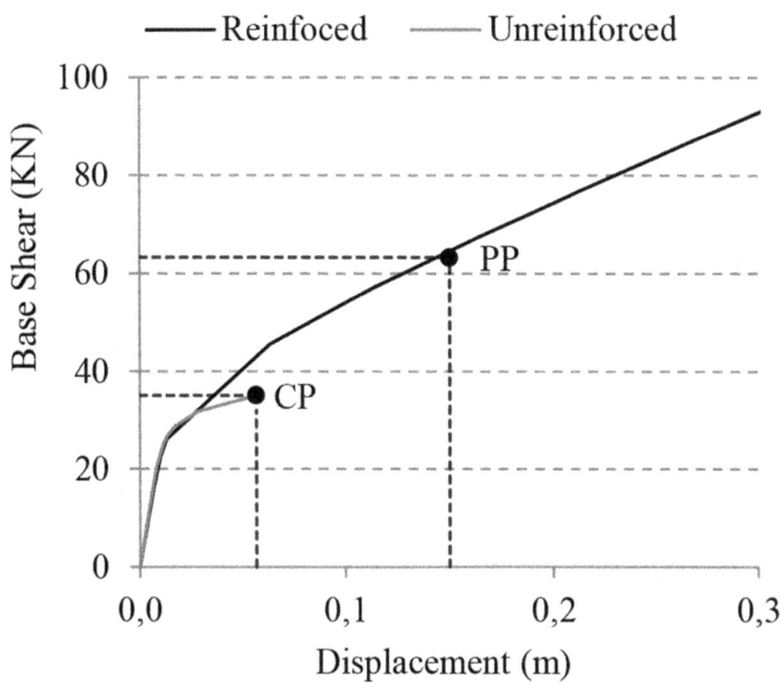

Figure 9: Nonlinear capacity curve for the mode 1.

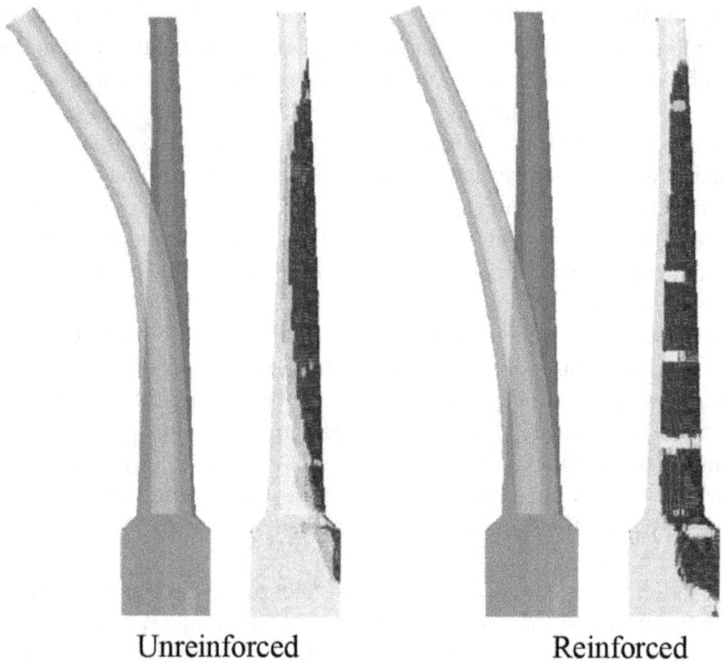

Figure 10: Cracking patterns for failure of pushover mode 1.

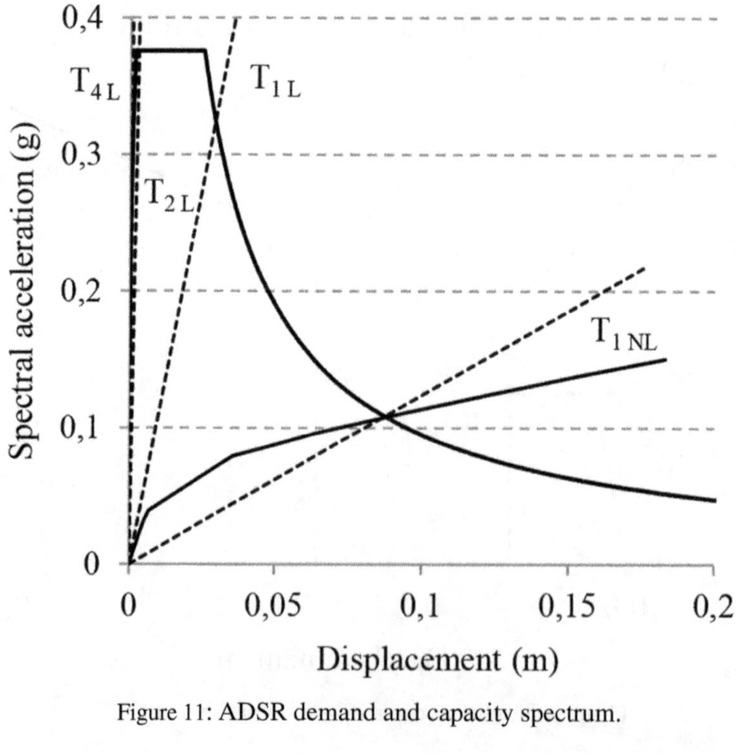

Figure 11: ADSR demand and capacity spectrum.

of highest modes to the full displacement is less important than the first mode contribution in nonlinear analysis.

$$d_n = \frac{a_n}{w_n^2} \tag{2}$$

where d_n = modal displacement; a_n = spectral acceleration.; and $w^2{}_n$ = natural frequency for mode n.

6 CONCLUSIONS

In this paper, a reinforcement scheme with composite materials was proposed for a 24.5m-high masonry chimney. This reinforcement comprised an internal textile reinforced mortar layer and local reinforcement with longitudinal carbon fiber bands. The results showed that the reinforcement prevents the structural collapse. Moreover, response spectrum analysis shows more seismic demand than pushover analysis, due to the effect of energy loss by cracking of the masonry chimney.

ACKNOWLEDGMENT

The authors express deep gratitude to Ministerio de Economia y Competitividad of the Spain's Government and MAPEI. This work was financed by them by means of the BIA2015-69952R research Project.

REFERENCES

[1] Magliulo, G., Ercolino, M., Petrone, C., Coppola, O. & Manfredi, G., The Emilia earthquake: seismic performance of precast reinforced concrete buildings. *Earthquake Spectra*, **30**(2), pp. 891–912, 2014. https://doi.org/10.1193/091012eqs285m

[2] Huang, W. & Gould, P.L., 3-D pushover analysis of a collapsed reinforced concrete chimney. *Finite Elements in Analysis and Design*, **43**(11–12), pp. 879–887, 2007. https://doi.org/10.1016/j.finel.2007.05.005

[3] Pallarés, F.J., Ivorra, S., Pallarés, L. & Adam, J.M., State of the art of industrial masonry chimneys: a review from construction to strengthening. *Construction and Building Materials*, **25**(12), 4351–4361, 2011. https://doi.org/10.1016/j.conbuildmat.2011.02.004

[4] Lourenço, P.B., Recommendations for restoration of ancient buildings and the survival of a masonry chimney. *Construction and Building Materials*, **20**(4), pp. 239–251, 2006. https://doi.org/10.1016/j.conbuildmat.2005.08.026

[5] Foti, D., Dynamic identification techniques to numerically detect the structural damage. *The Open Construction and Building Technology Journal*, **7**(1), pp. 43–50, 2013. https://doi.org/10.2174/1874836801307010043

[6] Peña, F., Lourenço, P.B., Mendes, N. & Oliveira, D.V., Numerical models for the seismic assessment of an old masonry tower. *Engineering Structures*, **32**(5), pp. 1466–1478, 2010. https://doi.org/10.1016/j.engstruct.2010.01.027

[7] Aoki, T. & Sabia, D., Structural characterization of brick chimney by experimental tests and numerical model updating. *Masonry International*, **19**, pp. 41–52, 2006.

[8] Pallarés, F.J., Ivorra, S., Pallarés, L. & Adam, J.M., Seismic assessment of a CFRP-strengthened masonry chimney. *Proceedings of the Institution of Civil Engineers - Structures and Buildings*, **162**(5), pp. 291–299, 2009. https://doi.org/10.1680/stbu.2009.162.5.291

[9] European Commitee for Standardization, Eurocode 8: Design of structures for earthquake resistance - Part 1 : General rules, seismic actions and rules for buildings. *European Committee for Standardization*, **1**(1), 2004.
[10] Ministerio de Fomento. Gobierno de España, "Norma de construcción sismorresistente: Parte general y edificación (NCSE-02)," *Real Decreto 997/2002, de 27 de Septiembre de 2002*. 2002.
[11] Chopra, A.K. & Goel, R.K., A modal pushover analysis procedure for estimating seismic demands for buildings. *Earthquake Engineering & Structural Dynamics*, **31**(3), pp. 561–582, 2002.
https://doi.org/10.1002/eqe.144
[12] Magalhães, A. & Veiga, R., Physical and mechanical characterisation of historic mortars. Application to the evaluation of the state of conservation [Caracterización física y mecánica de los morteros antiguos. Aplicación a la evaluación del estado de conservación]. *Materiales de Construcción*, **59**(295), pp. 61–77, 2009.
https://doi.org/10.3989/mc.2009.41907
[13] C. E. De Normalisation, "EN 1996-1-1:1995 Eurocode 6 - Design of masonry structures - Part 1-1: General rules for reinforced and unreinforced masonry structures," *Management*, pp. 1–123, 2005.
[14] CSI, "SAP2000. Analysis Reference Manual," *CSI: Berkeley (CA, USA): Computers and Structures INC*. p. 496, 2016.

A DIRECT PROCEDURE FOR THE SEISMIC DESIGN OF FRAME STRUCTURES WITH ADDED VISCOUS DAMPERS

STEFANO SILVESTRI, MICHELE PALERMO & TOMASO TROMBETTI
University of Bologna, Italy.

ABSTRACT

A direct procedure for the seismic design of building structures with added viscous dampers is described in this paper. The procedure is applicable to regular multi-storey frame structures, which are characterized by a period of vibration lower than 1.5 s. It aims at providing practical tools for a direct identification of the mechanical characteristics of the manufactured viscous dampers, which allow to achieve target levels of performances. Typically, the design philosophy is to limit the structural damages under severe earthquakes. In more detail, the procedure may be summarized as follows. First, a target damping reduction factor is selected to achieve a desired reduction in the peak structural response under earthquake excitation. Second, linear damping coefficients are calculated taking advantage of the properties of the modal damping ratios of classically damped systems. Then, analytical formulas allow the estimation of peak velocities and forces in the dissipative devices, and an energy criterion is used to identify the non-linear mechanical characteristics of the actual manufactured viscous dampers. Finally, the internal actions in the structural elements are estimated through the envelope of two equivalent static analyses (ESA), namely: ESA1 in which the naked structure is subjected to a given set of equivalent lateral forces, and ESA2 in which the structure, with rigid diagonal braces substituting the added viscous dampers, is subjected to a top floor lateral force. At this stage of the research, the procedure is suitable for the preliminary design phase, since correction factors for the higher modes contributions are necessary to improve its accuracy, especially for high-rise buildings. A numerical verification of the final behaviour of the system by means of non-linear time-history analyses is recommended. An applicative example is finally provided to highlight the simplicity of the proposed procedure.

Keywords: damping coefficient, five-step procedure, multi-storey frames, seismic design, viscous dampers

1 INTRODUCTION

For many years the seismic analysis and design of buildings have been carried out using methods based essentially on the concept of equivalent lateral forces. Nowadays, analysis and design procedures are mostly based on the use of non-linear dynamic analyses, which are available in most of common commercial software. The use of dynamic analyses was first introduced in the 1974 by the SEAOC Code [1] for major structures 'with highly irregular shapes, large differences in lateral resistance or stiffness between adjacent storeys'.

As a consequence, when energy dissipation and base isolation systems were first proposed for the mitigation of the seismic effects on the structural elements (1980s), the use of dynamic analyses was already well established as standard practice for the seismic design of building structures. At this point, according to most actual seismic codes (such as the Eurocode 8 [2] or the current Italian Code [3]), structures equipped with added viscous dampers can be analyzed and designed only by means of non-linear dynamic analyses. Indeed, only U.S. building codes (such as the ASCE 7-10 standard [4]) contain specific simplified procedures for analysis and design of buildings with passive energy dissipation systems. These procedures are grounded on the seminal research works carried out in the 1990s at the University at Buffalo [5–8] and summarized in the MCEER-01 report [9]. Nonetheless, none of these well-established procedures have been yet incorporated in Eurocode provisions.

DOI: 10.2495/SAFE-V7-N4-498-509

In the present work, a simplified procedure for the preliminary design of viscous dampers and structural elements of frame structures equipped with inter-storey viscous dampers is presented and exemplified for a three-storey frame building.

2 THE 'DIRECT FIVE-STEP PROCEDURE'

2.1 Procedure overview

A 'direct five-step procedure', synthetically schematized in the flow chart of Fig. 1a, is here described. It guides the professional engineer through the dimensioning of the non-linear viscous dampers to be inserted in the frame and the design of the structural members so that a given performance objective is achieved. It integrates some results of previous research works developed by the authors during the last 10 years [10–14]. Even though the procedure is also applicable to yielding frame structures (with an appropriate choice of the overall behaviour factor, see Ref. [13] for details), it is here presented assuming that the frames are designed in order to remain in the elastic phase. Thus, the design philosophy is to limit the structural damages under severe earthquakes.

In its current version, the 'direct five-step procedure' is applicable to regular multi-storey frame structures and to regular prefabricated pendular structures characterized by a period of vibration lower than 1.5 s. For larger periods, correction factors for the higher modes contributions are necessary to improve its accuracy, especially for high-rise buildings. For structures characterised by uniform mass and stiffness along the height and characterised by a period lower than 0.5 s, the procedure leads to conservative estimations of the internal actions in the structural members.

With reference to the seismic response along a given direction of an *N*-storey frame structure with uniform distribution of added viscous dampers along the height of the building, the steps of the procedures can be summarized as follows:

STEP 1: Identification of the performance objective, in terms of: (a) the desired x % reduction of the base shear due to the presence of the added viscous damper expressed in terms of damping reduction factor: $\eta = 1 - x/100$; (b) the equivalent damping ratio $\xi = 10/\eta^2 - 5$ (%),

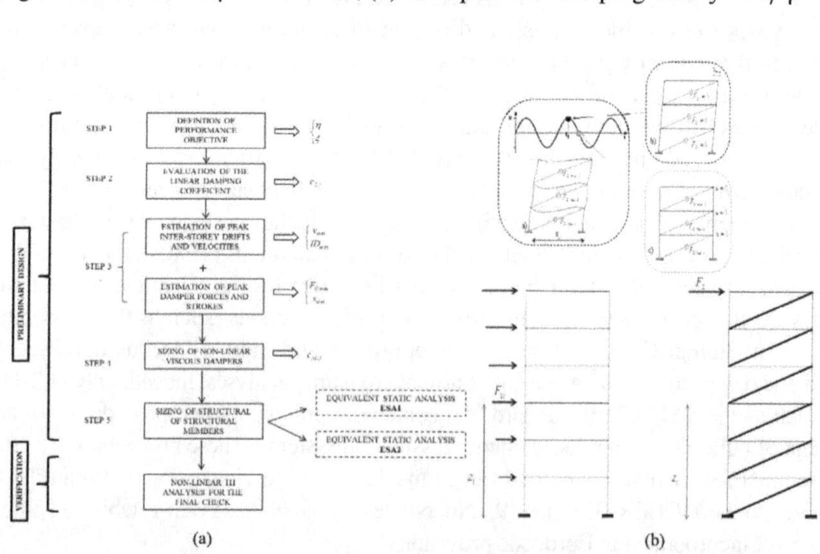

(a) (b)

Figure 1: (a) Flowchart of the proposed design procedure. (b) ESA1 and ESA2 analyses.

where $\xi = \xi_i + \xi_v$ is the sum of the inherent damping (ξ_i, conventionally set equal to 5 %) and the viscous damping provided by the added dampers (ξ_v).

STEP 2: Evaluation of the linear damping coefficient (c_L) of the single equivalent viscous damper characterised by a linear force-velocity relationship of the type $F_d = c_L \cdot sign(v) \cdot |v|$ (see derivation in Ref. [11]):

$$c_L = \xi \cdot \omega_1 \cdot m_{tot} \cdot \left(\frac{N+1}{n}\right) \cdot \frac{1}{\cos^2 \theta} \tag{1}$$

where ω_1 is the first mode circular frequency of the structure, m_{tot} is the total building mass, N is the total number of storeys of the building, n is the total number of dampers placed at each storey in each direction, and θ indicates the damper inclination with respect to the horizontal line.

STEP 3: Estimations of the peak damper velocity v_{max}, the peak inter-storey drift ID_{max}, the peak damper force $F_{d,max}$ and the peak damper stroke s_{max} under the design earthquake ($S_e(T_1,\eta)$ is the ordinate of the damped pseudo-acceleration elastic response spectrum at the fundamental period of the structure considering the effect of the dampers through factor η) assuming a linear first-mode shape (see derivation in Ref. [14]):

$$v_{max} = \frac{S_e(T_1,\eta)}{\omega_1} \cdot \frac{2}{N+1} \cdot \cos\theta \tag{2}$$

$$ID_{max} = \frac{S_e(T_1,\eta)}{\omega_1^2} \cdot \frac{2}{N+1} \tag{3}$$

$$F_{d,max} = 2 \cdot \xi \cdot m_{tot} \cdot \frac{S_e(T_1,\eta)}{\cos\theta} \tag{4}$$

$$s_{max} = \frac{S_e(T_1,\eta)}{\omega_1^2} \cdot \frac{2}{N+1} \cdot \cos\theta \tag{5}$$

STEP 4: Sizing of the 'non-linear' damping coefficient (c_{NL}) of the commercial damper characterized by a non-linear force–velocity relationship of the type $F_d = c_{NL} \cdot sign(v) \cdot |v|^{\alpha}$ where the α exponent is typically around 0.15 (see derivation in Ref. [11]):

$$c_{NL} = c_L \cdot (0.8 \cdot v_{max})^{1-\alpha} \tag{6}$$

In order to maintain high efficiency of the device, the axial stiffness (k_{axial}) of the dissipative brace(the stiffness due to the compressibility of the oil in the chamber of the damper in series with the stiffness of the supporting brace) should satisfy the following indication [11]:

$$k_{axial} \geq 10 \cdot c_L \cdot \omega_1 \tag{7}$$

STEP 5: Estimation of the internal actions in the structural elements through the envelope of two Equivalent Static Analyses (ESA, Fig. 1b), namely:

- Equivalent Static Analysis 1 (ESA1): the naked structure (e.g. the structure without the added viscous dampers) is subjected to the following set of equivalent lateral forces, with F_i indicating the lateral force to apply at the i-th floor:

$$F_i = m_{tot} \cdot S_e(T_1,\eta) \cdot \frac{z_i \cdot W_i}{\sum_{j=1}^{N} z_j \cdot W_j} \tag{8}$$

where W_i is the seismic weight of the i-th floor; the distribution of the lateral forces relates directly to the assumed fist-mode shape.

- Equivalent Static Analysis 2 (ESA2): the structure with rigid diagonal braces substituting the added viscous dampers is subjected to a top-storey lateral force:

$$F_{top-storey} = 0.8^{1-\alpha} \cdot 2 \cdot \xi \cdot m_{tot} \cdot \frac{S_e(T_1,\eta)}{n} \tag{9}$$

The application of a lateral force at the roof level only relates to the assumed first-mode shape and leads to the following estimation of the maximum axial force in the columns:

$$P_{i,\max} = (N - i + 1) \cdot 0.8^{1-\alpha} \cdot 2 \cdot \xi \cdot m_{tot} \cdot \frac{S_e(T_1,\eta)}{n} \cdot \tan\theta \tag{10}$$

Finally, non-linear time-history analyses are recommended to verify the actual behaviour of the structure and to bring little adjustments to the size of dampers and structural elements.

2.2 The rationale behind the Equivalent Static Analyses

During an earthquake, the internal actions in the structural members (such as columns and beams) achieve their maximum values at the instant of maximum lateral displacement. On the other hand, the damper forces are maximized at the instant of maximum inter-storey velocity (approximately coincident with the instant of zero lateral deformation). In case of diagonal dampers placed between two adjacent storeys (inter-storey placement), the forces exerted by the dampers transfer additional axial forces in columns and beams, which in some cases (see section 3) may govern the structural member sizing. For this reason, the maximum internal actions in the structural members can be estimated from the envelope of two following two equivalent static analyses:

- ESA1 is the static analysis of the naked structure subjected to a set of lateral forces producing the same lateral displacements as those developed at the instant of maximum lateral deformation.
- ESA2 is the static analysis of an appropriate structure schematization (e.g. the structure with diagonal rigid bracings replacing the diagonal viscous dampers) which provides the same axial forces in the structural members as those developed at the instant of maximum lateral velocity.

3 DISCUSSION

Figure 2 displays the base shear V_{base}, the maximum damper force $F_{d,max}$ and the maximum axial force at the ground floor $P_{base} = P_{1,max}$ as a function of the damping ratio, as obtained according to the predictive formulas of the direct five-step procedure (the cases of a 5-storey and 10-storey buildings are considered). The curves are normalized with respect to the base shear of the naked structure ($V_{base,\xi=5\%}$). As expected, with increasing damping ratio, the base shear V_{base} decreases at the expense of an increase in $F_{d,max}$ and P_{base}. It can be noted that the curves of V_{base} (blue curve) and $F_{d,max}$ (red curve) intersect at a ξ value of around 0.35. With increasing the total number of storeys, P_{base} increases as well and may become significantly large (for the 10-storey building even 4 times larger than $V_{base,\xi=5\%}$ for $\xi \cong 0.35$).

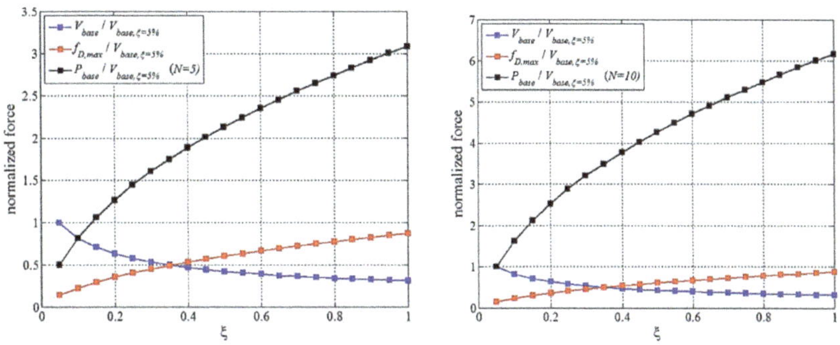

(a) (b)

Figure 2: Base shear, maximum damper force and base axial force: (a) N=5, (b) N=10.

It is of practical interest to evaluate the necessary amount of damper forces to obtain a target reduction in the base shear ($\Delta V_{base} = V_{base,\xi=5\%} - V_{base}$). By making use of eqn (4), it is possible to express directly the normalized versions of the maximum damper force as a function of the damping reduction factor:

$$\frac{F_{d,\max}}{V_{base,\xi=5\%}} = 2\frac{\left(10-5\eta^2\right)}{\eta \cdot \cos\theta} \tag{11}$$

$$\frac{F_{d,\max}}{\Delta V_{base}} = 2\frac{\left(10-5\eta^2\right)}{\left(1-\eta\right)\eta \cos\theta} \tag{12}$$

Figure 3 displays $\frac{F_{d,\max}}{V_{base,\xi=5\%}}$ versus η, and $\frac{F_{d,\max}}{\Delta V_{base}}$ versus 1-η, respectively (for the case of $\theta = 45°$). In detail, Fig. 3a illustrates the cost (in terms of maximum damper force normalized with respect to $V_{base,\xi=5\%}$) of achieving a prescribed performance (in terms of damping reduction factor η). Figure 3b illustrates the cost/benefit ratio (i.e. maximum damper force

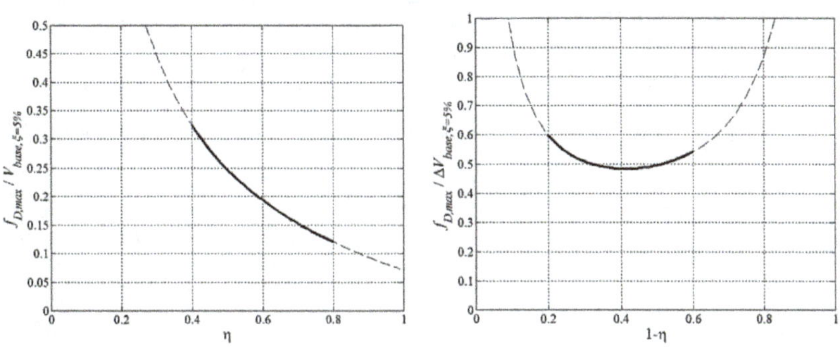

(a) (b)

Figure 3: (a) $\frac{F_{d,\max}}{V_{base,\xi=5\%}}$ vs. η; (b) $\frac{F_{d,\max}}{\Delta V_{base}}$vs. 1-$\eta$.

normalized with respect to ΔV_{base}) corresponding to a prescribed reduction in the response parameter (i.e. 1-η). For $1-\eta$ between 0.15 and 0.7 (e.g. η between 0.3 and 0.85, covering the range of added damping ratios between 10-30%), the ratio $\frac{F_{d,\max}}{\Delta V_{base}}$ is less than 1.0, which means that the benefit in terms of reduction of base shear is superior than the cost expressed by the maximum damper force.

4 APPLICATIVE EXAMPLE

The applicative example is referred to a 3-storey school building located in Bisignano (CS) in the Calabria region (Southern Italy). The reinforced-concrete frame structure has a 21.4 m x 15 m rectangular plan (Fig. 4a). The total height of the building is 9.9 m and the three inter-storey heights are around 3.3 m (Fig. 4b). The first two floors are school areas where people may congregate (live load 3 kN/m^2), while the third level represents an impracticable low attic and the roof (live and snow loads to to be not combined with the earthquake action, i.e. $\Psi_2 = 0$). The regular structural mesh is composed of four main frames placed along the longitudinal direction, each one characterized by five columns. The columns have 50 cm × 40 cm cross-section, whilst the longitudinal beams have 40 cm × 60 cm cross-section at each level. The main longitudinal frames are connected along the perimeter by 50 cm × 40 cm transversal beams.

Experimental tests on the materials have been carried out. The concrete is characterized by an average cubic compression strength around R_{cm} = 24.6 MPa, by a secant elastic modulus equal to $E_{c,sec}$ = 25000 MPa, and density mass of about 2500 kg/m^3. The average yielding strength of the reinforcement bar steel is equal to f_{ym} = 315 MPa.

Table 1 presents the load analysis for each floor. The masonry infills weight has been estimated equal to 4.00 kN/m^2. The total weight of the building in seismic conditions is equal to W_{tot} = 11900 kN.

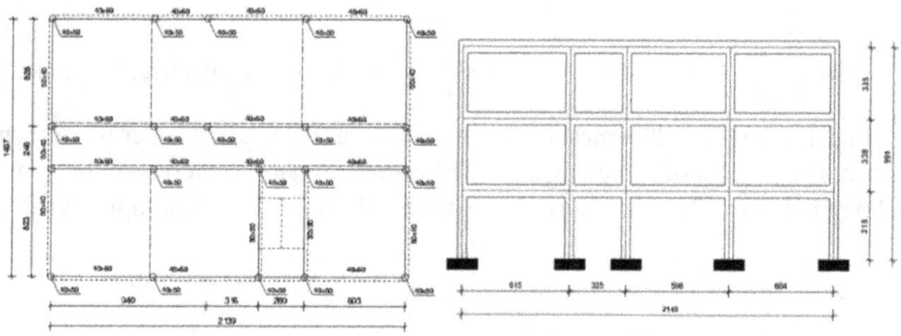

(a) (b)

Figure 4: (a) Building plan. (b) Longitudinal section of the structure.

Table 1: Load analysis.

Loads	*Floor 1*	*Floor 2*	*Floor 3 (attic + roof)*
Permanent *G1*	3.00 kN/m^2	3.00 kN/m^2	4.00 kN/m^2
Permanent *G2*	2.00 kN/m^2	2.00 kN/m^2	3.00 kN/m^2
Imposed Loads *Q*	3.00 kN/m^2 (Ψ_2= 0.6)	3.00 kN/m^2 (Ψ_2= 0.6)	2.50 kN/m^2 (Ψ_2= 0)
TOTAL in static conditions	8.00 kN/m^2	8.00 kN/m^2	9.50 kN/m^2
TOTAL in seismic conditions	6.80 kN/m^2	6.80 kN/m^2	7.00 kN/m^2

The frame structure is not capable of supporting the design 712 years-return period peak ground acceleration ($a_g \cdot S = 0.323g \cdot 1.23 = 0.40g$) of the Italian Code [3] provided for school buildings in the site of Bisignano. For the non-linear dynamic analyses, seven artificial accelerograms have been generated so as to match the elastic response spectrum shown in Fig. 5.

The three-dimensional Finite Element (FE) model of the structure is shown in Fig. 6.

Inter-storey viscous dampers have been added along the two main directions of the structure with the aim of keeping the structural elements (columns and beams) within the elastic phase by increasing the structural dissipative properties. In particular, the dampers have been positioned along the perimetrical frames in correspondence of the four corners of the building to reduce the plan rotational response and to obtain a symmetric configuration, as illustrated in Fig. 7.

The modal analysis of the FE model leads to the following results. Due to the absence of transversal beams, the first mode (fundamental mode along the transversal direction) is characterized by a period of vibration significantly larger (0.80 s) than the one (0.45 s) of the third mode (fundamental mode along the longitudinal direction). The second mode is characterized by a period equal to 0.52 s and is purely rotational. The first three modes excite more than 85% of the total mass of the building along both directions.

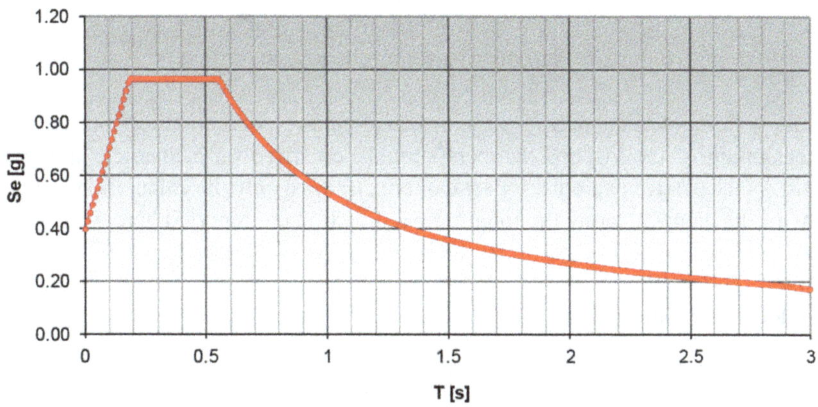

Figure 5: The horizontal pseudo-acceleration elastic response spectrum.

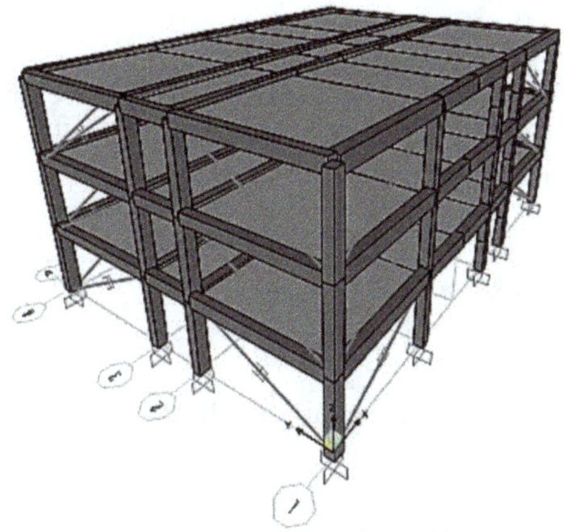

Figure 6: The FE model of the structure.

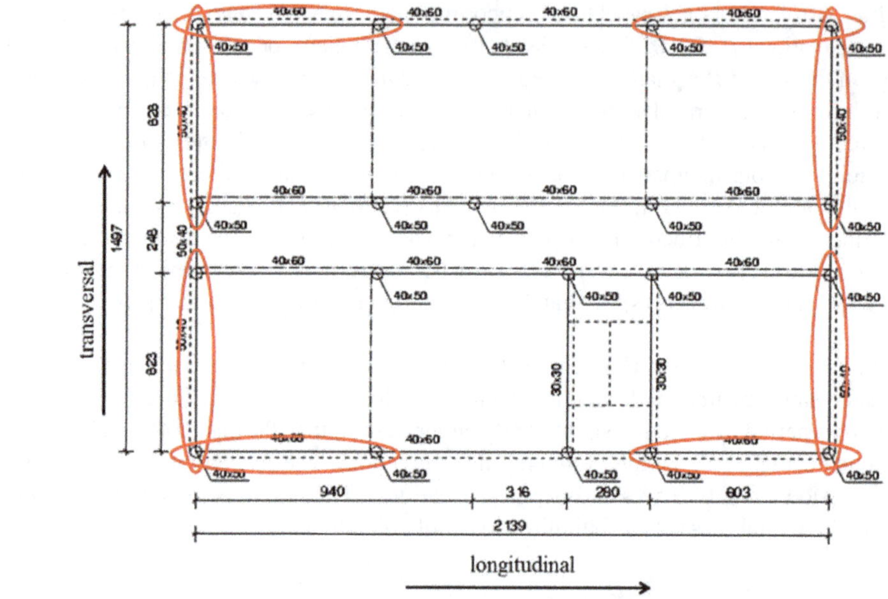

Figure 7: Plan distribution of the viscous dampers.

Since the difference in the periods of vibration along the two main directions is significant, the dimensioning of the viscous dampers should be carried out separately along the two directions. As illustrative example, for sake of conciseness, only the calculations necessary to the sizing of the dampers along the longitudinal direction are reported hereafter.

STEP 1

Assumed target damping ratio: $\xi = 0.30$

Corresponding damping reduction factor: $\eta = \sqrt{\dfrac{10}{5+\xi}} = \sqrt{\dfrac{10}{5+30}} = 0.53$

Fundamental period along the considered (longitudinal) direction: $T_1 = 0.45$ s

Spectral acceleration [3]: $S_e(T_1,\eta) = a_g \cdot S \cdot \eta \cdot F_o = 0.323g \cdot 1.23 \cdot 0.53 \cdot 2.43 = 0.52g$

STEP 2

Number of dampers per floor placed along the longitudinal direction: $n = 4$

Damper inclination with respect to the horizontal line: $\theta = 27°$

Linear damping coefficient, as per eqn (1):

$$c_L = \xi \cdot \omega_1 \cdot \frac{W_{tot}}{g} \cdot \left(\frac{N+1}{n}\right) \frac{1}{\cos^2\theta} = 0.30 \cdot \frac{2\pi}{0.45\text{s}} \cdot \frac{11900\ \text{kN}}{9.81\ \dfrac{\text{m}}{\text{s}^2}} \cdot \left(\frac{3+1}{4}\right) \cdot \frac{1}{\cos^2 27°} \cong 6400 \frac{\text{kN}\cdot\text{s}}{\text{m}}$$

STEP 3

Peak damper velocity estimation for the equivalent linear damper, as per eqn (2):

$$v_{\max} = \frac{S_e(T_1,\eta)}{\omega_1} \cdot \frac{2}{N+1} \cdot \cos\theta = \frac{0.52 \cdot 9.81\ \dfrac{\text{m}}{\text{s}^2}}{\left(\dfrac{2\pi}{0.45\text{s}}\right)} \cdot \frac{2}{3+1} \cdot \cos\theta \cong 0.16 \frac{\text{m}}{\text{s}}$$

Peak damper force estimation for the equivalent linear damper, as per eqn (4):

$$F_{d,\max} = 2 \cdot \xi \cdot \frac{W}{g} \cdot \frac{S_e(T_1,\eta)}{n \cdot \cos\theta} = 2 \cdot 0.30 \cdot \frac{11900 \text{ kN}}{g} \cdot \frac{(0.323g \cdot 1.23 \cdot 0.53 \cdot 2.43)}{4 \cdot \cos 27°} = 1025 \text{ kN}$$

Peak damper stroke estimation for the equivalent linear damper, as per eqn (5):

$$s_{\max} = \frac{S_e(T_1,\eta)}{\omega_1^2} \cdot \frac{2}{N+1} \cdot \cos\theta = \frac{\left(0.323 \cdot 981 \frac{\text{cm}}{\text{s}^2} \cdot 1.23 \cdot 0.53 \cdot 2.43\right)}{\left(\frac{2\pi}{0.45\text{s}}\right)^2} \cdot \frac{2}{3+1} \cdot \cos 27° \cong 1.15 \text{ cm}$$

STEP 4

α-exponent of the commercial damper: $\alpha = 0.15$

Non-linear damping coefficient of the commercial damper, as per eqn (6):

$$c_{NL} = c_L \cdot (0.8 \cdot v_{\max})^{1-\alpha} = 6400 \frac{\text{kN} \cdot \text{s}}{\text{m}} \cdot \left(0.8 \cdot 0.16 \frac{\text{m}}{s}\right)^{1-0.15} \cong 1115 \ \frac{\text{kN} \cdot \text{s}^{0.15}}{\text{m}^{0.15}}$$

Minimum axial stiffness of the device (non-linear damper + supporting brace), as per eqn (7):

$$k_{axial} \geq 10 \cdot c_L \cdot \omega_1 = 10 \cdot 6400 \frac{\text{kN} \cdot \text{s}}{\text{m}} \cdot \frac{2\pi}{0.45\text{s}} = 8.93 \cdot 10^5 \frac{\text{kN}}{\text{m}}$$

Peak damper force estimation for the 'non-linear' damper:

$$F_{d,\max,NL} = 0.8^{1-\alpha} \cdot F_{d,\max} = 0.8^{1-0.15} \cdot 1025 \text{ kN} \cong 848 \text{ kN}$$

STEP 5

ESA1 analysis:

Lateral forces to be applied at each floor for the whole structure, as per eqn (8):

$$F_h = \frac{W_{tot}}{g} \cdot S_e(T_1,\eta) = \frac{11900 \text{ kN}}{g} \cdot (0.323g \cdot 1.23 \cdot 0.53 \cdot 2.43) \cong 6090 \text{ kN}$$

$$F_1 = F_h \cdot \frac{z_1 \cdot W_1}{\sum_{j=1,2,\ldots,N} z_j \cdot W_j} = 6090 \text{ kN} \cdot \frac{3.18 \text{ m} \cdot 3928 \text{ kN}}{(3.18 \text{ m} \cdot 3928 \text{ kN} + 6.56 \text{ m} \cdot 3928 \text{ kN} + 9.91 \text{ m} \cdot 4044 \text{ kN})}$$

$$\cong 6090 \text{ kN} \cdot 0.16 = 975 \text{ kN}$$

$$F_2 = F_h \cdot \frac{z_2 \cdot W_2}{\sum_{j=1,2,\ldots,N} z_j \cdot W_j} = 6090 \text{ kN} \cdot \frac{6.56 \text{ m} \cdot 3928 \text{ kN}}{(3.18 \text{ m} \cdot 3928 \text{ kN} + 6.56 \text{ m} \cdot 3928 \text{ kN} + 9.91 \text{ m} \cdot 4044 \text{ kN})}$$

$$\cong 6090 \text{ kN} \cdot 0.33 = 2016 \text{ kN}$$

$$F_3 = F_h \cdot \frac{z_3 \cdot W_3}{\sum_{j=1,2,\ldots,N} z_j \cdot W_j} = 6090 \text{ kN} \cdot \frac{9.91 \text{ m} \cdot 4044 \text{ kN}}{(3.18 \text{ m} \cdot 3928 \text{ kN} + 6.56 \text{ m} \cdot 3928 \text{ kN} + 9.91 \text{ m} \cdot 4044 \text{ kN})}$$

$$\cong 6090 \text{ kN} \cdot 0.51 = 3116 \text{ kN}$$

Assuming that each floor is infinitely stiff in its own plane and considering that the four longitudinal frames are equal to each other, the storey forces are evenly distributed along the four frames. Thus, with reference to a single longitudinal frame, Fig. 8 illustrates the static scheme to be solved for ESA1 to obtain the maximum bending moments and shear forces in the structural elements (columns and beams).

ESA2 analysis:

Lateral force to be applied at the roof level for each single frame in which the dampers are present, as per eqn (9):

$$F_{top-storey} = 0.8^{1-\alpha} \cdot 2 \cdot \xi \cdot m_{tot} \cdot \frac{S_e(T_1,\eta)}{n} =$$

$$= 0.8^{1-0.15} \cdot 2 \cdot 0.30 \cdot \frac{11900\ \text{kN}}{g} \cdot \frac{(0.323 \cdot g \cdot 1.23 \cdot 0.53 \cdot 2.43)}{4} = 756\ \text{kN}$$

With reference to one of the two perimetrical frames, Fig. 9 illustrates the static scheme (in which dampers are replaced by rigid braces) to be solved for ESA2 to obtain the maximum axial forces in the columns.

Estimated maximum axial force at the base (i=1) of columns B and D, as per eqn (10):

$$P_{base} = P_{1,\max} = N \cdot 0.8^{1-\alpha} \cdot 2 \cdot \xi \cdot m_{tot} \cdot \frac{S_e(T_1,\eta)}{n} \cdot \tan\theta =$$

$$= 3 \cdot 0.8^{1-0.15} \cdot 2 \cdot 0.30 \cdot \frac{11900\ \text{kN}}{g} \cdot \frac{(0.323 \cdot g \cdot 1.23 \cdot 0.53 \cdot 2.43)}{4} \cdot \tan 27° = 1155\ \text{kN}$$

It is recommended that, after this preliminary dimensioning of the viscous dampers and the structural elements, with the aim of verifying the obtained seismic performances (e.g. the achieved value of η) and of bringing the convenient adjustments, non-linear time-history dynamic analyses are performed to check the actual seismic behaviour of the structure under earthquake ground motions.

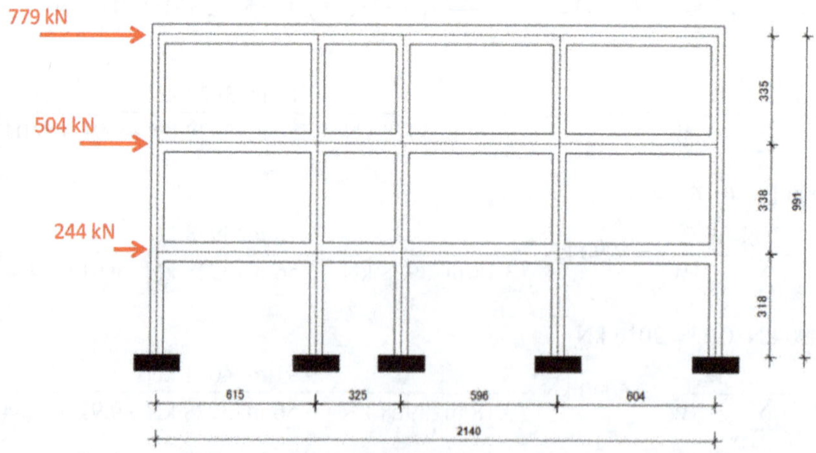

Figure 8: Static scheme to be solved for ESA1.

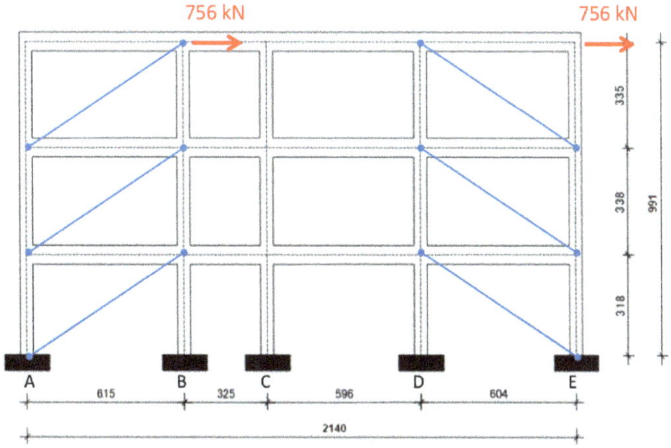

Figure 9: Static scheme to be solved for ESA1.

5 CONCLUSIONS

A direct design procedure for frame buildings equipped with inter-storey viscous dampers has been presented. The procedure is aimed at guiding the structural engineer from the choice and sizing of the added viscous dampers to the dimensioning of the structural elements. It allows to obtain analytical estimations/predictions of peak displacements, peak inter-storey drifts and velocities, maximum forces in the dampers and maximum internal actions in the structural elements. Although the procedure can be further improved through the introduction of appropriate correction coefficients accounting for the higher modes contribution, it is simple to apply and produces results of sufficient accuracy for the purpose of preliminary design of regular moment-resisting frames (characterised by period of vibration lower than 1.5 s).

ACKNOWLEDGEMENTS

Financial supports of Department of Civil Protection (DPC-Reluis 2014–2018 Grant—Research line 6: 'Seismic isolation and dissipation') is gratefully acknowledged.

REFERENCES

[1] Seismology Committee, *Recommended lateral force requirements and commentary, structural engineers association of California*, San Francisco, 1974.

[2] CEN, Eurocode 8. *Design of structures for earthquake resistance – Part 1: general rules*. Seismic Actions and Rules for Buildings, Brussels, 2003.

[3] Norme Tecniche per le Costruzioni, Italian building code, adopted with D.M. 14/01/2008, published on S.O. n. 30 G.U. n. 29 04/02/2008, 2008.

[4] American Society of Civil Engineers ASCE. *Minimum design loads for buildings and other structures*, ASCE 7–10. Reston, VA, 2010.

[5] Constantinou, M.C. & Symans, M.D., Seismic response of structures with supplemental damping. *Structural Design of Tall Buildings*, **2**, pp. 77–92, 1993. https://doi.org/10.1002/tal.4320020202

[6] Ramirez, O.M., Constantinou, M.C., Whittaker, A.S., Kircher, C.A. & Chrysostomou C.Z., Elastic and inelastic seismic response of buildings with damping systems. *Earthquake Spectra*, **18**(3), pp. 531–547, 2002. https://doi.org/10.1193/1.1509762

[7] Ramirez, O.M., Constantinou, M.C., Whittaker, A.S., Kircher, C.A., Johnson, M.W. & Chrysostomou, C.Z., Validation of the 2000 NEHRP provisions' equivalent lateral force and modal analysis procedures for buildings with damping systems. *Earthquake Spectra*, **19**(4), pp. 981–999, 2003.
https://doi.org/10.1193/1.1622392

[8] Whittaker, A.S., Constantinou, M.C., Ramirez, O.M., Johnson, M.W. & Chrysostomou, C.Z., Equivalent lateral force and modal analysis procedures of the 2000 NEHRP Provisions for buildings with damping systems. *Earthquake Spectra*, **19**(4), pp. 959–980, 2003.
https://doi.org/10.1193/1.1622391

[9] Ramirez, O.M., Constantinou, M.C., Kircher, C.A., Whittaker, A.S., Johnson, M.W., Gomez, J.D. & Chrysostomou, C.Z., *Development and evaluation of simplified procedures for analysis and design of buildings with passive energy dissipation systems.* Report MCEER-00-0010, State University of New York, Buffalo, 2001.

[10] Trombetti, T. & Silvestri, S., On the modal damping ratios of shear-type structures equipped with Rayleigh damping systems. *Journal of Sound and Vibration*, **292**(1), pp. 21–58, 2006.
https://doi.org/10.1016/j.jsv.2005.07.023

[11] Silvestri, S., Gasparini, G. & Trombetti, T., A five-step procedure for the dimensioning of viscous dampers to be inserted in building structures. *Journal of Earthquake Engineering*, **14**(3), pp. 417–447, 2010.
https://doi.org/10.1080/13632460903093891

[12] Palermo, M., Muscio, S., Silvestri, S., Landi, L. & Trombetti, T., On the dimensioning of viscous dampers for the mitigation of the earthquake-induced effects in moment-resisting frame structures. *Bulletin of Earthquake Engineering*, **11**(6), pp. 2429–2446, 2013.
https://doi.org/10.1007/s10518-013-9474-z

[13] Palermo, M., Silvestri, S., Trombetti, T. & Landi, L., Force reduction factor for building structures equipped with added viscous dampers. *Bulletin of Earthquake Engineering*, **11**(5), pp. 1661–1681, 2013.
https://doi.org/10.1007/s10518-013-9458-z

[14] Palermo, M., Silvestri, S., Landi, L., Gasparini, G. & Trombetti, T., Peak velocities estimation for a direct five-step design procedure of inter-storey viscous dampers. *Bulletin of Earthquake Engineering*, **14**(2), pp. 599–619, 2016.
https://doi.org/10.1007/s10518-015-9829-8

EARTHQUAKE SHAKE: A REGULATORY RECIPE FOR QUAKE STRICKEN KAIKOURA, NEW ZEALAND

KATE MADSEN
Paua Planning Ltd, Auckland.

ABSTRACT

Earthquakes and their aftermath have played a big part in the lives of many New Zealander's over the past five years – and while Kiwis are resilient, the tourism sector (as our largest single export industry in terms of foreign exchange earnings) has suffered, particularly in the quake-hit areas of Christchurch and Kaikoura. An integrated approach between local and central government, tourism bodies, communities and civil defence is required in forming both pre-emptive and recovery methodologies.

Through the lens of the New Zealand regulatory planning framework, we will discuss the current situation (three months post-earthquake) in Kaikoura. In addition, we will be exploring strategies and tools to assist rather than constrain tourism operators and operations during, and subsequent to, any future natural disasters. Although seismic activity seems to have increased over the past ten years, it could be assumed that little has been done to refine or instigate regulatory measures to respond to the specific issue of maintaining and/or improving sustainable tourism in NZ.

A series of photographs by a top NZ social media photographer and environmental advocate are included to illustrate the scope and impact of the recent 7.8 magnitude Kaikoura earthquake (14 November 2016). We will outline problems encountered by tourism operators, the existing regulatory local and central government measures in place to avoid/remedy/mitigate effects of natural disasters on the tourism industry, and respond with potential long term strategic options to overcome these identified obstacles.

Keywords: civil defence, local government, natural hazard, regulatory planning, resilience, sustainability, tourism.

1 INTRODUCTION

Kaikoura is one of New Zealand's tourist meccas, famous for whale watching, seal swimming and other marine related activities, all experienced through a lens of breath-taking scenery. It is also the first community in New Zealand and second in the world to embrace the 'EarthCheck' program, which is a global tourism certification standard birthed from the principles of Agenda 21 (the United Nations action for climate change). This has included initiatives relating to zero waste, energy efficiency, tree planting by tourists, water quality projects, marine protection and biodiversity.

Continuing to set the scene for discussion, it is noted that visitor arrivals to New Zealand set a new record to 3.5 million in the December 2016 year, a 12% increase from 2015, and

Figure 1: Kaikoura, New Zealand. (*Source: Wings over Whales flight, 11 Feb 2017*).

DOI: 10.2495/SAFE-V7-N4-510-518

this is forecast to expand to 4.5 million by 2022 [1]. At present, according to David Simmons (Professor of Tourism, Lincoln University), this equates to a city of tourists the size of Dunedin (120,000 or 2.8% of New Zealand's population) [2] moving across the landscape daily. The impacts of this cannot be under-estimated.

Tourism directly contributes $12.9 billion (5.6%) to New Zealand total GDP, with a further $9.8 billion (4.3%) indirectly contributed. This is a significant industry for New Zealand, and for Kaikoura, where $109 million tourist dollars are estimated to have been spent in 2016 [3].

Unfortunately, New Zealand also sits precariously astride an active boundary between two large tectonic plates of the Earth's crust. Regional geological mapping has previously identified 14 areas of known or suspected active faults and/or folds within the Kaikoura District [4]. As a natural consequence of this, the tourism sector is vulnerable to the impacts of large disasters, and '*...may be subjected to prolonged reduction in visitation, damage to critical infrastructure and negative media attention.*' [5]. Even before the 7.8 magnitude earthquake on 14 November 2016, experts at the Institute of Geological and Nuclear Sciences Ltd (GNS Science) were recommending that specific vulnerabilities within the Kaikoura District be identified (such as berthing facilities at the marina, the major coastal highway/railway routes and low intensity residential development areas) and that contingency plans for first response emergency management and disaster recover in the event of an earthquake be prioritised by regional and territorial local authorities.

Figure 2: Kaikoura Township. *(Source: Wings over Whales flight, 11 Feb 2017).*

Figure 3: Fur seal up close. *(Source: Excursion with 'Kaikoura Kayaks', 12 Feb 2017).*

Figure 4: Ruptured Local Landscape – Entire hillside collapse. *(Source: Wings over Whales flight, 11 Feb 2017).*

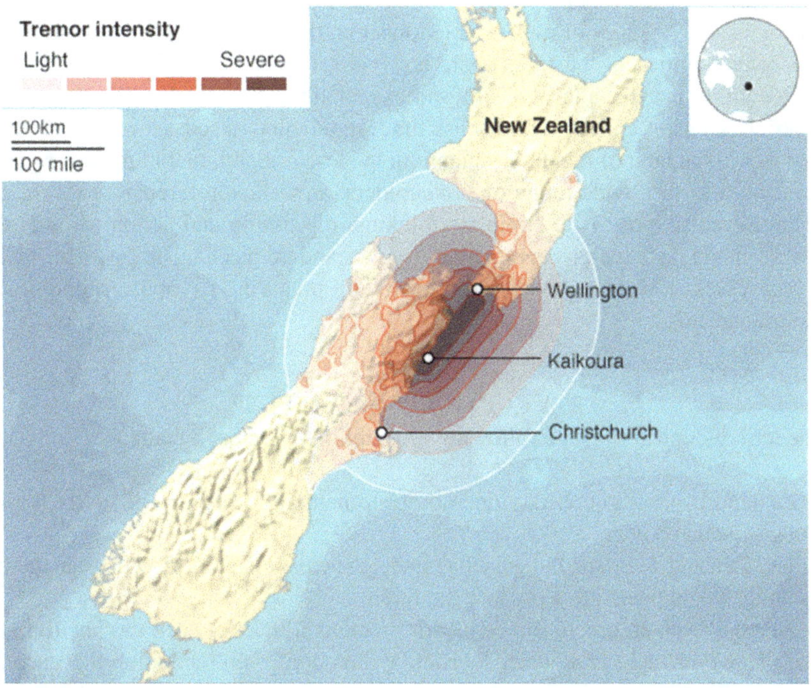

Figure 5: Location map identifying severity and epi-centre of Kaikoura earthquake. *(Source: http://www.bbc.com/news/).*

Further research has revealed that rather than five faults rupturing, as previously thought, at least 21 fault lines ripped apart – possibly a world record for the number of faults rupturing in a single earthquake. Fourteen of these faults were so violent, the land moved by more than a metre [6].

The final ingredient in this regulatory recipe is legislative. Currently, New Zealand has a strong environmental statutory framework, the backbone being the Resource Management Act 1991 (RMA), which seeks to promote the sustainable management of natural and physical resources in a way that enables people/communities to provide for their social, economic and cultural well-being (and for their health and safety), whilst avoiding/remedying/mitigating effect on the environment. This is supported by the Local Government Act 2002 (LGA), The Land Transport Management Act 2003 (LTMA), the Building Act 2004 (BA), the Conservation Act 1987 (CA), and the Marine Mammals Protection Act 1978 (MMPA), amongst others. Various supporting national, regional and district documents outwork the key strategies of the legislative framework.

The Civil Defence Emergency Management Act 2002 (CDEMA) aims to improve and promote the sustainable management of hazards in a way that contributes to the social, economic, cultural and environmental well-being and safety of the public and encourage/enable communities to achieve acceptable levels of risk.

The New Zealand Tourism Board Act 1991 (NZTBA) is enacted by a Tourism Board, whose job it is to create tourism strategies and advise the government and tourism industry on matters related to these strategies.

Whilst independently of each other these industries and legislative/regulatory bodies are active and relevant within their sphere of influence, collaboration seems to be thin, with no easily recognisable linkages or transferable tools that the wider users of the system (i.e. 'the public' or 'communities' or 'business operators') can either rely upon for continuity, or expect enforcement of, or provide a level of uniform expectation.

This article seeks to collectively discuss the 'ingredients' raised above, based on experience working in the field of regulatory planning in New Zealand, including past and ongoing consultancy work for a wide range of stakeholders, as well as reference to recent publications. The overall purpose is to stimulate thinking, discussion and action towards utilising currently available resource management tools as a standard mechanism for ensuring tourism related businesses and communities are proactive in preparing for and responding to large scale natural hazards.

2 A CUP OF COMMON SENSE

From the introduction above, the following assumptions have been made:

- New Zealand cannot ignore its vulnerability to natural hazards, particularly the risks of ongoing seismic activity.
- The New Zealand Tourism Industry has undergone exponential growth over the past ten years and is not expected to slow into the future.
- New Zealand's 'clean green' image clearly markets a link between the environment and tourism, however there is no clear legislative direction or collaboration that incorporates natural hazard risk in a cohesive response.

As an exercise in curiosity, the legislations referred to earlier (environmental, civil defence and tourism related) were explored using basic word searches, including 'tourists/tourism', 'visitors' and 'natural hazards/earthquakes'. The basis of this was to determine how many obvious linkages/specific references occur in New Zealand's statutory framework.

The environmental backbone of New Zealand's legal system (RMA, LGA & LTMA) does not refer to 'tourists', 'tourism' or 'visitors' at all. A single cursory reference in the LTMA refers to tourism within a definition for 'excluded passenger service', which has no direct relevance to the context of this paper. Neither does the CDEMA make any reference to tourism. Whilst one may argue that the use of more generalised nouns ('public' 'people' and 'communities') are inclusive of visiting populations, the lack of any direct acknowledgement of visitors to our shores may be likened to a home without hospitality – uninviting and unprepared.

Out of ten different Acts reviewed, only four mention tourism, and one mentions visitors, whilst six refer to natural hazards and seven refer to earthquakes. One presumption is that while New Zealand may be informed of (and prepared for) the risks related to natural hazards within the natural environment, they do not, as yet, inform (or prepare) the tourism industry via environmentally focussed legislation. In contrast, there is the Conservation Act and Reserves Act, which do have a focus on tourism, but few references to natural hazards.

Laws are passed to make improvement and produce better outcomes and legislation is used as an instrument to change behaviour and shape society in various ways, whether it be the economy, the environment, health, housing, education or crime. The RMA in particular, has a central purpose of 'sustainable management' – that is, growth within the constraints of the environment. Yet how is it that the growth in tourism and associated potential effects on the environment (and the impact of the environment on people by way of natural hazards) is severely underrepresented within NZ environmental legislation?

The mechanism of National Environment Standards (NES) and National Policy Statements (NPS) under the RMA umbrella allow for government statements of intent regarding matters of national significance. These contain objectives and policies that councils must 'give effect to' in their regional and territorial plans. Currently there are only five NPS – Coastal Policy (required by law), Freshwater Management, Renewable Electricity Generation, Electricity Transmission and most recently, Urban Development Capacity. Only five NES have been set in regulation since 1991, a fact which Sir Geoffrey Palmer (one of the original architects of the RMA) suggests is a gross underutilisation of a potentially very effective tool [7].

In addition, the New Zealand Planning Institute and the Productivity Commission have independently commented that: 'There are just too many plans and consequently too many varied planning responses to the same issues across region[s]. This has resulted in ineffective and inefficient planning and resource management outcomes' [8]. And: 'the lack of integration between the RMA, the Local Government Act 2002 and the Land Transport Management Act 2002 causes difficulties and should be addressed. The legislation makes it too difficult to integrate decisions about land use, transport and housing… [there is also] insufficient responsive infrastructure provision, a sluggish planning system and some in-built incentives to oppose the growth of cities' [9].

Recent proposed amendments to the RMA in 2015 recommend including 'natural hazards' within Part 2, which is seen by planning practitioners as common sense: that councils must 'recognise and provide for…the management of significant risks from natural hazards' [10].

Taking this a step forward, introducing a National Policy Statement that outlines objectives and policies related to effects of tourism on the environment in addition to the effects of the environment on tourism would allow a gateway for territorial authorities to create and strengthen new links between tourism, natural hazards and local government processes. This would improve cooperative governance for community sustainability and resilience before, during and after such an event as the Kaikoura earthquake presented.

For example, the Kaikoura earthquake resulted in significant landslides which cut off road and rail access completely, isolating the township and outlying district from the rest of New Zealand. The Hurunui/Kaikoura Earthquakes Recovery Act 2016 [11] fast-tracked legislative powers to rebuild infrastructure and aid with economic and community recovery, however if an NPS related to tourism existed, local government (in this case Canterbury Regional

Figure 6: Access to Kaikoura from the north along the main State Highway was completely devastated. *(Source: Wings over Whales flight, 11 Feb 2017).*

Authority and Kaikoura District Council) may have had more strategic goals in place relating to such a scenario. Informal talks with residents/tourism operators during a recent visit identified that some would prefer the coastal road to remain just that – a road serving tourists and cyclists only, rather than being reconstructed into the main route south. This would enable Kaikoura to establish as a destination in its own right, rather than a drive by attraction. However, the urgency to re-open the road does not allow the luxury of such an option to be explored, and any potential opportunity for creative and collaborative planning is restricted.

3 A TEASPOON OF TOOLS

There are a wide range of planning tools at the disposal of local government that could be utilised to coordinate the approach between tourism and natural hazards. The direction for these may be set by the NPS process discussed above, and include: territorial plans, the resource consent process, the building consent process, Land Information Memorandum (LIM's) under the Local Government Official Information and Meetings Act 1987 (LGOIMA) and hazard registers. Wider tools available include structure plans, growth strategies, IWI management plans, tourism and visitor action plans and community response plans (managed under the CDEMA). While there is no shortage of tools that have been implemented at a local level as a direct response to environmental or economic issues within a given community (and the list above is by no means complete), this paper suggests that there is no focus – the recipe now has too many ingredients and too many cooks, and is too complex.

More research and findings should be written in plain language and in a format familiar to planning practitioners and decision makers. Practical tools that are already available need to be explored for additional capacity and application. One such tool it the humble resource consent process. In brief, a resource consent provides permission to carry out an activity that would otherwise contravene certain sections of the RMA (that relate to land-use, subdivision, coastal marine, rivers and lakes, water, discharge of contaminants, noise and general adverse effects) so long as it complies with any conditions attached to the consent. Rules contained within regional and district plans determine whether an activity is considered permitted, controlled, restricted discretionary, discretionary, non-complying or prohibited (Section 87A, RMA).

While some natural hazards such as land stability, flooding and inundation are well registered in NZ and therefore taken into account during the consent process, other less measurable and unpredictable hazard risks (such as seismic or volcanic) are generally ignored (from the

Figure 7: Repair to the Kaikoura Harbour, which rose 1–2 m during the earthquake – emergency works are occurring under section 330 of the RMA. (*Source: T Madsen, Feb 2017*).

writer's experience in assessing resource consents). As New Zealand is located entirely within the 'ring of fire' band of fault lines, adverse effects of tourism activities and development on the environment in this regard is difficult to predict or plan for. Whilst the resource consent process has been utilised in the response and recovery phases of the Kaikoura earthquake (and previous Christchurch earthquakes) by way of 'fast-tracking' consents to repair infrastructure such as roading, rail and access to the marina – it has not been exploited as a tool to link tourism operators and communities that experience high visitor numbers directly to the civil defence framework prior to any large scale natural disaster occurring.

Applying a standard suite of natural hazard management conditions to every resource consent granted would allow Council to cross-pollinate and streamline training, education and information and business continuity planning/emergency response plans, requiring the monitoring of such conditions be coordinated with the Ministry of Civil Defence and Emergency Management and the Ministry of Tourism. In the context of this paper, this would include every application to undertake a tourism-related activity; however, in the context of New Zealand, this could eventually include every application for land-use development. While traditionally the RMA has been utilised to avoid, remedy and/or mitigate adverse effects of development on the environment, future application might see the RMA seek to enable people and communities to proactively provide for their social, economic and cultural well-being and for their health and safety by preparing for potential adverse effects of large scale natural hazards. Resource consents are a familiar and well utilised tool within the tourism industry and utilising the process to network civil defence, tourism and planning in a single spoon could simplify the regulatory recipe profoundly.

4 A TRUCKLOAD OF COOPERATION

As mentioned earlier, the fundamental building blocks to interlock planning, natural hazards and pressures associated with the growth of tourism have been duly established and are well grounded within New Zealand's legal system. However, mechanisms to link these separate processes could be considered stunted at best. While recent seismic events have put pressure on central and local government to refine and strengthen responses to and between natural hazards and planning, the economy and business of tourism remains on an island of its own. If the Kaikoura earthquake had occurred during the day, rather than at night, the local community of 3500 and visiting population of approximately 1000 may have experienced more serious consequences in terms of injury and death. While some would argue that local government is not directly responsible for the health and safety of people and communities, section 11A (d) of the LGA 2002 considers the avoidance or mitigation of natural hazards as

Figure 8: The coastline has been thrust up to 2 m, exposing Paua and Koura (Crayfish) above the high tide mark in some areas. *(Source: Facebook/Anna Redmond).*

a core service of local authorities. Fundamentally, natural hazards in populated areas cannot be considered independently of the adverse impact they have on people – as much as we utilise the resource consent process to minimise adverse effects on the environment, the end result always includes the health, safety and well-being of the public – both local communities and visitors alike.

In the immediate aftermath of the Kaikoura Earthquake, it was a combined response between Te Rūnanga o Ngāi Tahu (local Maori community in Kaikoura), Kaikoura District Council, Civil Defence, tourist operators and community members that ensured the stranded tourists were accounted for and hosted until they could be evacuated. Crayfish and Paua were served by local fishermen and Te Rūnanga o Ngāi Tahu, and tourists slept at the local Marae (Maori meeting house/grounds). Tourist operators (such as 'Wings over Whales') provided air transport for emergency services, reconnaissance flights and access to check on remote communities; civil defence coordinated with other government response groups to check public safety and organise supplies and access, along with local and regional council involvement.

While collaboration was apparent in the immediate response to the Kaikoura earthquake, (as is often the case when urgent action is required), it is this level of cooperation that communities and authorities should engage in during times of non-conflict. As a planning practitioner, too often it becomes apparent that there is limited involvement of key stakeholders either due to misconception regarding RMA processes or the lack of follow-through from policy to practice.

In summary, a recipe is only as good as the quality of the ingredients. Whilst New Zealand has a strong tourism economy and environmental legislation, the mix as yet is not working and it is considered that available tools, such as National Policy Statements (at the national level) and the resource consent process (at the local level) are underutilised. Both central and local government need to cooperate with each other, and with external organisations (both statutory and non-statutory) to achieve the perfect recipe before the next big shake-up down-under.

ACKNOWLEDGEMENTS

The author would like to acknowledge discussions held with Professor David Simmons, Professor of Tourism/Director of Research Strategy and Development, Lincoln University and Mr. Vernon Rive, Senior Lecturer in Law and Associate Head of School – External Relations, Auckland University of Technology, and the research papers they provided for recommended reading. In addition, thanks to Talman Madsen Photography for the majority of photos included, unless otherwise referenced.

REFERENCES

[1] International visitor arrivals commentary; Ministry of Business, Innovation & Employment (Hīkina Whakatutuki), available at http://www.mbie.govt.nz/info-services/sectors-industries/tourism/tourism-research-data/international-travel/international-visitor-arrivals-commentary (accessed 18 February 2017).

[2] 2013 Census QuickStats about a place: Dunedin City; Statistics New Zealand, available at: http://www.stats.govt.nz/Census/2013-census/profile-and-summary-reports/quickstats-about-a-place.aspx?request_value=15022&tabname (accessed 19 February 2017).

[3] Tourism Data & Statistics, available at http://www.tourismnewzealand.com/about/about-the-industry/ (accessed 19 February 2017).

[4] Barrell, D.J.A., *General distribution and characteristics of active faults and folds in the Kaikoura District, North Canterbury.* GNS Science Consultancy Report 2014/210 & Environment Canterbury Report No. R15/23, p. 3, 2015. available at: www.kaikoura.govt.nz/assets/Docs/Council/Publications/GNS-CR-2014-210-Kaikoura-District-active-faults-web-version.pdf (accessed 18 February 2017).

[5] Orchiston, C., Tourism business preparedness, resilience and disaster planning in a region of high seismic risk: the case of the Southern Alps, New Zealand. *Current Issues in Tourism,* **16**(5), pp. 477–494, 2013.
https://doi.org/10.1080/13683500.2012.741115

[6] https://www.nzgeo.com/audio/kaikoura-quakes-21-faultlines-could-be-world-record/.

[7] Ruminations on the problems with the Resource Management Act, Sir Geoffrey Palmer QC, *Keynote address to the Local Government Environmental Compliance Conference 2 – 3 November 2015.* available at: https://www.planning.org.nz/Attachment?Action=Download&Attachment_id=3538

[8] Wellington Region Local Government Review Panel, Future Wellington-Proud, Prosperous and Resilient (Wellington October 2012) para. 515.

[9] New Zealand Productivity Commission, Using Land for Housing (Wellington October 2015). available at: http://www.productivity.govt.nz/inquiry-content/2060?stage=4

[10] New Zealand Legislation, Parliamentary Counsel Office, Resource Legislatioin Amendment Bill. available at: http://www.legislation.govt.nz/bill/government/2015/0101/latest/whole.html

[11] New Zealand Legislation, Parliamentary Counsel Office, Hurunui/Kaikōura Earthquakes Recovery Act, 2016. available at: http://www.legislation.govt.nz/act/public/2016/0102/latest/whole.html

MODEL UPDATING BASED ON THE DYNAMIC IDENTIFICATION OF A BAROQUE BELL TOWER

MARIELLA DIAFERIO[1], DORA FOTI[1], NICOLA IVAN GIANNOCCARO[2] & SALVADOR IVORRA[3]
[1]Polytechnic of Bari, Department of Civil Engineering and Architecture, Bari, Italy.
[2]University of Salento, Department of Innovation Engineering, Lecce, Italy.
[3]University of Alicante, Department of Civil Engineering, Alicante, Spain.

ABSTRACT

The sanctuary of Santa Maria di Loreto (Mola di Bari, Italy) was built in in the sixteen century and it is constituted by a church with a simple hut façade characterized by a fine calcareous rose window and on its left side a high bell tower with heavy bells.

In the present study, a non-destructive dynamic identification of the modal characteristics of the structure has been carried out by firstly organising an experimental setup of environmental measurements with several accelerometers placed on different levels of the bell tower. Secondly, the data acquired by the accelerometers have been elaborated by using modern statistic techniques of operational modal analysis (OMA). Finally,these results have been used for calibrating a complete finite element (FE) model of the bell tower structure that may permit to obtain important information about the state of integrity of the medieval structure. This calibrated model was developed to obtain a preliminary evaluation of the seismic vulnerability of the bell tower by performing a nonlinear static analysis.

Keywords: Environmental vibrations, Masonry Tower, Operational Modal Analysis, Seismic Analysis, Structural identification

1 INTRODUCTION

The analysis of the seismic vulnerability of historical structures has achieved numerous attentions in order to answer to the public request of preserving the wide building heritage, which in many cases, as for Italy, lays on sites with a high seismic level. However, this analysis requires the definition of a reliable finite element model which is able to estimate the structural response also taking into account the existence of cracks and/or damage. In order to define such model a great amount of the researches have performed dynamic tests on such structures and updated a refined finite element model in order to match the experimental results [1–12].

The present study deals with the dynamic identification and the evaluation of the seismic performances of the historical masonry bell tower of Santa Maria di Loreto in Mola di Bari (Italy). The church was erected in the 1587 at the southern outskirts of the town. The church is located on the seafront. The bell tower (shown in Fig. 1) is aligned with the front wall of the church on its Eastern side and it has a total height of 38.3 m. It has a square plan with a side of about 6.3 m and it connected to the church for about 9 m. Its brick stone masonry walls have a thickness of about 2 m at the first body (see Fig. 2), at the second body, where the bells are located, the wall thickness becomes equal to about 1.6 m, while at the third body the thickness is about 0.85 m. The tower shows good preservation conditions at the first body; however, at the second and third bodies (Fig. 2) it presents damages due to the atmospheric exposition, as it stands near the sea. The present paper deals with the seismic vulnerability of the bell tower of the sanctuary. For this purpose, a refined 3D finite element model of the bell tower has been defined, which has been updated and validated by comparing the first natural frequencies with the experimental ones. A nonlinear static analysis has been performed on the aforementioned finite element model of the bell tower in order to evaluate the capacity

DOI: 10.2495/SAFE-V7-N4-519-531

Figure 1: Bell tower of Santa Maria di Loreto (Mola, Bari, Italy). (a) North view. (b) Main façade (c) Plan view of the church with the bell tower.

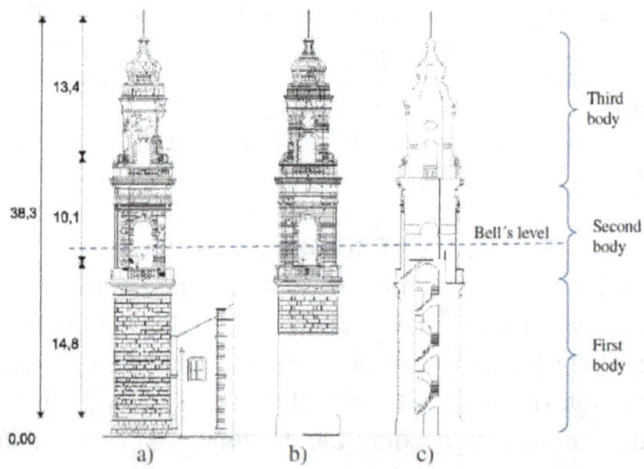

Figure 2: Frontal view of the bell tower of Santa Maria di Loreto (Mola, Bari, Italy). (a) facade in front of sea -North; (b) South facade; (c) Vertical section.

curves along the principal directions. The vulnerability of the bell tower is also evaluated according to the building Italian guidelines using the results of the performed nonlinear static analysis.

This bell tower has the classical structure of three bodies. The first body, resting on an ashlar basement, arrives to the belfry, and it is characterized by four walls practically orthogonal each other, with a thickness of 2 m without any window or opening. The second body houses the belfry. It is characterized by the opening of four windows with rounded arches, with the base of molded balustrades; pillars projecting creates a very lively language pattern and ends with a sequence of moldings alternating straight and curved profiles.

The third body is characterized by pillars at the corners with swirls of basic pedestal. Each prospect has an opening with rounded arches, this one with a balustrade in front, topped by an oval punch. The bell tower has four bells situated at the height of 18 m. They are situated on a box steel structure not directly anchored to the walls of the bell tower. It is directly supported by the floor of the bell room. Figure 3 shows a general view of the west, north

Figure 3: Bells of the tower.

and south bells. The steel structure and the swinging mechanism are very deteriorated by corrosion effects.

2 THE EXPERIMENTAL CAMPAIGN AND DYNAMICAL IDENTIFICATION

a) Experimental campaign

A dynamic experimental analysis has been performed on the bell tower of Santa Maria di Loreto church on 21st April 2016. The data have been acquired during a sunny day and with a reduced level of wind and environmental vibrations. Four tests of 15 minutes have been recorded in environmental conditions in order to have a statistical analysis of the identified modal parameters. A reference coordinate system *Oxyz* has been introduced (see Fig. 4) in such a way that the *x* axis is parallel to the church façade (see also Fig. 1b).

Moreover, additional tests (two) were carried out by considering the effect of the bells. The first test has been carried out considering the effect of the bells oriented in the *x* axis direction, the second test considering the bells in y direction.

All the experimental tests have been carried out with 1024 Hz sampling frequency.

The monitoring system consisted of 10 seismic accelerometers ICP PCB 393B31 with a NI data acquisition system. The DAQ is positioned with the laptop on the base of the tower connected with the accelerometers by flexible shielded cables. The ten uniaxial accelerometers have been connected in couples with opportune blocks in such a way to measure orthogonal components (parallel to *xy* axis, as shown in Fig. 5) in five different point of the structure. The five points have been selected at three different levels of the bell as shown in Fig. 6 where for each level, the map of the section the accelerometers nomenclature and the reference system are reported. The accelerometers 1, 2, 3, 4 at the higher level (height of 28.6 m up the bell level), the accelerometer 5, 6, 7, 8 at an average level close to the bell

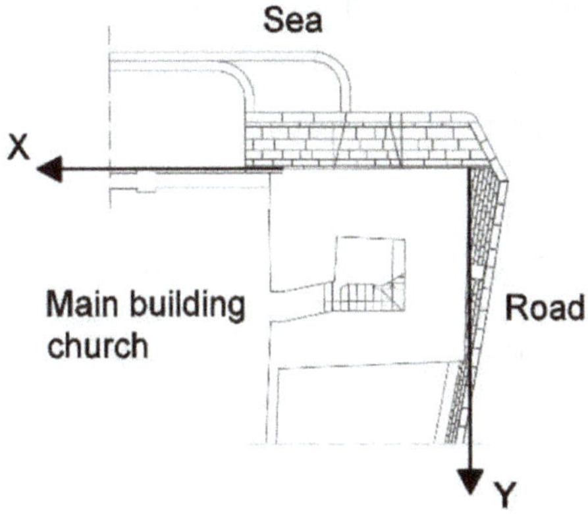

Figure 4: The considered reference system with respect to the geographical bell location.

Figure 5: The considered couples of uniaxial accelerometers joined by a block.

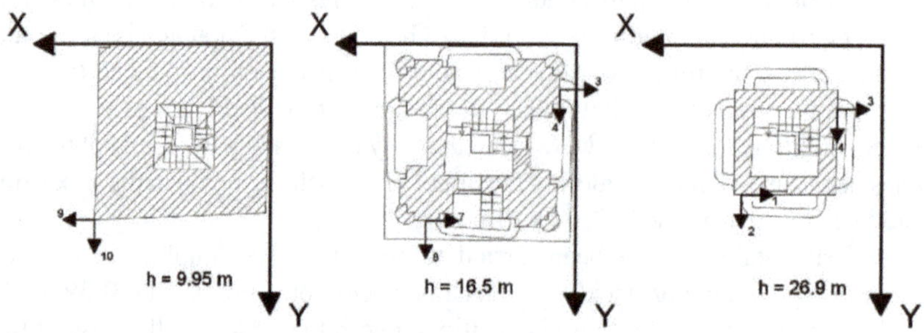

Figure 6: The position of the accelerometers used in the tests.

(height of 16.5 meters) and the accelerometers 9 and 10 are placed in the first body, just upper the connection with the church. Accelerometers 9 and 10 may be considered reference accelerometers. The complete nomenclature, the exact coordinates and the directions (positive and

Table 1: Code and coordinates of the accelerometers used in the test.

Accelerometer	Code	Direction	x[m]	y[m]	z[m]
1	35063	x- →	4.68	5.74	26.64
2	35098	y+ ↓	4.68	5.74	26.64
3	35211	x- →	0.591	1.87	26.76
4	35208	y+ ↓	0.591	1.87	26.76
5	35089	x- →	0.04	1.38	16.88
6	33995	y+ ↓	0.04	1.38	16.88
7	35205	x- →	5.13	6.26	16.88
8	35087	y+ ↓	5.13	6.26	16.88
9	35206	x+ ←	6.65	6.28	10.10
10	35201	y+ ↓	6.65	6.28	10.10

negative with respect to the axis) of the ten accelerometers are reported in a unifying manner in Table 1.

b) Dynamical identification

The four tests developed under environmental conditions have been considered for the dynamical identification of the structure: Test 1, Test 2, Test 3 and Test 4, respectively. A preliminary analysis of the experimental data has been carried out for determining eventual unexpected events on the registered time histories. Figure 7 shows the time histories of accelerometers 1, 2, 3, 4, 5, 6, 7, 8 in Test 1.

The preliminary check noted unexpected peaks registered for accelerometer 2 in Test 2, accelerometer 6 in Test 3, accelerometer 3 in Test 4. For this reason, these signals have not been considered in the following dynamical identification.

The dynamical identification has been carried out by using two different and well known techniques, one implemented in the frequency domain (the Curve-fit Frequency Domain Decomposition Modal Analysis, CFDD method) and the second in the time domain (the Stochastic Subspace Identification Modal Analysis, SSI method with UPC (Unweighted Principal Component)).

The analysis has been carried out by searching the first clear experimental frequencies and their corresponding modes for all the tests by using both the methods in such a way to have a statistical overview that guarantees the estimation of the frequencies.

The model realized in an opportune software [13] is shown in Fig. 8, where the arrows indicate the placement and orientation of the accelerometers, the red arrows indicate the accelerometers whose signal is used for the frequencies identification. In Fig. 9 the results of the CFDD identification (for Test 3) and in Fig. 10 the results of SSI identification (Test 1).

The identification results of the four experimental tests in environmental condition are summarized in Table 2, where an analysis about the statistical repeatability is also reported. It is evident the excellent repeatability of the first three identified frequencies for almost all the tests with both the considered techniques. The average value of 1.69 Hz may be considered for the first frequency, 1.75 for the second one and 5.33 for the third one, with a very low standard deviation between the different tests.

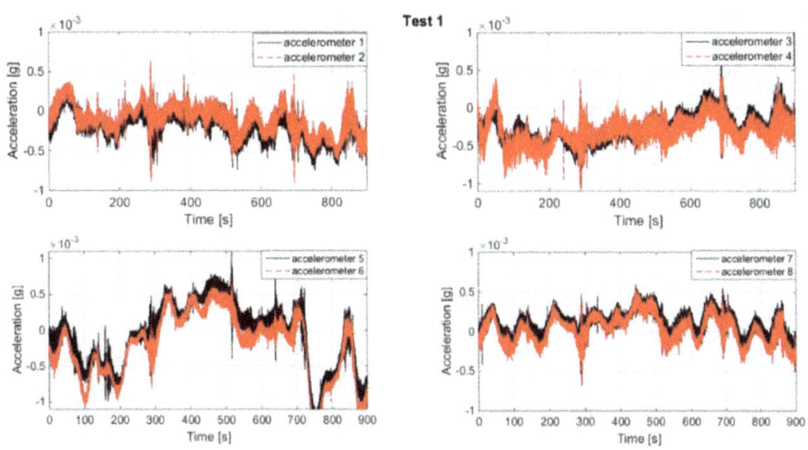

Figure 7: Time histories for Test 1.

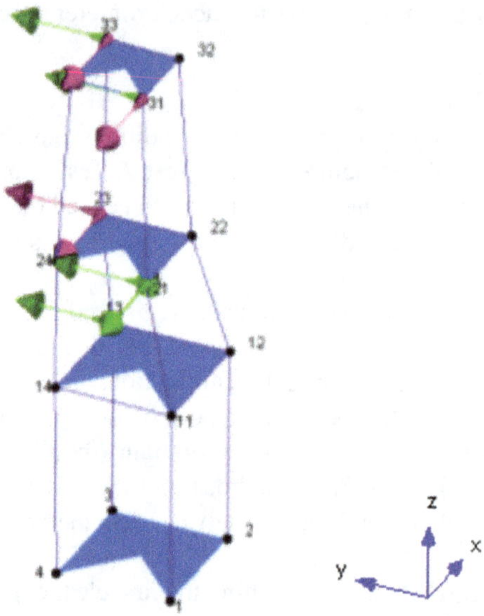

Figure 8: Model used for analysing the experimental data.

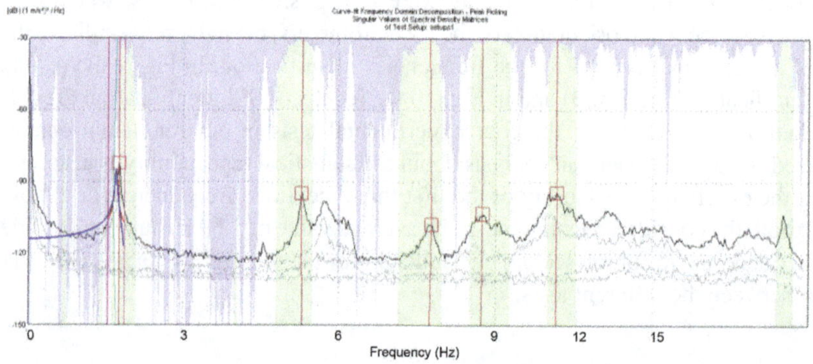

Figure 9: CFDD identification results (Test 3). Automatically detected modes: Mode 1: 1.696 Hz, Mode 2: 1.752 Hz, Mode 3: 5.309 Hz, Mode4: 7.748 Hz, Mode 5 (8.738 Hz, Mode 6: 10.16 Hz).

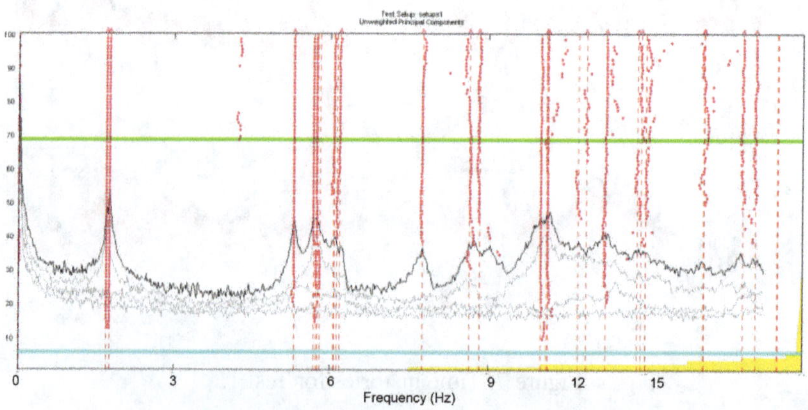

Figure 10: SSI identification results (Test1).

Table 2: Identified frequencies [Hz] for the environmental tests and statistical repeatability.

Freq	Test1 CFDD	Test1 SSI	Test2 CFDD	Test2 SSI	Test3 CFDD	Test3 SSI	Test4 CFDD	Test4 SSI	Aver. value	Stand. Dev.
1	1.67	1.70	1.70	1.71	–	1.70	1.69	1.69	1.694	$1.2\ 10^{-3}$
2	1.75	1.75	1.75	1.75	1.76	1.76	–	1.76	1.754	$5.3\ 10^{-4}$
3	–	5.38	5.31	–	5.30	5.34	5.32	5.33	5.330	$2.8\ 10^{-3}$

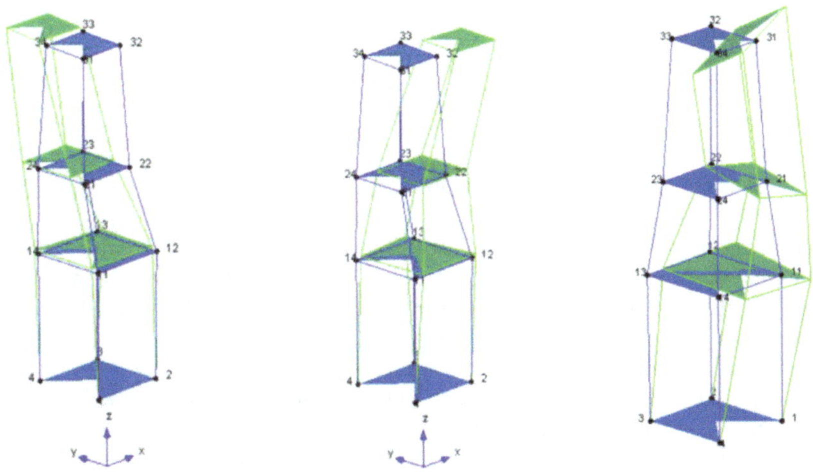

(a) (b) (c)

Figure 11: Experimental mode shapes. (a) First mode, (b) second mode and (c) third mode.

Other higher frequencies are also evident in the CFDD and SSI diagram shown in Figs 9 and 10; but for the validation of a sufficiently accurate FE model of the bell tower, the first three frequencies have been considered enough for guaranteeing a good matching. Moreover, the higher frequencies do not have a so clear repeatability in all the tests.

The mode shapes associated to the identified frequencies are depicted in Fig. 11 and may be clearly considered purely flexional (first flexional) for the first two frequencies (along y axis the first mode, along x axis the second mode) and torsional for the third identified frequency.

3 THE NUMERICAL MODEL

The aim of the performed structural identification was the evaluation of the Young modulus E and density w of the masonry, and of the degree of constraint due to the adjacent church to the bell tower.

This evaluation is performed, as a first approximation, modelling the tower walls by means of shell elements having the same geometrical characteristics of the tower and a constant Young's modulus. Moreover, the tower has been assumed fixed at the base and a preliminary analysis has been performed assuming E = 1080 MPa and w = 1630 kg/m^3 [14]. The model

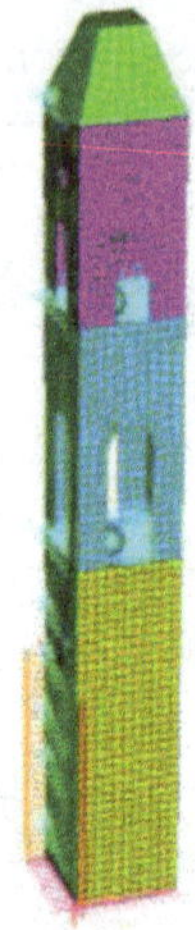

Figure 12: Finite element model of the bell tower of Santa Maria di Loreto (Mola, Bari, Italy).

is defined by means of 3498 shell elements of the type QUAD4 which are defined by means of 4 nodes each one with six degrees of freedom [15]. The model (Fig. 12) has 3654 nodes and three structural masses located at the centre of each floor; moreover, the stairs and the bells have been considered as non-structural masses (Fig. 12).

Two different hypotheses have been adopted: (i) the church is not able to constrain the tower, which behaves as an isolated tower; (ii) the tower is fully constrained by the church [14]. The comparison between the natural frequencies of the abovementioned model and the experimental ones allows concluding that the adjacent church affects the level of constraint of the tower and cannot be neglected in the structural analysis. Thus, here a new model has been adopted which is been obtained from the one discussed in Ref. [14] by updating the Young modulus, the density and the constraints due to the church. In detail, the church is modelled by means of elastic springs acting along the directions of the church walls.

After performing a sensitivity analysis, which for sake of conciseness is not here discussed, it has been adopted a Young's modulus equal to 1500 MPa and a density of 1500 kg/m^3, while the stiffness of the elastic springs is varied in order to match the experimental

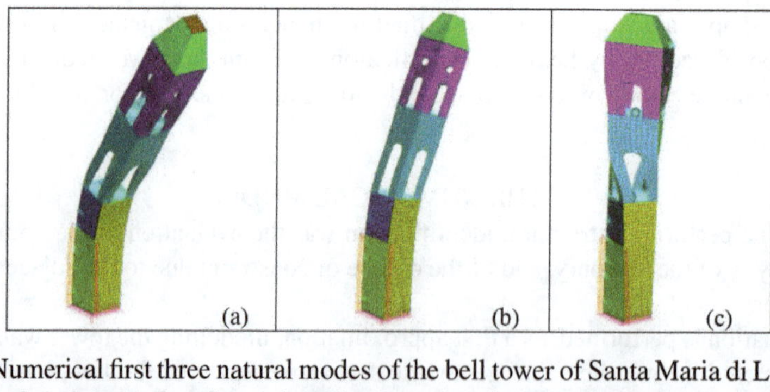

Figure 13: Numerical first three natural modes of the bell tower of Santa Maria di Loreto (Mola, Bari, Italy): (a) first flexural mode along the direction normal to the main façade of the church: $f_1 = 1.698$ Hz; (b) second flexural mode along the direction of the main façade of the church: $f_2 = 1.751$ Hz; and (c) third torsional mode: $f_3 = 5.310$ Hz.

Table 3: Comparison between experimental and numerical frequencies.

≠ MODE	Experimental frequency [Hz]	Numerical frequency [Hz]	Difference [%]	MAC Experimental/ Numerical
1	1.694	1.698	0.2	80.7 %
2	1.754	1.751	0.2	79.5 %
3	5.330	5.310	0.4	18.1 %

frequency. The model has elastic springs along the connection with the main façade of the church and acting along this direction whose stiffness is equal to 10^6 KN/m, and elastic springs acting along the direction orthogonal to the previous one and located along the connection with the church whose stiffness is equal to 10^6 KN/m. In Fig. 13, the first three natural modes of the bell tower model are shown.

As it can be easily verified in Table 3, there is a good agreement between experimental (average value) and numerical frequencies. The comparison allows considering the structural model reliable to prevent the dynamic response of the structure. In Table 3, the Modal Assurance Criterion (MAC) between the experimental and the numerical modes is also reported. The MAC values are satisfying for the first two flexural modes and this result allows considering the model reliable for the following nonlinear static analysis.

4 THE NONLINEAR STATIC ANALYSIS

The area of Mola di Bari (Italy) is a seismic prone zone. According to the actual Italian building code [16] the site is characterized by the response horizontal spectrum, with behaviour factor q = 2.25, reported in Fig. 14 in terms of Peak Ground Acceleration (PGA) at the bedrock a_g versus the return period T_R. The PGA at the bedrock with a 475 years return period is $a_g = 0.091$ g.

In order to evaluate the seismic vulnerability of the examined tower, a pushover analysis has been performed on the validated finite element model. In accordance with the Italian building code [16], the tower has been subjected to horizontal forces acting along the principal axes and concentrated at the lumped structural masses. The forces are established in accordance with the eqn (1):

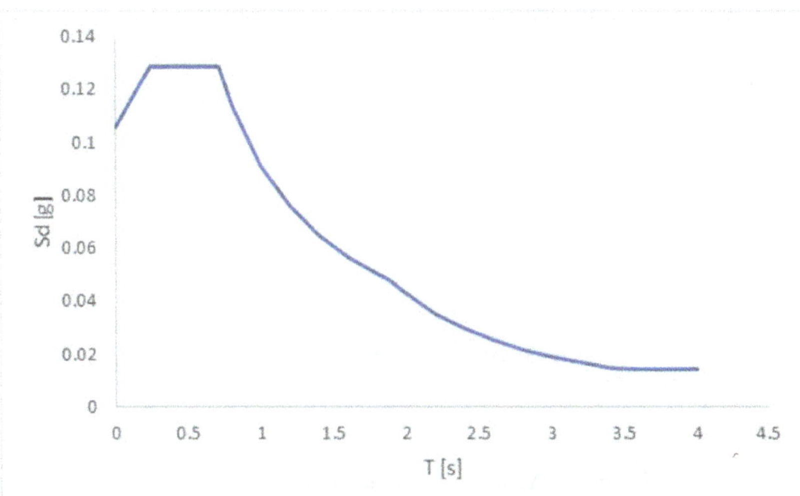

Figure 14: Design horizontal spectrum for Mola di Bari (Italy) [16].

$$F_i = F_h \frac{z_i W_i}{\sum_{j=1}^{n} W_j z_j} \text{ with } F_h = S_d(T_1)\frac{W\lambda}{g} \tag{1}$$

where F_i is the i-th concentrated force located at the i-th lumped mass; W_i is the weight of the i-th mass, z_i is the altitude of the i-th mass respetct to the foundation, $S_d(T_1)$ is the design horizontal spectral value for the first natural period of the structure, W is the total weight of the structure, $\lambda = 1$ and g the gravity acceleration. Figure 15 shows the direction and position of forces presented in eqn (1) on the structural model of this bell tower.

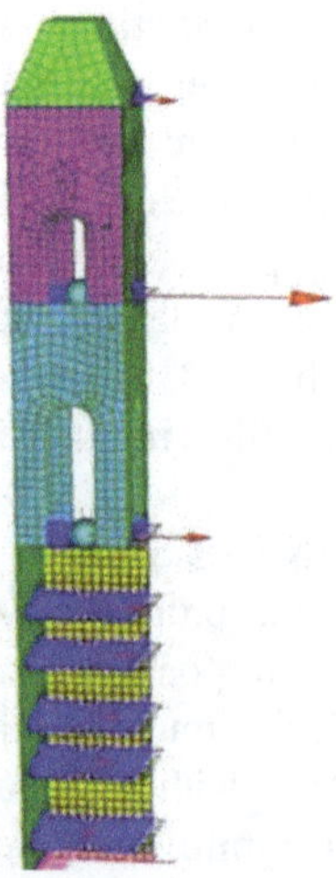

Figure 15: View of the horizontal forces for the pushover analysis of the bell tower of Santa Maria di Loreto (Mola di Bari, Italy).

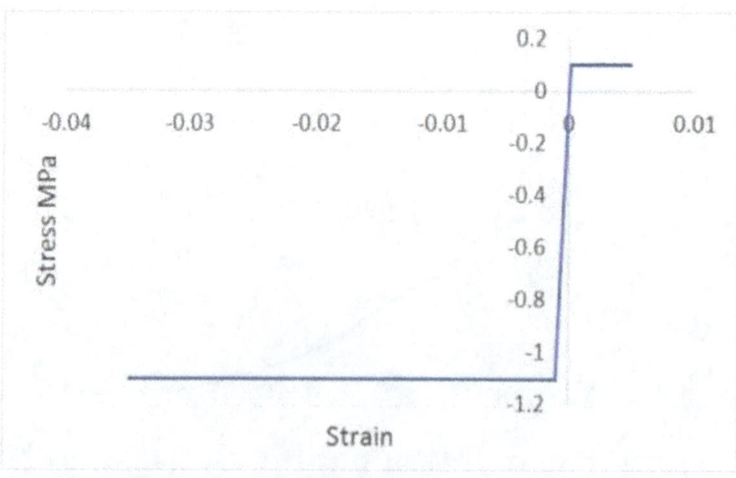

Figure 16: Adopted constitutive law for masonry.

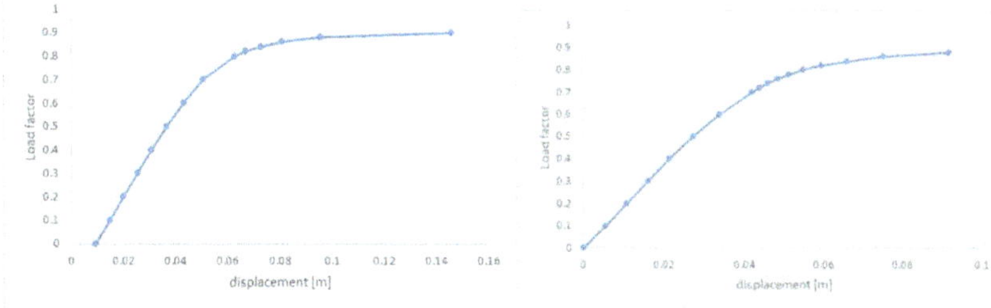

(a) (b)

Figure 17: Capacity curves of the bell tower of Santa Maria di Loreto (Mola di Bari, Italy): (a) forces acting along the x axis and (b) forces acting along the y axis.

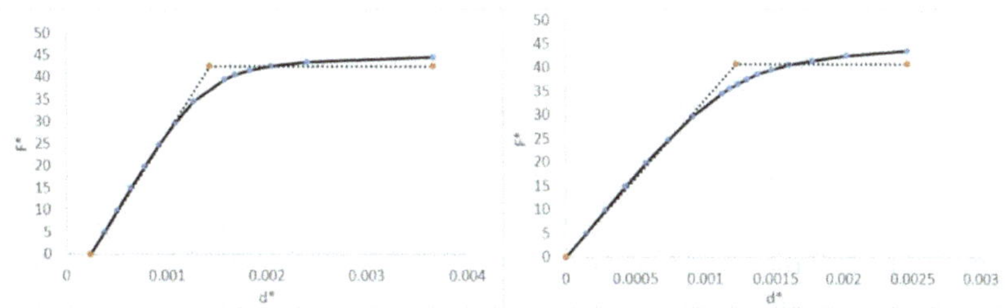

(a) (b)

Figure 18: Equivalent capacity curves and bilinear curves of the bell tower of Santa Maria di Loreto (Mola di Bari, Italy): (a) forces acting along the x axis and (b) forces acting along the y axis.

The nonlinear static analysis has been performed adopting for the masonry the constitutive law in Fig.16 where the negative values are related to compression forces (stress) while the positive values represent tensions.

In Fig. 17 the capacity curves evaluated along the x direction and the y direction are plotted.

In order to evaluate the seismic vulnerability, the equivalent Single Degree of Freedom (SDF) system has been obtained. In detail, the capacity curve has been modified by introducing the modifications indicated in eqn (2):

$$F^* = \frac{F}{\Gamma} \quad d^* = \frac{d}{\Gamma}, \text{ where } \Gamma = \frac{\sum m_i \Phi_i}{\sum m_i \Phi_i^2} \tag{2}$$

Figure 18 shows the equivalent bilinear capacity curves of the SDF equivalent system for each one of the principal directions.

The bilinear capacity curves evidence maximum displacement values equal to 0.031 m along the x axis and equal to 0.021 m along the y axis, while the maximum displacement demand, according with Ref. [16], is equal to 0.42 m and 0.37 m along the x and y axis, respectively.

5 CONCLUSION

This paper presents a preliminary analysis of the structural behaviour of Santa Maria di Loreto bell tower. The structural numerical model was calibrated with the dynamic experimental results with a good level of approximation for the complex behaviour of masonry structures. The non-linear analysis developed by the application of a pushover analysis on the calibrated numerical model applying the Italian Standards concludes the high seismic vulnerability of this structure and the probably collapse under the proposed seismic loads for Mola di Bari area.

REFERENCES

[1] Lourenço, P.B., Computations of historical masonry constructions. *Pro Structural Engineering*, **4**(3), pp. 301–319, 2002.
https://doi.org/10.1002/pse.120

[2] Gentile, C. & Saisi, A., Ambient vibration testing of historic masonry towers for structural identification and damage assessment. *Construction and Building Material*, **21**(6), pp. 1311–1321, 2007.
https://doi.org/10.1016/j.conbuildmat.2006.01.007

[3] Russo, G., Bergamo, O., Damiani, L. & Lugato, D., Experimental analysis of the ``Saint Andrea'' Masonry Bell Tower in Venice. A new method for the determination of ``Tower Global Young's Modulus E. *Engineering Structures*, **32**, pp. 353–360, 2010.
https://doi.org/10.1016/j.engstruct.2009.08.002

[4] D'Ambrisi, A., Mariani, V. & Mezzi, M., Seismic assessment of a historical masonry tower with nonlinear static and dynamic analyses tuned on ambient vibration tests. *Engineering Structures*, **36**, pp. 210–219, 2012.
https://doi.org/10.1016/j.engstruct.2011.12.009

[5] Foti, D., Diaferio, M., Giannoccaro, N.I. & Ivorra, S., Structural identification and numerical models for slender historical structures. *Handbook of Research on Seismic Assessment and Rehabilitation of Historic Structures*, pp. 674–703, 2015.
https://doi.org/10.4018/978-1-4666-8286-3.ch023

[6] Ivorra, S., Brotóns, V., Foti, D. & Diaferio, M., A preliminary approach of dynamic identification of slender buildings by neuronal networks. *International Journal of Non-Linear Mechanics*, **80**, pp. 183–189, 2016.
https://doi.org/10.1016/j.ijnonlinmec.2015.11.009

[7] Ramos, L., Marques, L., Lourenco, P., De Roeck, G., Campos-Costa, A. & Roque, J., Monitoring historical masonry structures with operational modal analysis: two case studies. *Mechanical Systems and Signal Processing*, **24**(5), pp. 1291–1305, 2010.
https://doi.org/10.1016/j.ymssp.2010.01.011

[8] Diaferio, M., Foti, D., Gentile, C., Giannoccaro, N.I. & Saisi, A., *Dynamic testing of a historical slender building using accelerometers and radar.* Prooceedings of 6th International Operational Modal Analysis Conference, IOMAC 2015; Abba HotelGijon; Spain; 12–14 May 2015.

[9] Betti, M., Orlando, M. & Vignoli. A., Static behaviour of an Italian medieval castle: damage assessment by numerical modelling. *Computers and Structures*, **89**(21–22), pp. 1956–1970, 2011.
https://doi.org/10.1016/j.compstruc.2011.05.022

[10] Diaferio, M., Foti, D. & Giannoccaro, N.I., *Identification of the modal properties of a squat historic tower for the tuning of a FE model.* Proocedings 6th International Operational Modal Analysis Conference, IOMAC 2015; Abba Hotel Gijon; Spain; 12–14 May 2015.

[11] Gentile, C. & Saisi, A., Operational modal testing of historic structures at different levels of excitation. *Construction and Building Materials*, **48**, pp. 1273–1285, 2013. https://doi.org/10.1016/j.conbuildmat.2013.01.013
[12] Tomaszewska, A. & Szymczak, C., Identification of the Vistula Mounting tower model using measured modal data. *Engineering Structures*, **42**, pp. 342–348, 2012. https://doi.org/10.1016/j.engstruct.2012.04.031
[13] Artemis Modal Pro v4.5. Structural Vibrations Solutions. Aalborg, Denmark. 2016.
[14] Ivorra, S., Foti, D., Diaferio, M. & Carabellese, I., *Preliminary OMA results on a soft calcarenite stone bell-tower in Mola di Bari (Italy)*. Prooceedings of 7th International Operational Modal Analysis Conference, IOMAC 2017; Ingolstadt - Germany; 10–12 May 2017.
[15] Strauss 7 Finite element analysis system. Strand7 Pty Ltd, available at: www.strauss7.com
[16] NTC. Norme tecniche per le costruzioni. D.M. Ministero Infrastrutture e Trasporti 14 gennaio 2008, G.U.R.I. February 4th 2008, Roma, Italy.

USING PBEE TO ASSESS AND IMPROVE PERFORMANCE OF DIFFERENT STRUCTURAL SYSTEMS FOR LOW-RISE STEEL BUILDINGS

VESNA TERZIC[1] & STEPHEN A. MAHIN[2]
[1]Civil Engineering and Construction Management, California State University Long Beach, California, USA.
[2]Civil and Environmental Engineering, University of California Berkeley, California, USA.

ABSTRACT
In recent earthquakes in Chile, New Zealand, and Japan modern buildings were generally quite safe. However, there was a tremendous variability in the economic and social consequences associated with damage repair and loss of occupancy. Code-compliant structures are generally designed to provide safety and prevent collapse at minimum costs, but when severe ground shaking occurs, the damage to contents, nonstructural components and the structural system can result in loss of function, which can have a dramatic impact on the occupants, owners, and community. Such situations can be avoided by mitigating future seismic damage through a better structural design. Although structural enhancements may likely increase the initial cost of a structure they should be compensated through benefits realized over its lifetime. This paper presents the results of repair cost and repair time analyses of conventional and high-performance code-compliant low-rise commercial steel buildings designed for the same location. The buildings are located at a site with high seismic hazard representative of western North America. The conventional lateral load-resisting systems include: special moment resisting frame, special concentrically braced frame, and buckling restrained braced frame; and high-performance systems include: seismically isolated intermediate moment resisting frame, seismically isolated ordinary concentrically braced frame, seismically isolated buckling restrained braced frame, and viscously damped moment frame. In addition, enhancements of the fixed, seismically isolated, and viscously damped moment frames beyond the code minimum and their effect on the repair cost and repair time are studied. The analysis reveals significant damage savings for code-compliant seismically isolated systems relative to their conventional counterpart, with seismically isolated ordinary concentrically braced frames yielding the largest savings. It also shows that enhancements of the isolation and viscously damped system beyond code minimum standards results in significant reduction in damage-induced losses, while the enhancement of the SMRF does not yield desirable loss reduction.
Keywords: Base isolation, braced frames, dampers, moment resisting frames, repair loss, repair time.

1 INTRODUCTION

The aim of building codes is to design structures that achieve functional requirements with adequate safety but at a minimum cost. As evidenced by the M8.8 27 February 2010 Chile earthquake, the tremendous variation in consequences associated with damage, repair, and loss of occupancy of the modern buildings was observed. While some of the buildings had minimal damage, great number of buildings experienced considerable damage to nonstructural and structural components, resulting in not only a large economic loss in terms of repair costs but also incurred an additional financial burden associated with the loss of function due to necessity of performing these extensive repairs [1].

Earthquake hazards and risks can be systematically characterized through performance-based earthquake evaluation (PBEE) methodologies. This paper has employed PBEE methods developed by the Pacific Earthquake Engineering Research Center (PEER) [2] adopted for general use in FEMA P-58 [3] to evaluate the repair cost and repair time of conventional, high-performance, and enhanced (i.e., beyond minimum requirements) structural designs for a low-rise commercial steel building. The conventional steel lateral load-resisting

DOI: 10.2495/SAFE-V7-N4-532-544

systems include: special moment resisting frame (SMRF), special concentrically braced frame (SCBF), and buckling restrained braced frame (BRBF); high-performance systems include: base-isolated intermediate moment resisting frame (BI-IMRF), base-isolated ordinary concentrically braced frame (BI-OCBF), base-isolated buckling restrained braced frame (BI-BRBF), and viscously damped moment frame (VDMF); enhanced systems include beyond the code minimum designs of: SMRF, BI-IMRF, and VDMF, designated as SMRFe, BI-IMRFe, and VDMFe. The conventional and high-performance buildings are designed to satisfy close to minimum code requirements for a site with high seismic hazard representative of western North America. The performance of the considered building designs is evaluated and compared for three seismic hazard levels: 2%, 10%, and 50% probabilities of exceedance in 50 years (referred to as 2% in 50-year, 10% in 50-year, and 50% in 50-year).

This paper illustrates the ability of simplified PBEE procedures to identify cost-effective strategies for reducing earthquake-induced costs during the preliminary stages of seismic design. To aid understanding of the relative performance of the considered systems, median values of key engineering demand parameters (EDPs) (i.e., maximum and residual interstory drifts and floor accelerations) are compared at different hazard levels. The paper further highlights the factors that contribute significantly to the economic losses and the ability of high-performance or enhanced structural systems in reducing these losses.

2 WORKFLOW ANALYSIS WITHIN PBEE FRAMEWORK

The workflow analysis within PBEE framework necessary for evaluation of repair cost and repair time is presented herein. The hazard model and ground motion selections for the considered site were adopted from Baker *et al.* [4]. These ground motions were used in conjunction with nonlinear numerical models of different building configurations that simulated the full range of responses that the facility might experience throughout its lifetime. The computer software OpenSees [5] was utilized to perform the numerical simulations. Engineering demand parameters were monitored during the numerical simulations and used in conjunction with PACT [6] to calculate cost and time needed to repair each damaged component (structural and non-structural) for each realization of Monte Carlo simulations at every considered hazard level. The repair costs for all components of the building at one hazard level were summed for each realization to obtain the cumulative distribution function of the repair cost for that hazard level. To estimate the cumulative distribution function of the repair time for a specified hazard level, the repair times of components along with their damage states and number of damaged units were used in conjunction with the repair time model developed by Terzic *et al.* [7]. Cost and time needed for a building repair is presented utilizing the median and 90th percentile outcomes.

2.1 Building configurations

The study considered a three-story steel office building located on a site representative of the high seismic hazard characteristic of western North America. The basic building plan dimensions are 120 ft × 180 ft, with a bay spacing of 30 ft in each direction and a typical story height of 15 ft, with the exception of the first story height of 17 ft for SMRFs and VDMFs. The building is located on stiff soil (site class D with reference shear wave velocity = 180 to 360 m/s). Code spectral accelerations were selected to be $S_s = 2.2g$ for short periods and $S_1 = 0.74g$ at a period of 1 sec, which are representative of many locations in California.

All systems were designed by professional engineers to meet minimum code standards for design according to the Equivalent Lateral Force Method [8]. Building configurations and specifications are provided in Fig. 1 and Table 1. The fixed base and base-isolated moment resisting and concentrically braced frames were designed by Forell/Elsesser Engineers, Inc [9], Miyamoto International designed VDMFs, Ugljesa Terzic S.E. designed fixed base and base-isolated BRBF, and the authors designed the enhanced versions of SMRF and BI-IMRF.

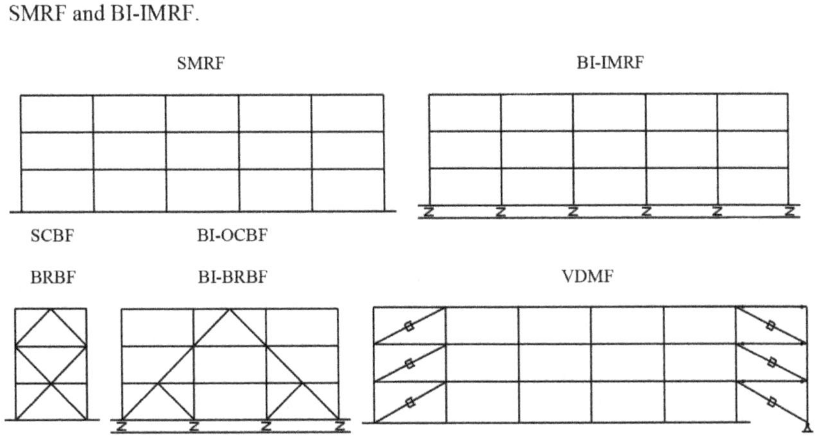

Figure 1: Configurations of different lateral load-resisting systems.

Table 1: Beam, column, and brace sizes of different lateral load-resisting systems.

Systems	Elements	Base	Story 1	Story 2	Roof
SMRF	Beams	–	W33x141	W33x130	W102x102
	Columns	–	W14x370	W14x370	W14x211
SMRFe	Beams	–	W14x500	W14x500	W14x370
	Columns	–	W36x182	W36x182	W33x130
BI-IMRF	Beams	W24x94	W24x86	W24x76	W18x60
	Columns	–	W14x176	W14x176	W14x109
VDMF	Beams	–	W18x86	W18x65	W18x60
	Columns	–	W14x145	W14x145	W14x99
SCBF	Beams	–	W27x84	W30x99	W36x150
	Columns	–	W14x176	W14x176	W14x109
	Braces	–	HSS12x12x0.625	HSS10x10x0.625	HSS8x8x0.5
BI-OCBF	Beams	W24x94	W24x84	W24x76	W18x60
	Columns	–	W12x72	W12x72	W12x72
	Braces	–	HSS8x8x0.625* HSS6x6x0.375**	HSS8x8x0.5	HSS6x6x0.625
BRBF	Beams	–	W18x86	W18x86	W18x86
	Columns	–	W12x120	W12x120	W12x96
	Braces	–	400 kips	300 kips	200 kips
BI-BRBF	Beams	W24x94	W18x86	W18x86	W18x86
	Columns	–	W12x96	W12x96	W12x96
	Braces	–	250 kips* 150 kips**	200 kips	150 kips

*External Brace, **Internal Brace

The SMRF was designed with a response modification factor (R) of 8 and an interstory drift limit of 2.5%, and utilized reduced beam sections (prequalified beam-to-column connections). The enhanced design of SMRF (i.e, SMRFe) utilized larger sections for beams and columns to limit interstory drifts to 1%. Compared to SMRF, an isolated IMRF was designed utilizing lower R factor (1.69) and drift limit (1.5%). It uses simpler connection details and does not require a strong column-weak girder design approach. The isolation system utilized triple pendulum (TP) friction bearings, 24 in. in diameter, designed to accommodate displacements expected to occur at Maximum Considered Earthquake (MCE)-level shaking. Characteristics of these isolators, designated as ID, are given in Table 2. The enhanced BI-IMRF (i.e., BI-IMRFe) utilized larger isolators (30 in. in diameter) which reduced design forces by about 33% through the increase of the effective period of the isolation system by 40% (designated as IDe in Table 2). The SCBF and BRBF utilized chevron configuration of braces and were designed with R factors of 6 and 8, respectively, and interstory drift limit of 2.5%. Compared to a SCBF and BRBF, the isolated OCBF and BRBF were designed utilizing lower R factors (1.22 and 2, respectively) and drift limit (1.5%). To increase the robustness of the braces under the near-fault ground shaking, the braces of the OCBF were selected to have 1.5 to 2 times larger area than what is minimally needed. An alternative configuration of braces was used for the isolated OCBF and BRBF in an attempt to minimize tensile forces and uplift in the bearings. They utilized the same isolation system as BI-IMRF. The VDMF was designed for 0.75% of the seismic base shear and R factor of 8. To limit interstory drifts to 2.5%, dampers of VDMF had a stiffness of 2000 kips/in., viscous damping of 32 kip/(in./s), and velocity exponent of 0.5. The enhanced design of VDMF utilized larger dampers that had a viscous damping of 135 kip/(in./s) at first two stories and 35 kip/(in./s) at the third story to limit interstory drifts to 1%, which is a common engineering practice.

2.2 Ground motion selection

The ground motions used in the analyses are adopted from Baker *et al.* [4]. Forty ground motion records were selected for a site in Oakland, CA, to represent the ground motion hazard at each of three hazard levels: 2%, 10%, and 50% probabilities of exceedence in 50 years (referred to as 2% in 50-year, 10% in 50-year, and 50% in 50-year). These ground motions were selected and scaled to match the uniform hazard spectrum and its variance associated with the causal events for the Oakland site. Although uniform hazard spectrum cannot be interpreted as the response spectrum associated with any single ground motion, such selection of ground motions has the advantage that their amplitude at any given period has approximately the same probability of exceedence. For this comparative study, this is a useful property as a single set of ground motions can be used to analyze structures sensitive to

Table 2: Isolation system parameters.

	*DBE		**MCE	
Isolator properties	ID	IDe	ID	IDe
Effective period	2.77 sec	3.95 sec	3.07 sec	4.35 sec
Effective damping	24.2 %	22.9 %	15.8 %	15.1 %
Isolator displacement	12.7 in.	16.1 in.	24.3in.	30 in.

*Design basis earthquake, **Maximum considered earthquake

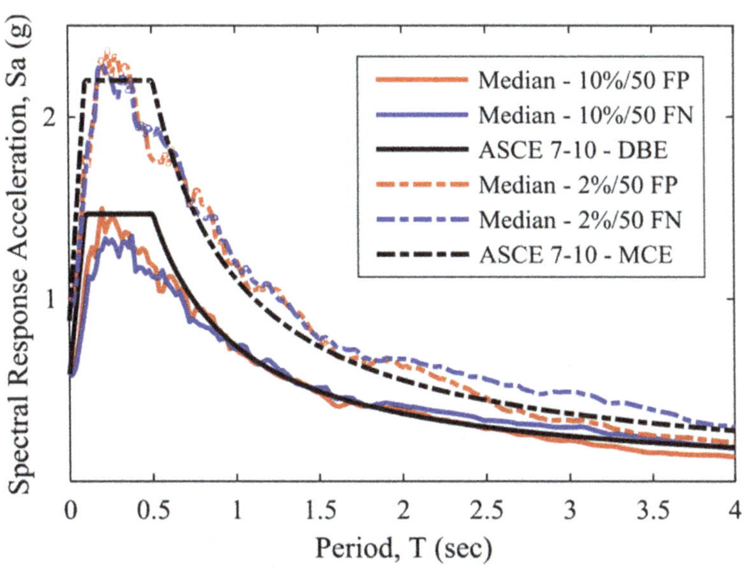

Figure 2: Comparison of code DBE and MCE spectra with the median pseudo-accelerations for the fault-normal and fault-parallel components of the 10% in 50-year and 2% in 50-year hazard level events used in analysis.

excitation at differing periods with ground motions that are comparably 'intense' in their excitation of each building. Figure 2 highlights agreement between spectra for the code design basis earthquake (DBE) and MCE and the median pseudo-acceleration response spectra for the fault parallel (FP) and fault normal (FN) components of the 10% in 50-year and 2% in 50-year hazard level events.

2.3 Numerical models and methods

To simplify the analysis for this study, time history analyses were performed on appropriately modeled two-dimensional (2D) frames. The lateral load resisting frames described above are located only on the perimeter of the building. Gravity-load-only type connections were used elsewhere in the structure. Numerical modeling assumptions are described below:

1. Half of the lateral floor mass was assigned at each floor of the 2D frame, equally distributed among all nodes of that floor.
2. Floor slabs were assumed to be axially inextensible.
3. P-Δ effects from the gravity columns were taken into account by using a single leaning column [10].
4. The effects of large deformations of beam and column elements were accounted for utilizing P-Δ nonlinear geometric transformation.
5. The frames were subjected to two horizontal components of ground motion (one at a time).

The numerical model was implemented in OpenSees [5]. Centerline models were used for all systems. Beams of the SMRF utilized a reduced beam section (RBS) and were modeled using a series of elastic elements and concentrated plastic hinges, with stiffness of the elastic beam and initial stiffness of the plastic hinges modified as suggested by Ibarra and Krawinkler [11]. Plastic hinges were defined using zero-length elements (rotational springs). The hinge moment-rotation relationship was defined using *Hysteretic Material* of OpenSees. The moment-rotation relationships for RBS connections were developed based on recommendations from PEER/ATC [12], calibrated to closely resemble experimental results for RBS connection tests performed by Uang [13].

Columns of the SMRF as well as beams and columns of all other systems considered in this study were modeled utilizing two element types. The portions of beams and columns along the clear element length were modeled utilizing force-based beam-column elements of OpenSees, which accounts for the distribution of plasticity along the element. Sections were discretized into fibers. The Menegotto-Pinto hysteretic model was used to model behavior of steel fibers. Low-cycle fatigue failure of beams and columns was accounted for by using an OpenSees fatigue model that has been well calibrated for I-sections [14]. Panel-zone regions of beam-column connections were modeled with elastic elements. Braces of the SCBF, OCBF, and BRBF were modeled as suggested by Uriz and Mahin [14]. Gusset plate regions of braced frame connections were modeled with elastic elements. Viscous dampers of VDMF were modeled utilizing two node link elements in conjunction with *ViscouseDamper Material* of OpenSees.

For the isolated systems, isolators were modeled with zero-length elements (horizontal springs), one beneath each column of the structural frame. The bearing model represented the sum of the properties of all building isolators divided by two (number of lateral-load resisting frames in one direction) and divided by the number of columns in one frame. Vertical displacements and rotations at the top and the bottom of isolators were assumed to be fixed. To represent hysteretic behavior of bearings, tri-linear models were used. Yield points and equivalent periods of the tri-linear bearing models for the two isolation designs are given in Table 3.

Damping was assigned to the frames based on PEER/ATC [12] recommendations. The damping was modeled utilizing Rayleigh damping model and damping ratio of 3% for all lateral load-resisting systems. For the fixed-base building, mass and tangent stiffness proportional Rayleigh coefficients were calculated based on two periods. The first (T_1) and third (T_3) periods were selected for the 50% in 50-year hazard level, and $1.5T_1$ and T_3 for the 10% in 50-year and 2% in 50-year hazards. The first period was elongated 1.5 times to account for the change in period due to the nonlinear deformations of the system. For the isolated building, the damping was assumed to be proportional only to the tangent stiffness of the structure as suggested by Hall [15]. The stiffness-proportional damping was calculated from the fundamental period of the structure T_1 for the 50% in 50-year hazard, and T_{eff} for the 10% in 50-year and 2% in 50-year hazards, where T_{eff} is the effective fundamental period of the seismically isolated structure (Table 2). Viscous damping contributions associated with

Table 3: Yield points and equivalent periods of the tri-linear bearing models of ID and IDe designs.

Bearing	1st branch			2nd branch			3rd branch
Design	d_1 [in.]	V_1/W	T_1 [sec]	d_2 [in.]	V_2/W	T_2 [sec]	T_3 [sec]
ID	2.18	0.08	1.75	6.64	0.13	3.09	4.0
IDe	0.82	0.03	2.0	5.08	0.06	3.82	5.0

stiffness were based on the stiffness of elements used to model the lateral-load resisting frames except for: (1) elements of the leaning column; (2) highly rigid truss elements that link the frame and leaning column; and (3) zero-length elements used to model beam plastic hinges of SMRF and bearings of BI-IMRF. Because the stiffness of the elastic beam elements of SMRF was modified, the stiffness proportional damping coefficients used with these elements were also modified (following Zareian and Medina [16]).

2.4 Loss models

Two loss metrics were used to assess the performance of considered structural systems: repair cost as a percentage of the building replacement cost and repair time. The probable repair costs and repair time were determined for each structural system at each hazard level utilizing the *FEMA P-58* computer software Performance Assessment Calculation Tool (PACT) [6]. To account for the many uncertainties affecting calculation of seismic performance, the *FEMA P-58* methodology uses a Monte Carlo procedure [3]. Loss modeling with PACT consists of three stages:

1. The user first inputs basic building information, such as replacement cost, occupancy type, footprint, and story height;
2. The user then defines the type and quantity of structural and non-structural components; and
3. Computed engineering demand parameters (EDPs) (maximum story drifts, maximum absolute floor horizontal acceleration, and peak residual story drifts) are input for each ground motion at each hazard level in each planar direction.

The type and quantities of structural components stem from the structural design (Table 1), while the type and quantities of non-structural components for a commercial building were determined using the normative quantities recommended by *FEMA P-58* [3]. In PACT, each building component is associated with a fragility curve that correlates EDPs to the probability of that item reaching a particular damage state. The component's damage is then related to a repair cost and repair time utilizing consequence functions. The repair costs for all components of the building at one hazard level are then summed for each realization to obtain the cumulative distribution function of the repair cost for that hazard level. To estimate the cumulative distribution function of the repair time for a specified hazard level, the repair times of components along with their damage states and number of damaged units were used in conjunction with the repair time model developed by Terzic *et al.* [7]. Repair schedule and labor allocations were adopted from a case study of a three-story building by Yoo [17].

3 COMPARISON OF STRUCTURAL RESPONSES

While numerous parameters need to be considered to fully evaluate structural response, it is common to correlate performance to EDPs based on interstory drifts (ISD), floor accelerations, and residual interstory drifts. By comparing the medians of the absolute maximum responses of these EDPs for the three hazard levels considered, the relative performance characteristics of the systems studied herein can be assessed.

Compared to the SMRF, the VDMF and BRBF designed based on code-minimum requirements are effective in reducing acceleration demand at all considered hazard levels but have significantly higher median ISD demands at higher hazard levels than any other system considered in this study (Fig. 3). With the median ISD approaching 4% at the 2% in 50-year

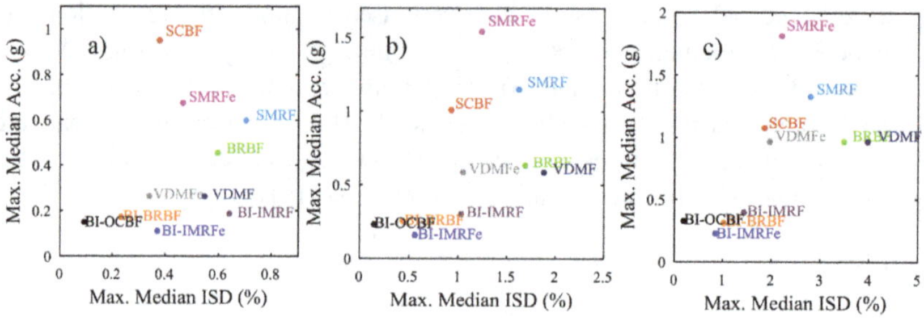

Figure 3: Maximum median interstory drift (ISD) vs. maximum median floor acceleration for the considered systems for three hazard levels: a) 50% in 50-year, b) 10% in 50-year, and c) 2% in 50-year.

hazard level, VDMF and BRBF may not be able to provide structural integrity after this level of shaking. By comparison, the enhanced viscously damped moment frame, VDMFe, achieves 30–50% ISD reduction relative to SMRF across different hazard levels through utilization of larger dampers, therefore anticipating smaller repair losses. The enhanced configuration of the moment frame, SMRFe, although effective in reducing median ISD (21–37%), amplifies floor accelerations (30–50%), not necessarily resulting in loss reduction. Lastly, the SCBF has smaller drifts and accelerations (except for 50% in 50-year hazard) than SMRF which should result in less pronounced damage to non-structural components but more structural damage due to the early yielding of braces (at a median ISD of 0.35%) relative to the beams of a moment resisting frame (at a median ISD of 1.00%).

Base-isolation substantially reduces acceleration and also drifts in both, moment resisting and braced frames at all hazard levels (Fig. 3). From the systems considered in this study, the BI-OCBF has the smallest EDPs as the isolation system substantially reduces maximum median ISDs (63%, 78%, and 83%) and maximum median floor accelerations (84%, 79%, and 70%) across all three hazard levels. The BI-BRBF follows next with the comparable level of efficiency in reducing both, the ISDs and the floor accelerations. By comparison, BI-IMRF is equally effective in reducing the floor accelerations (69%, 73%, and 69%) but not as effective in reducing the ISDs (9%, 35%, and 47%). The enhanced system, BI-IMRFe, achieves more pronounced reduction of the ISDs (47%, 63%, and 67%) as it utilizes larger isolators relative to BI-IMRF.

Figure 4 shows median residual ISDs, demonstrating none or negligible residual drifts at 50% in 50-year hazard level. While the isolated systems do not experience any residual drift at the 10% in 50-year hazard level, all fixed-based system except from BRBF (0.48%) have small median residual drifts, permitted by the code of the construction practice (<0.2%). At the 2% in 50-year hazard level, the BI-OCBF and BI-IMRFe are the only systems with no residual ISD. At this level of excitation, the median ISD is very high for VDMF (1.22%) and BRBF (0.95%) requiring building replacement, and high for SMRF (0.59%) requiring structural realignment or strengthening [3].

4 LOSS EVALUATION

Two loss metrics used to assess the performance of considered structural systems: repair cost as a percentage of the building replacement cost (i.e., loss ratio) and repair time are presented in Figs 5–8. The fixed base systems have significantly greater median repair cost and repair time than the isolated systems at all considered hazard levels. The best performing fixed-base

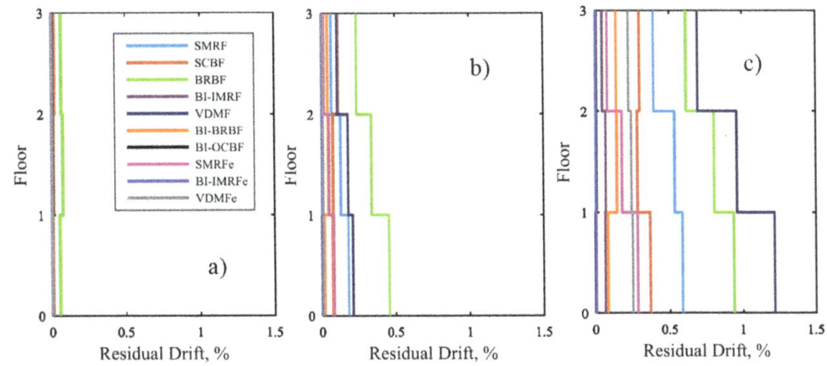

Figure 4: Median residual interstory drifts for the considered systems for three hazard levels: a) 50% in 50-year, b) 10% in 50-year, and c) 2% in 50-year.

system is VDMFe with the median loss ratio of 10% and median repair time of 60 days under the strongest hazard considered. In case of the base isolated systems, the best performer is BI-OCBF with the median loss ratio of 0.8% and median repair time of 2 days under the strongest hazard considered. The median loss ratio (Fig. 5) of the fixed base systems is in the range from 0.3% (VDMFe) to 3% (SCBF) at 50% in 50- year hazard, 3% (VDMFe) to 6% (SMRFe) at 10% in 50-year hazard, and 10% (VDMFe) to 14% (SMRF) at 2% in 50-year hazard with VDMF and BRBF requiring replacement. The median loss ratio of the base-isolated systems is very small at all hazard level, being smaller than 0.15%, 1%, and 5% for 50%, 10%, and 2% in 50-year hazards, respectively.

90th percentile loss ratios (Fig. 6) of the fixed base systems are several magnitudes higher than their median losses reaching 6%, 13%, and 100% (replacement of all systems) for the

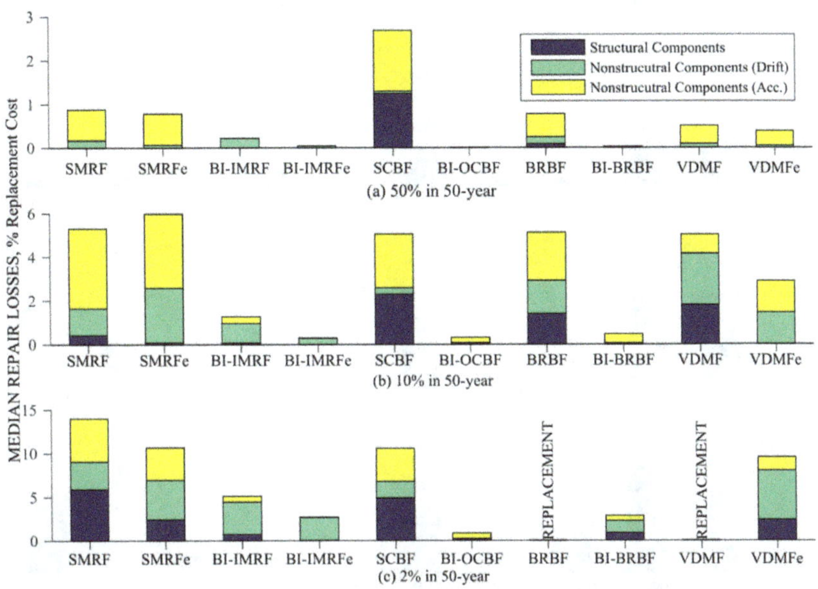

Figure 5: Median repair losses as a % of replacement cost for the considered systems and the three hazard levels.

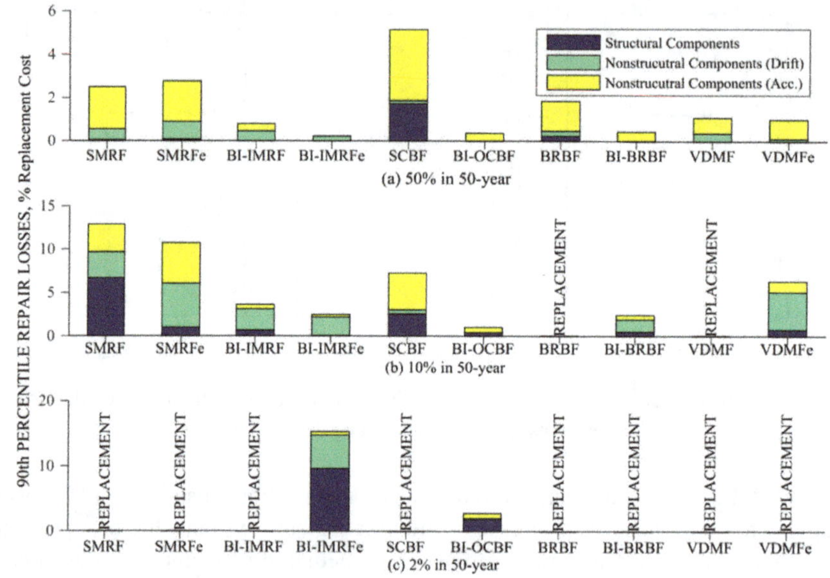

Figure 6: 90th percentile repair losses as a % of replacement cost for the considered systems and the three hazard levels.

50%, 10%, and 2% in 50-year hazards, respectively, with BRBF and VDMF requiring replacement even at 10% in 50-year hazard. For the base isolated systems, these losses are smaller reaching 0.8%, 4%, and 12% for the 50%, 10%, and 2% in 50-year hazards, respectively, with BI-IMRF and BI-BRBF requiring replacement at 2% in 50-year hazard.

The median repair time (Fig. 7) of the fixed base systems has a larger spread than the loss ratio and is in the range from 3 days (VDMFe) to 55 days (SCBF) at 50% in 50-year hazard,

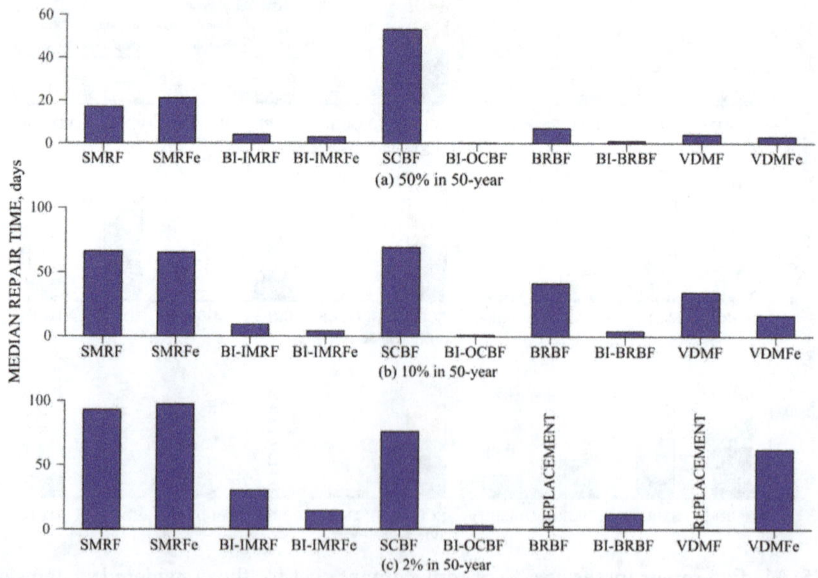

Figure 7: Median repair time for the considered systems and the three hazard levels.

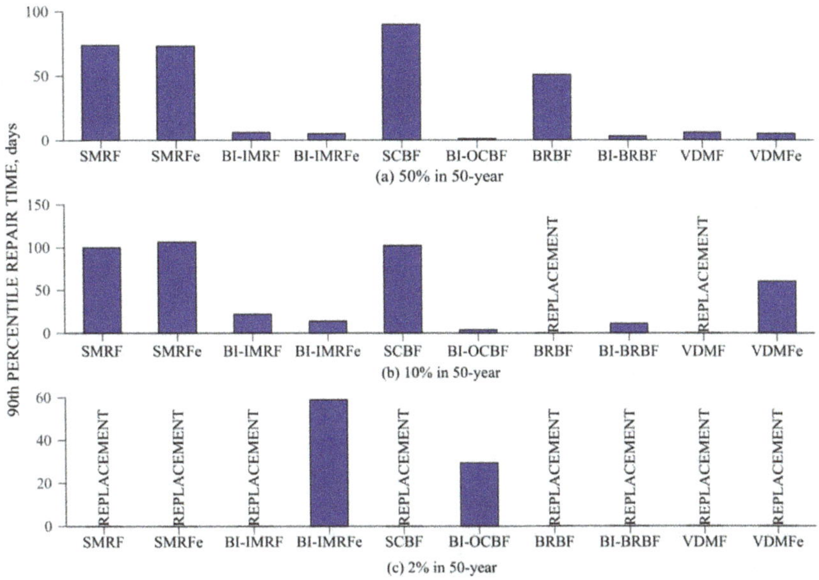

Figure 8: 90th percentile repair time for the considered systems and the three hazard levels.

20 days (VDMFe) to 70 days (SCBF) at 10% in 50-year hazard, and 75 days (VDMFe) to 100 days (SMRFe) at 2% in 50-year hazard with the exception of systems that require replacement. The median repair time of the base-isolated systems is very small at all hazard level, being smaller than 3, 8, and 20 days for 50%, 10%, and 2% in 50-year hazards, respectively.

90th percentile repair times (Fig. 8) of the fixed base systems are several magnitudes higher than their median values reaching 95, 110, and 720 days (replacement time) for the 50%, 10%, and 2% in 50-year hazards, respectively. For the base isolated systems, these losses are smaller reaching 8, 20, and 60 days for the 50%, 10%, and 2% in 50-year hazards, respectively, with BI-IMRF and BI-BRBF requiring replacement at 2% in 50-year hazard.

CONCLUSIONS

Current U.S. building codes are oriented toward preserving lives and preventing collapse rather than minimizing economic loss. As shown in this study that considers three-story steel commercial building, performance of the seven considered structural systems designed to meet the minimum and beyond code requirements ranged from poor to superior. Considering the range of hazard levels, from the systems designed to meet the minimum code requirements the least desirable are the special concentrically braced frame with 55 days of repair time at 50% in 50-year hazard level associated with the brace buckling; and the buckling restrained braced frame along with the viscously damped moment frame which at stronger intensities of shaking exhibit substantial residual interstory drifts resulting in a great probability of deeming the building irreparable. Overall, the fixed base systems had significantly greater median repair cost and repair time than their isolated configurations at all considered hazard levels. The best performing fixed-base system is an enhanced viscously damped moment frame reaching the median loss ratio of 10% and median repair time of 60 days under the 2% in 50-year hazard level. In case of the base isolated configurations, the best performer is the ordinarily concentrically braced frame reaching the median loss ratio of 0.8% and median repair time of 2 days under the 2% in 50-year hazard level. The study also demonstrated that an investment into a stronger and stiffer moment frame may not result in loss savings, while an investment into larger dampers or isolators can yield substantial savings. It is thus recommended to utilize holistically oriented performance-based earthquake evaluation methodologies in seismic design to identify the proportions and details of the system that can improve the overall performance of a structure.

ACKNOWLEDGEMENTS

I authors acknowledge financial support for this work from the Pacific Earthquake Engineering Research Center under the projects 2002-NCEEVT. We are grateful for the Applied Technology Council providing the latest release of the PACT software application. The conclusions, observations, and findings presented herein are those of the authors and not necessarily those of the sponsors.

REFERENCES

[1] Miranda, E., Mosqueda, G., Retamales R. & Pekcan, G., Performance of nonstructural components during the 27 February 2010 Chile earthquake. *Earthquake Spectra*, **28**(S1), pp. S453–S471, 2012.
https://doi.org/10.1193/1.4000032

[2] Miranda, E. & Aslani, H., *Probabilistic response assessment for building specific loss estimation*. Pacific Earthquake Engineering Research Center, University of California, Berkeley, CA, USA, 2003. Report PEER 2003/03.

[3] FEMA. *Next-generation methodology for seismic performance assessment of buildings*. Prepared by the Applied Technology Council for the Federal Emergency Management Agency, Report No. FEMA P-58, Washington, DC., 2012.

[4] Baker, J.W., Lin, T., Shahi, S.K. & Jayaram, N., *New ground motion selection procedures and selected motions for the PEER transportation research program*. Pacific Earthquake Engineering Research Center, University of California, Berkeley, CA, USA, 2011. Report PEER 2011/03.

[5] McKenna, F. & Fenves, G.L. *Open system for earthquake engineering simulation (OpenSees)*. PEER, University of California, Berkeley, CA, 2004.

[6] Applied Technology Council (ATC). *Performance assessment computation tool (PACT v3.0.3)*. Applied Technology Council, Redwood City, CA, 2012.

[7] Terzic, V., Yoo, D. & Aryan, A. H., *Repair time model for buildings considering the earthquake hazard*. SEAOC Convention, Maui, HI, 2016.

[8] American Society of Civil Engineers. *Minimum design loads for buildings and other structures*. ASCE 7-10, ASCE, Reston, VA, 2010.

[9] *Structural Design Calculations and Drawings, Forell/Elessesser Engineers*, available at: http://www.neng.usu.edu/cee/faculty/kryan/NEESTIPS/PBEE_study.html. (accessed March 15, 2015).

[10] Gupta, A. & Krawinkler, H., *Seismic demands for performance evaluation of steel moment resisting frame structures*. Stanford University, Stanford, CA, pp. 26–28, 1999.

[11] Ibarra L.F. & Krawinkler, H., *Global collapse of frame structures under seismic excitations* (Rep. No. TB 152), The John A. Blume Earthquake Engineering Center, Stanford University, Stanford, CA, 2005.

[12] PEER/ATC. *Modeling and acceptance criteria for seismic design and analysis of tall buildings* (PEER/ATC-72-1). prepared by the Applied Technology Council in cooperation with PEER, Redwood City, CA, 2010.
[13] Uang, C.M., *Cyclic response of RBS moment connections: loading sequence and lateral bracing effects*. University of San Diego, San Diego, California, 2000. (Report No. SSR-99/13)
[14] Uriz, P. & Mahin, S.A., *Toward earthquake-resistant design of concentrically braced steel-frame structures*. Pacific Earthquake Engineering Research Center, University of California, Berkeley, CA, USA, 2008. (Report PEER 2008/08.)
[15] Hall, J.F., Problems encountered from the use (or misuse) of Raleigh damping. *Earthquake Engineering and Structural Dynamics,* **35**(5), pp. 525–545, 2005. https://doi.org/10.1002/eqe.541
[16] Zareian, F. & Medina, R.A., A practical method for proper modeling of structural damping in inelastic plane structural systems. *Computers & Structures*, **88**(1–2), pp. 45–53, 2010.
https://doi.org/10.1016/j.compstruc.2009.08.001
[17] Yoo, D.Y., *Repair time model for different builidnig sizes considering the earthquake hazard*. Master Thesis, California State University Long Beach, CA, 2016.

CODE INADEQUACIES DISCOURAGING THE EARTHQUAKE-BASED SEISMIC ANALYSIS OF BUILDINGS

MARIA CRISTINA PORCU
Department of Mechanical, Chemical & Materials Engineering, University of Cagliari, Italy.

ABSTRACT
Despite being the more accurate method to design earthquake-resistant buildings, the non-linear dynamic analysis still remains the least used in practice. The paper shows that some code inadequacies may even contribute to ward off the designer from this kind of seismic analysis. To this purpose, non-linear analyses under different sets of spectrum-compatible earthquakes and by considering two distinct plastic-hinge models were carried out for a multistory reinforced concrete (r/c) frame designed according to the Eurocode 8 provisions. It has been shown that owing to the inadequate limits imposed to the spectrum-compatible earthquakes, the European codes of practice may lead to significant inconsistencies in the seismic design of structures through the time-history non-linear analysis.
Keywords: earthquake-resistant buildings, EC8 inadequacies, spectrum-compatible earthquakes, time-history non-linear seismic analysis.

1 INTRODUCTION

The seismic design philosophy typically requires that buildings may withstand events of moderate severity deforming in the elastic range, without suffering any irreversible structural damage. Conversely, during strong earthquakes structures are required to behave in a dissipative ductile way, even getting damaged severely but without collapsing. These achievements have to be reached regardless to which type of code-based analysis (linear or non-linear) is chosen. If a linear analysis is performed, in fact, a reduced response spectrum can be adopted to design the structure so as it can withstand moderate earthquakes in the elastic range, while compliance with the detailed code provisions (local ductility and strength hierarchy rules) ensure that the plastic deformation capacity of the system can satisfy the demand under stronger earthquakes. If a non-linear analysis is carried out instead, the elastic and post-elastic behaviour of the structure can be straight monitored. Both a non-linear static analysis (often referred to as 'pushover') and a non-linear time-history analysis (NLTHA) are allowed by codes. The latter is expected to be the more powerful since it obtains the deformation time-history of the system through a numerical integration of the differential equations of motion under the considered earthquake.

Despite its indubitable potentials, the NLTHA is still the least applied method of seismic analysis. This may be due to the difficulties that even an experienced designer has to overcome when (i) modelling the non-linear behaviour of the system, (ii) defining the suite of spectrum-compatible earthquakes, (iii) carrying out the numerical calculations and even (iv) interpreting the results and choosing the requirements to be met. To help the designer, some codes of practice, i.e. FEMA-356 [1] and ATC-40 [2], give default properties for r/c plastic hinges, which are implemented by finite element structural programs such as SAP2000 [3]. User-friendly software tools, as for instance REXEL [4], are also available to obtain code-based selections of accelerograms.

Nevertheless, the NLTHA is still not widely adopted by practitioners and even the scientific community is sceptical about the adoption of this kind of analysis in the design practice

DOI: 10.2495/SAFE-V7-N4-545-556

due to the risks that its inexpert use may involve. A warning against these possible risks was given for instance in Ref. [5], where some of the difficulties that the non-linear dynamic analysis of reinforced concrete buildings may involve were put in evidence. In the same paper the likely dependence of the results of a NLTHA on the chosen structural model (fibre discretization model or plastic-hinge model) and on the adopted computer program (SAP2000 or SeismoStruct) were also highlighted.

With the aim to help make the non-linear dynamic analysis less difficult for the designer, the present paper further investigated on the reasons of its unattractiveness. To this end, a three-story reinforced concrete (r/c) frame was considered and assumed to be built in a high-seismicity Italian area. The frame was preliminarily designed through the modal response spectrum analysis (MRSA) according to Eurocode 8 (EC8) [6], so as to comply with all the code requirements ensuring adequate local ductility and strength hierarchy rules. In a second stage, the same building was spatially modelled in SAP2000 through a lamped plasticity idealization to perform a NLTHA. Two different mechanical models of plastic hinges and four suites of seven spectrum-compatible earthquakes were considered in the numerical investigation. The difficulties to be faced up by the designer during both the modelling and the calculation phases were evidenced.

A key question raises from the present investigation: are buildings designed through the MRSA unable to withstand earthquakes compatible with the design response spectrum or rather are the earthquakes that EC8 assumes as spectrum-compatible too strong to be considered as such? Answering to this question highlights some code inadequacies, which should be overcome to allow a wider use of the NLTHA in the engineering practice.

2 CASE STUDY

A three-story r/c frame, regular both in plan and in elevation, supposed to be built in a high seismicity Italian area (Messina), was considered. The frame was initially designed through a MRSA, so as to be consistent with the no-collapse and damage limitation requirements as well as with all the ductility and strength hierarchy rules provided by EC8. The assumption of a high ductility class was made. Geometrical data, materials' properties, loads at each level of the longitudinal frames and some reinforcement details are collected in Fig. 1.

The elastic and the design response spectra adopted in the present investigation are provided in Fig. 2. They are obtained by referring to the no-collapse requirement and assuming soil type B, no topographic amplification effects and a building lifetime of 50 years. The values of all the parameters defining the shape of the response spectrum are listed in Fig. 2. They were determined following the Italian standard [7] in accordance with the EC8 provisions [6; 3.2.2.2]. The spectra in Fig. 2 are relevant to the horizontal components of the earthquake, since the vertical component may be neglected in this case, according to EC8.

It should be noted that the design spectrum in Fig. 2 was obtained by reducing the elastic spectrum through the behaviour factor q. Such a spectrum has to be adopted when performing a MRSA. On the contrary, the elastic (unreduced) spectrum plotted in the same figure should be considered when defining the suite of spectrum-compatible earthquakes to be used in the NLTHA.

3 PERFORMING THE NLTHA ACCORDING TO EC8

The r/c frame illustrated in Fig. 1 was designed through a MRSA with reference to the design spectrum given in Fig. 2 and to all the detailing rules provided by EC8. In accordance with the code design philosophy this frame is expected to be able to withstand any earthquake

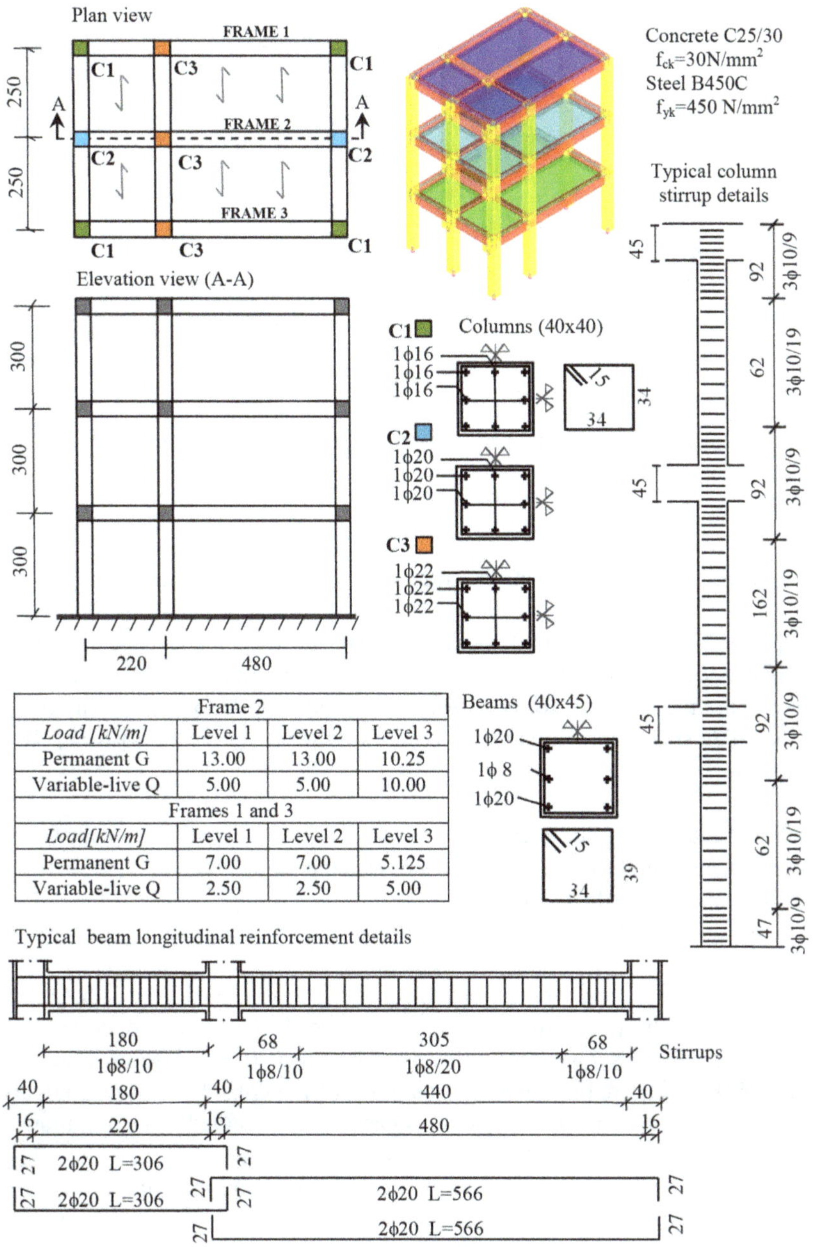

Frame 2			
Load [kN/m]	Level 1	Level 2	Level 3
Permanent G	13.00	13.00	10.25
Variable-live Q	5.00	5.00	10.00
Frames 1 and 3			
Load[kN/m]	Level 1	Level 2	Level 3
Permanent G	7.00	7.00	5.125
Variable-live Q	2.50	2.50	5.00

Figure 1: The r/c three-story building considered in the study.

compatible with the elastic spectrum given in Fig. 2. The criteria that decide if an earthquake is compatible with a given spectrum are stated by codes. To check whether the building concerned was able or not to withstand earthquakes that EC8 considers spectrum-compatible, a series of non-linear analyses were performed. The steps the designer has to follow to perform a NLTHA are firstly examined. The general-purpose structural program SAP2000 [3] and the software REXEL [4] were adopted to this end.

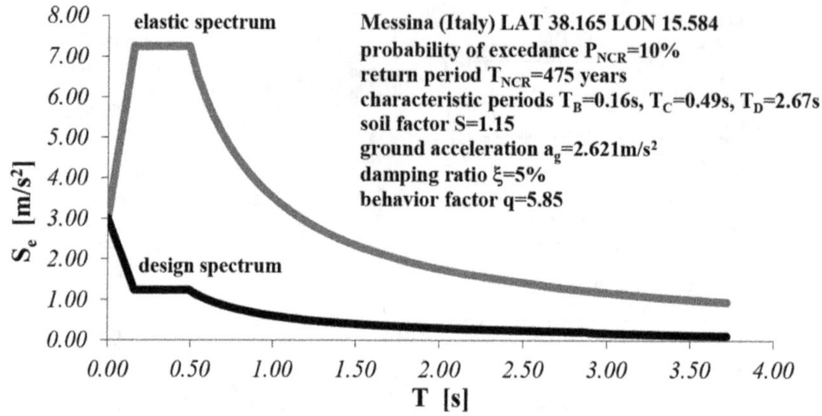

Figure 2: Elastic and design response spectra for the no-collapse requirement (horizontal component).

3.1 Step 1: defining a suite of spectrum-compatible real accelerograms

According to EC8 [6; 3.2.3.1.2, (4)], the designer should define in advance a suite of at least three ground acceleration time histories consistent with the reference spectrum. Some requirements, not recalled here for brevity, are provided by EC8 to achieve this goal. To find suites of recorded spectrum-compatible acceleration time histories the software REXEL [4] requires the user:

a) to define the target elastic spectrum;
b) to select an earthquake database and some related parameters;
c) to set how many records for the suite;
d) to give the spectrum matching parameters.

In the present case study, (a) the elastic spectrum of Fig. 2 was considered as target spectrum; (b) reference to the European Strong Motion Database (ESD) [8] was done by assuming a magnitude range $6 \le M_w \le 7$ and an epicentre distance $\Delta \le 20km$; (c) suites of seven records were considered, as required by EC8 (section 4.3.3.4.3) to consider the mean effects, rather than the maxima, on the structure. On the other hand, it can be noted that a minimum of seven records for suite is prescribed by the Italian Code [7].

As for issue (d), both an upper and a lower tolerated deviation of the average spectrum from the target spectrum should be specified. EC8 states a lower tolerance value (no more than 10%), but does not provide any indication about the upper bound. The designer is left alone in making this choice despite the fact that it may drastically affect the final results of the analysis.

To evaluate the influence of the designer choices, two upper tolerance values, namely 30% and 20%, were considered in the investigation. Since REXEL [4] allows to obtain combinations of accelerograms whose average meets the lower and upper bound tolerance being not manipulated (original records) or if linearly scaled in amplitude (scaled records), both the options of original and scaled records were set. Table 1 summarizes the main options set to obtain four suites of spectrum-compatible earthquakes to use in the investigation.

Table 1: Suites of earthquakes compatible with the elastic spectrum of Fig. 2.

suite	number of histories	earthquake database	M_w	Δ [km]	lower tolerance	upper tolerance	records scaled
Suite-1	7	ESD	6÷7	>20	10%	30%	no
Suite-2	7	ESD	6÷7	>20	10%	30%	yes
Suite-3	7	ESD	6÷7	>20	10%	20%	no
Suite-4	7	ESD	6÷7	>20	10%	20%	yes

For each case in Table 1, REXEL returned hundreds of suites of adequately qualified records, sorted for dispersion of the spectra with respect to the target spectrum. The first set of each list (relevant to the lowest dispersion), was always chosen in this study. Tables 2 and 3 give the characteristics of Suite-1 and Suite-2. For the sake of brevity, the tables with the data relevant to Suite 3 and Suite 4 are not provided herein. It can be observed that the same waveforms may appear in different suites, scaled by factors SFx and SFy (see Table 2 and Table 3). Figures 3a–d plot the spectra of the records of the four suites, together with the average, the lower and the upper tolerance spectra and the reference spectrum. Note that two acceleration histories (by the directions x and y) are actually given for each earthquake by REXEL and, therefore, 14 record spectra are given in each diagram of Fig. 3.

Table 2: Suite-1 (original records) as obtained from REXEL.

Waveform ID	Earthquake ID	Earthquake Name	M_w	Δ [km]
146	65	Friuli (aftershock)	6	14
291	146	Campano Lucano	6.9	16
1715	474	Ano Liosia	6	14
230	108	Montenegro (aftershock)	6.2	8
6334	2142	South Iceland (aftershock)	6.4	11
199	93	Montenegro	6.9	16
6263	1635	South Iceland	6.5	7

Table 3: Suite-2 (scaled records) as obtained from REXEL.

Waveform ID	Earthquake ID	Earthquake Name	M_w	Δ [km]	SFx	SFy
232	108	Montenegro (aftershock)	6.2	20	5.2458	5.4141
1713	474	Ano Liosia	6	18	2.7018	3.5011
1711	474	Ano Liosia	6	20	3.4359	3.863
535	250	Erzincan	6.6	13	0.77013	0.58428
230	108	Montenegro (aftershock)	6.2	8	2.5057	1.1195
594	286	Umbria Marche	6	11	0.57168	0.64726
291	146	Campano Lucano	6.9	16	1.9254	1.7032

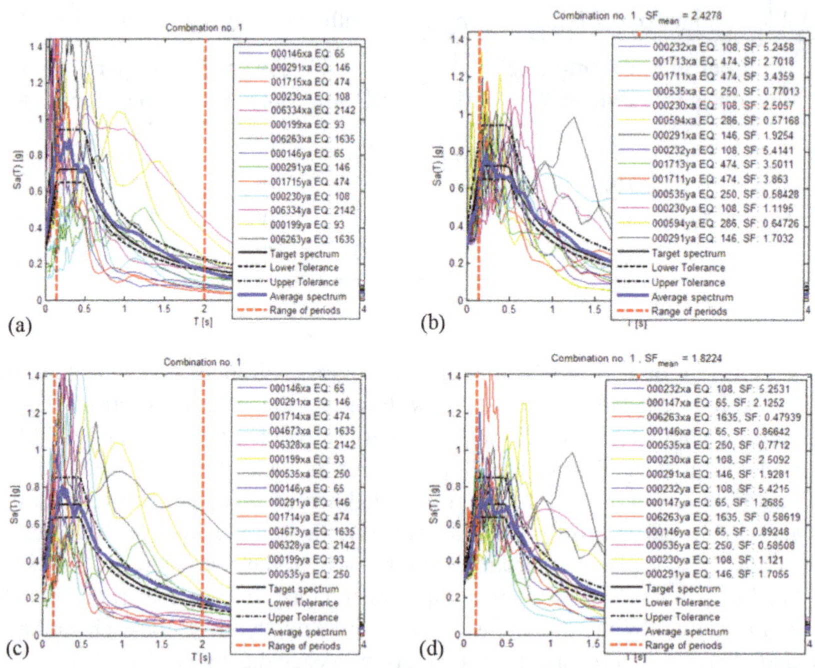

Figure 3: Response spectra for (a) Suite-1; (b) Suite-2; Suite-3; (d) Suite-4, from REXEL.

3.2 Step 2: modelling the nonlinear behaviour of the structure

A concentrated plasticity three-dimensional model was adopted to describe the seismic behaviour of the building in Fig. 1. Linear-elastic *frame elements* were adopted to model beams and columns while non-linear *hinge elements* were put at the end of all elements to account for the post-elastic rotation of the extreme sections. Slab masses (from loads given in Fig. 1) were assigned to beams while *rigid diaphragm constraints* accounted for the slab stiffening effects.

The default properties provided by SAP2000 for r/c *hinge elements* are based on American codes FEMA-356 [1] and ATC-40 [2]. Although relieving modelling work considerably, default hinges of SAP2000 should be used cautiously for design guidelines different from the ones adopted in the USA, so as to prevent unreasonable results. The possible effects of the misuse of the default-hinge properties assumed by SAP2000 were also investigated in Ref. [9], with reference to the pushover analysis. It was shown that some parameters (meeting the American detailing rules in the SAP default hinge model), such as the plastic hinge length and the transverse reinforcement spacing, may play a significant role on the displacement capacity of r/c frames and, therefore, the adoption of user-defined plastic hinges is generally preferable to catch the building non-linear behaviour [9].

It can be observed that EC8 does not suggest any specific plastic-hinge model to be implemented in the finite element analysis. To evaluate the impact of the choice of the plastic-hinge model on the results of the code-based NLTHA, both a Default-Hinge (DH) model, as assigned by SAP2000, and a User-Hinge (UH) model, based on the Dolšek–Fajfar moment–rotation relationship [10], were considered. The moment-curvature diagrams of the two models, simplified as required by SAP2000, are given in Fig. 4. The values to be assigned to

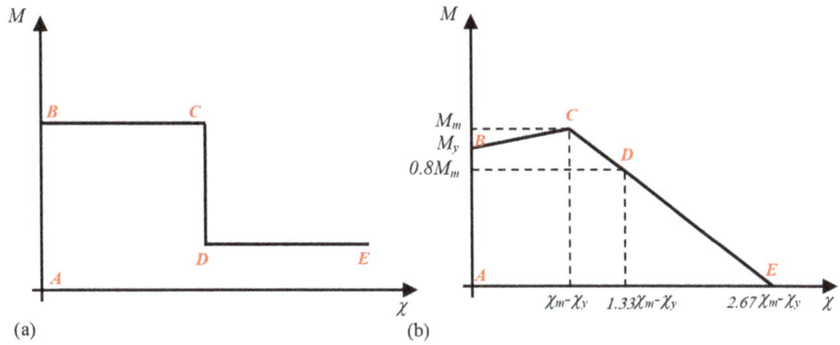

Figure 4: Moment–curvature diagram of (a) Default-Hinge and (b) User-Hinge.

points B, C, D and E vary depending on type of element, kind of materials, longitudinal steel percent and axial load level. They were automatically assigned for the DH, while a complete moment-curvature analysis was needed for the UH, involving also a preliminary evaluation of axial loads in columns (due to dead loads and 30% of live loads). This step required the use of an external application and a significant expenditure of time to compute the large amount of hysteretic relationships (as also evidenced in Ref. [5]).

A kinematic hardening hysteretic model for unloading was implemented for the considered both models. According to EC8, the plastic hinge length should be at least:

$$\begin{aligned} &beams \quad && l_{crB} = h_b \\ &columns \quad && l_{crC} = \max\{h_c; l_{cl}/6; 0.45\} \quad \text{[in metres]}, \end{aligned} \tag{1}$$

where h_b and h_c are the section height of beams and columns, while l_{cl} is the column clear length. Cautiously, $l_{crB} = 1.5h_b = 0.675m$ and $l_{crC} = l_{cl}/6 = 0.50m$ was assumed for both models, also in accordance with the Italian code [7]. Note, that the DH elements needed to be converted into user-editable to allow modifying length and unloading law.

A Rayleigh classic damping was supposed. The damping matrix was thus obtained as a linear combination between the mass and the stiffness matrixes:

$$[d] = a[m] + b[k]. \tag{2}$$

The damping factor of the jth mode can be expressed, in this case, as:

$$\xi_j = \frac{1}{2}\left(\frac{a}{\omega_j} + b\omega_j\right). \tag{3}$$

A constant value $\xi_j = \xi = 0.05$ was assumed for each mode. The values of the coefficients appearing in eqn (2) in this case are then a = 1.2526 s^{-1} and b = 1.995E-3 s.

3.3 Step 3: setting the parameters for the numerical analysis with SAP2000

Before starting a transient analysis some parameters should be set in SAP2000. After assigning the initial conditions under gravitational loads, the time step and the number of steps has to be chosen. Of course, shorting the time step generally leads to much precise results but it can also highly increase the computing effort.

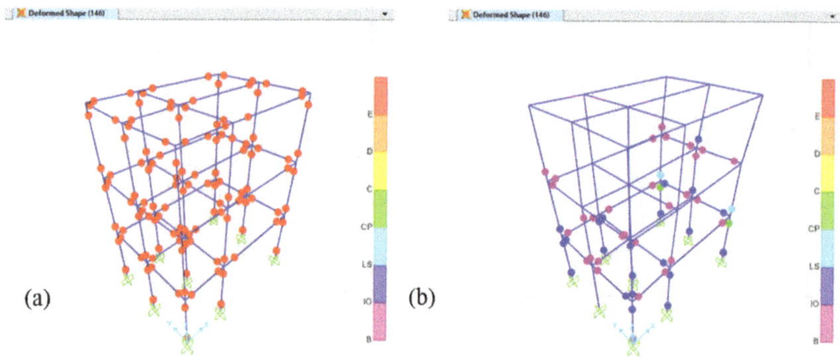

Figure 5: Deformation envelope for (a) $\alpha = 0$ and (b) $\alpha = -0.2$ (waveform 146 of Suite 1; UH model).

The Hilber–Hughes–Taylor (HHT) method (also known as alpha method) [11] is adopted by default by SAP2000 for the direct integration of the motion equations. Based on the Newmark formulation, it defines the two parameters β and γ as a function of a single parameter α:

$$\beta = \frac{(1-\alpha)^2}{4}, \gamma = \frac{1}{2} - \alpha. \tag{4}$$

The range of practical interest for α is $-1/3 \leq \alpha \leq 0$, [11]. For α = *0* the method coincides with the Newmark formulation. On the other hand, values of α lower than zero (negative) may facilitate the convergence of the solution by reducing the contribution of the higher modes, which acts as a sort of numerical dissipation. However, this may affect the accuracy of the final results, as the instances provided by Fig. 5 put in evidence.

To avoid such an accuracy problem, $\alpha = 0$ was initially set for every transient analysis of the present investigation. However, this value being set, a significant number of analyses, typically the ones relevant to the strongest events of each suite, failed the convergence. This required lowering the value of α till the convergence was reached. In many of the considered cases, the convergence occurred only when the limit value $\alpha = -1/3$ was chosen. In some other cases this was not enough, thus leading to some considerations about the violence of the suites of earthquakes considered and about the ductile resources of the building, as discussed in the following section.

4 CONVERGENCE PROBLEMS AND EXTRA DUCTILITY DEMAND

A great amount of non-linear analyses were performed, by considering the earthquakes of the four suites of Table 1, the two plastic hinge models and different values of the integration parameters. The solution failed in several cases, particularly when the earthquakes of Suite 1 and Suite 3 were involved. Although matching in average the same tolerance bounds, the earthquakes of Suite 1 (or of Suite 3) are in fact stronger than those of Suite 2 (or Suite 4). In addition, for each given earthquake, the success or failure of the solution depended also on the plastic-hinge model and on the HHT alpha parameter.

Setting the lowest admissible value $\alpha = -0.33$ leaded to solve most of the convergence problems, but in some cases, this was not enough. For instance, under the 6263 earthquake of

Table 4: Solution convergence map (Y = success; N = failure).

SUITE 1														
waveform	**146**		**199**		**230**		**291**		**1715**		**6263**		**6334**	
hinge / α	UH	DH	UH	DH	UH	DH	UH	DH	UH	DH	UH	DH	UH	DH
0	Y	Y	**N**	Y	Y	Y	Y	Y	**N**	Y	**N**	**N**	**N**	Y
-0.33	Y	Y	**N**	Y	Y	Y	Y	Y	Y	Y	**N**	**N**	**N**	Y
SUITE 2														
waveform	**230**		**232**		**291**		**535**		**594**		**1711**		**1713**	
hinge / α	UH	DH	UH	DH	UH	DH	UH	DH	UH	DH	UH	DH	UH	DH
0	Y	Y	**N**	Y	**N**	Y	Y	Y	Y	Y	**N**	Y	**N**	Y
-0.33	Y	Y	Y	Y	Y	Y	Y	Y	Y	Y	Y	Y	Y	Y
SUITE 3														
waveform	**146**		**291**		**1714**		**4673**		**6328**		**199**		**535**	
hinge / α	UH	DH	UH	DH	UH	DH	UH	DH	UH	DH	UH	DH	UH	DH
0	Y	Y	Y	Y	Y	Y	**N**	**N**	**N**	**N**	**N**	**N**	**N**	**N**
-0.33	Y	Y	Y	Y	Y	Y	**N**	**N**	**N**	**N**	**N**	Y	**N**	**N**
SUITE 4														
waveform	**232**		**147**		**6263**		**146**		**535**		**230**		**291**	
hinge / α	UH	DH	UH	DH	UH	DH	UH	DH	UH	DH	UH	DH	UH	DH
0	**N**	Y	Y	Y	**N**	**N**	Y	Y	**N**	Y	Y	Y	**N**	Y
-0.33	Y	Y	Y	Y	Y	Y	Y	Y	Y	Y	Y	Y	Y	Y

Suite 1 the convergence was never reached for both the plastic hinge models, see Table 4. Although spectrum-compatible for EC8 criteria, the 6263 earthquake has a response spectrum even five times higher than the target elastic spectrum, see Fig. 3a. Some numerical tests indicated that the convergence problems were due to the excessive violence of the earthquake with respect to the building design features, which implied a ductility demand greater than was available. As an example, Figs 6a and b show that the 6263 y-acceleration should be scaled by a factor 0.2 to lead the solution to converge (even if all the plastic hinges in the UH model reached the maximum curvature). Conversely, to reach the convergence under the original 6263 earthquake, the ductility capacity should be increased tenfold in both the UH and DH models. Figures 6c and 6d plot the x and y dispacements of the first floor target point as obtained by increasing tenfold the ultimate curvature of all end sections. Note, however, that the two plastic-hinge models behave differently under the same earthquake. Larger displacements are in fact, demanded to the UH model in both directions.

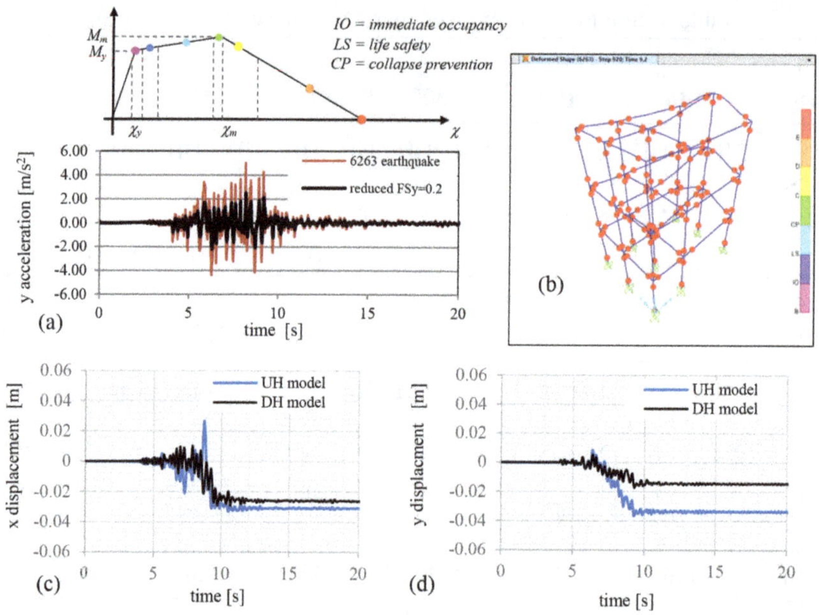

Figure 6: (a) Original and reduced 6263 y-acceleration; (b) deformed shape under the reduced earthquake; (c-d) x and y displacements of the first-floor target point under the original 6263 earthquake, as obtained by increasing tenfold the ductility of end sections.

To obtain the solution convergence under any earthquake of the considered suites, the ductility capacity of the plastic hinges was fictitiously increased every time it was necessary (this required, of course, a very lengthy user job). An extra ductility demand ranging from two times (record 535 of Suite 3) to ten times (record 6263 of Suite 1) the capacity was found for the earthquakes that caused convergence problems. Moreover, to compare homogeneous results, $\alpha = -0.33$ was set for all the events and for both plastic-hinge models. A comparison between the averaged y-displacements at storey levels, as obtained from the NLTHA and from the MRSA, is given in Fig. 7. The diagrams show that when scaled records are taken into account (Suites 2 and 4), the MRSA gives, as expected, more conservative results than the NLTHA. On the contrary, as far as original records are taken into account (Suites 1 and 3), the MRSA may be less conservative. In particular, the averaged displacements of the UH model may greatly exceed those of the MRSA. This result contrasts with the fact that, being more simpler (and less reliable) the MRSA should be more conservative than the NLTHA (similar concerns are discussed in Ref. [12]). It is to note that the response of the two plastic-hinge models is very similar under less violent events (Suites 2 and 4), while it may largely differ under stronger earthquakes (Suites 1 and 3).

5 CONCLUDING REMARKS

Although meeting all the ductility and strength hierarchy rules given by EC8, reinforced concrete frames designed through a code-compliant linear analysis (MRSA), may be unable to resist earthquakes that EC8 considers compatible with the design response spectrum. The present investigation showed in fact that some of such earthquakes are so strong that the ductility demand they involve can be even ten times the building capacity. Such an inconsistency

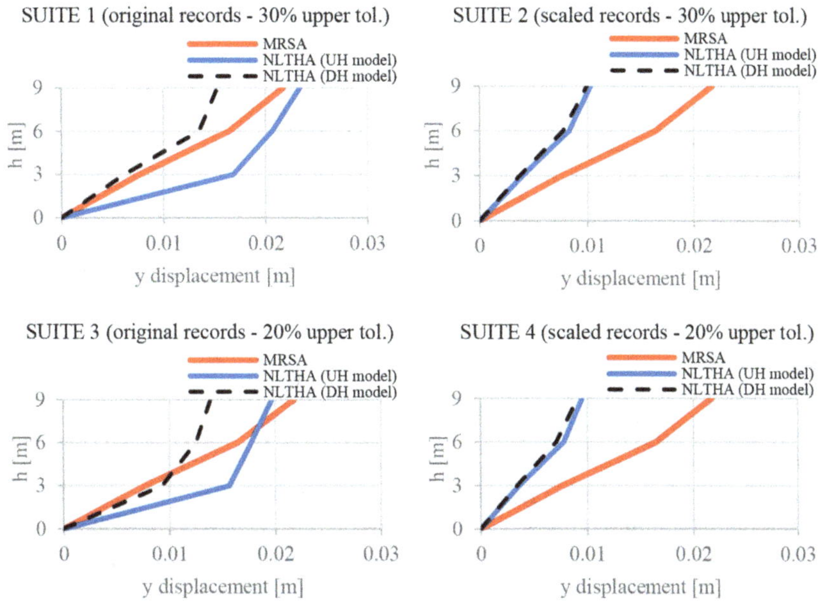

Figure 7: Comparison of the averaged y-displacements of the floors of the building obtained from the NLTHA and the MRSA.

is mainly due to the lack (in the EC8 provisions) of an upper tolerance bound for the average spectrum of the suite of compatible earthquakes. Setting an upper limit of 30% or even only of 20% may lead to convergence problems, at least under the strongest records of the suite. On the other hand, refer to a suite of earthquakes scaled to meet the upper tolerance bound, may significantly increase the chances of success of the analysis than to refer to original records. In addition, the paper showed that adopting the default hinge model given by SAP (based on the American codes) might lead to underestimate the actual ductility demand. However, assuming a more suitable user-defined hinge model requires a large amount of preliminary work, involving also the use of external applications to obtain the moment-curvature relationship for each critical section. In any case, no indications are given by EC8 on which plastic hinge model is better to adopt. Such difficulties can discourage the designer by using the more accurate non-linear dynamic analysis for seismic design. Some revisions of the EC8 provisions might help to avoid the inconsistencies evidenced by the present investigation.

AKNOWLEDGEMENTS

The author gratefully acknowledges Eng. Dario Cannaos and Eng. Peiman Novin for the numerical simulations made during their master's thesis.

REFERENCES

[1] FEMA-356, Prestandard Commentary for the Seismic Rehabilitation of Buildings. Federal Emergency Management Agency, Washington, DC, 2000.

[2] ATC-40, Seismic evaluation and retrofit of concrete buildings. Applied Technology Council, report ATC-40. Redwood City, 1996.

[3] CSI, SAP2000 v.17. Integrated Finite Element Analysis and Design of Structures Basic Analysis Reference Manual. Computers and Structures, Inc., Berkeley, California, USA, 2014.
[4] Iervolino, I., Galasso, C. & Cosenza, E., REXEL: computer aided record selection for code-based seismic structural analysis. *Bulletin of Earthquake Engineering*, **8**(2), pp. 339–362, 2010.
https://doi.org/10.1007/s10518-009-9146-1
[5] Carvalho, G., Bento, R. & Bhatt, C., Nonlinear static and dynamic analyses of reinforced concrete buildings-comparison of different modelling approaches. *Earthquake and Structures*, **4**(5), pp. 451–470, 2013.
https://doi.org/10.12989/eas.2013.4.5.451
[6] CEN. Eurocode 8: Design of Structures for Earthquake Resistance - Part 1: General Rules, Seismic Actions and Rules for Buildings. European Comitee for Standardization, Brussels, p. 232., 2004.
[7] NTC 2008, Norme tecniche per il calcolo, l'esecuzione ed il collaudo delle strutture in cemento armato, normale e precompresso e per le strutture metalliche. In Italian, 2008.
[8] Ambraseys, N., Smit, P., Douglas, J., Margaris, B., Sigbjörnsson, R., Olafsson, S., Suhadolc, P. & Costa, G., Internet site for European strong-motion data. *Bollettino di Geofisica Teorica ed Applicata*, **45**(3), pp. 113–129, 2004.
[9] Inel, M. & Ozmen, H.B., Effects of plastic hinge properties in nonlinear analysis of reinforced concrete buildings. *Engineering Structures*, **28**(11), pp. 1494–1502, 2006.
https://doi.org/10.1016/j.engstruct.2006.01.017
[10] Dolšek, M. & Fajfar, P., The effect of masonry infills on the seismic response of a four-storey reinforced concrete frame—a deterministic assessment. *Engineering Structures,* **30**(7), pp. 1991–2001, 2008.
https://doi.org/10.1016/j.engstruct.2008.01.001
[11] Hilber, H.M., Hughes, T.J.R. & Taylor, R.L., Improved numerical dissipation for time integration algorithms in structural dynamics. *Earthquake Engineering and Structural Dynamics*, **5**, pp. 283–292, 1977.
https://doi.org/10.1002/eqe.4290050306
[12] Paglietti, A., Porcu, M.C. & Pittaluga, M., A loophole in the Eurocode 8 allowing for non-conservative seismic design. *Engineering Structures,* **33**, pp. 780–785, 2011.
https://doi.org/10.1016/j.engstruct.2010.12.001

HANDS-ON-EXPERIENCE ON SEISMIC RETROFIT IN FOUR DIFFERENT COUNTRIES

HAZIM YILMAZ & THOMAS HACHMANN
OBERMEYER Planen + Beraten GmbH, Munich, Germany.

ABSTRACT
There are various different seismic vulnerability assessment procedures and seismic retrofit methods that have been applied to existing buildings in seismic regions. After the analytical assessment and design phase, it is of critical importance that the retrofit design is properly applied on-site. Conventional parties such as local authorities, construction culture, construction companies, quality of workmanship and availability of materials play a crucial role in the construction of seismic resistant buildings and in the selection of retrofit method and application. Furthermore there is a lack of experience on the performance of buildings subjected to earthquakes. Authors assessed and retrofitted eight reinforced concrete buildings and one masonry building from one to six stories in Nepal, Djibouti, Turkmenistan and Haïti, respectively, in 2011, 2013, 2015 and 2016. Retrofitted buildings in Nepal were subjected to 7.8 magnitude earthquake in April 2015, which gave authors the opportunity to document the seismic performance. This paper summarizes the hands-on-experience gained from four different seismic assessment and retrofitted projects conducted in four different countries. Performance of the retrofitted buildings subjected to a 7.8 magnitude earthquake and difficulties in the application of retrofit are present.
Keywords: implementation of retrofit, retrofit design, Seismic assessment, seismic retrofit, site survey

1 INTRODUCTION

Structural engineering has learned a lot from previous earthquake hazards. Seismic codes have been improved and reviewed after each earthquake event, which leads to investigation of earthquake safety of existing buildings. This need is more pronounced in seismic active regions. Obermeyer was commissioned to carry out the seismic assessment of nine buildings worldwide. This paper presents the practical experience gained in four different countries and the related findings. The project locations can be seen in Fig. 1.

Effect of various aspects like characterization of seismic hazards, performance of non-structural systems, costs of retrofit and public policy is thoroughly discussed and summarized in Holmes [1]. Owner's investment strategies, construction cost, disruption to the building users, aesthetics are the important parameters need to be considered during the whole evaluation process [1]. Review of various building codes, different structural materials, building techniques and seismic retrofit methods in three different countries such as USA, Italy and New Zealand are reported in Ref. [2].

2 ASSESSED BUILDINGS

2.1 Available documentation

The degree of information level about the geometry, details and material properties is very important for the seismic assessment and strongly dependant on the construction year. Detailed drawings and calculations are available for recently built structures, but there is often a lack of information for older buildings. Nonetheless, drawings and calculations are only unconfirmed information and need to be checked on site. Especially as-built drawings were, according to experience, often inaccurate, incomplete, and only helpful in preparation

DOI: 10.2495/SAFE-V7-N4-557-567

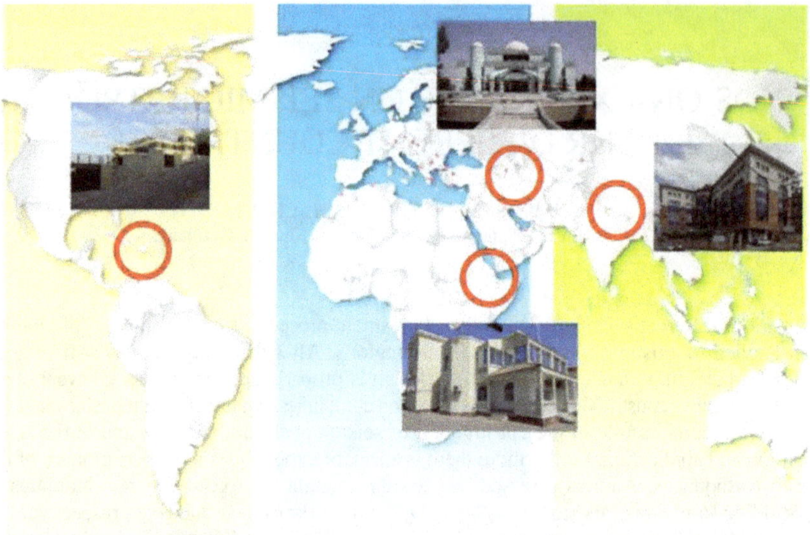

Figure 1: Project locations.

of the site survey. Digital copies of drawings in, for example, dwg-Format, are the exception. Standard are hardcopies of existing plans, which are often hard to read because of their inadequate quality.

Availability of floor plans, reinforcement plans, structural calculations are one of the most important information sources. Table 1 gives an overview of the available information for the assessed buildings.

Table 1: Available documentation.

Location	Available documentation	Comments
Nepal – Kathmandu (5 buildings built between 1979–2002)	• Former seismic assessment report • Geotechnical report for foundation design • As-built drawings • Architectural drawings • Structural drawings and details partly available	In general an adequate number and quality of documentation is available. Information has to be verified during the site survey.
Djibouti – Djibouti (built in 2008)	• Architectural drawings • Structural drawings • Description of construction type	Reinforcement information available but only in principle.
Turkmenistan – Aşgabat (built in 1993)	• Architectural drawings	Architectural drawings in deficient quality. Major differences between architectural drawings and as-built situation.
Haïti – Port-au-Prince (2 buildings built in between 1985 and 2013)	• Architectural drawings • Structural drawings partly available • Soil report partly available • Photo documentation of construction period	Architectural drawings in partly deficient quality.

Table 2: Structural type.

		Existing Buildings			
		Construction Type	Construction Year	No. of Stories	Design Code
Nepal	Building 1	RC Frame	1979	3	Nepalese Building Code NBC (based on Indian Building Code) [3]
	Building 2	RC Frame	1979	2	
	Building 3	RC Frame	2002	3	
	Building 4	RC Frame	2001	1	
	Building 5	RC Frame	1996	6	
Djibouti [4]	Building 6	RC Frame	2008	3	Code parasismique 69, PS69 (French building code 1969) [5]
Turkmenistan	Building 7	RC Frame	1993	5	SNIP [6]
Haïti	Building 8	RC Frame	2013	5	ASCE [7]
	Building 9	URM	1985	2	N.A.

2.2 Buildings structural system

Except of one unreinforced masonry building in Haïti, all buildings were reinforced concrete structures where the seismic loads are resisted by frames. More detailed information about 9 buildings can be seen in Table 2. Assessed buildings have a storey range of one to six. Year of construction varies between 1979 and 2013.

3 ASSESSMENT PROCEDURE (GENERAL)

3.1 Site survey

On the basis of the available information the assessment starts with a site survey. The survey program for a building is developed after the detailed review of all documents.
The objective of the survey efforts is to be able to set-up a most realistic simulation model of the buildings in order to rate their behaviour under earthquake loads and to determine necessary strengthening measures.
The major constraint is often that no information of the structural system is available. The investigation program in general is:

1. Assess the building's geometry, using digital measuring equipment and cross check the provided plans to identify the level of accurateness and reliability.
2. Identify the building's structural elements like concrete columns and beams using visual inspection, thermal imaging and magnetic field scanning to understand the overall load bearing system and the connection points.
3. Identify and rate possible shear walls, using visual inspection, thermal imaging and magnetic field scanning, to identify the three-dimensional bracing of the construction.
4. Non-destructive testing by means of reinforcement scanners at representative structural parts of the building to identify and to quantify the reinforcement inside the structural elements.

5. Destructive testing by means of opening representative structural parts to verify the reinforcement scan results.
6. Schmidt hammer concrete tests to verify the provided data of the former assessment reports.

A reinforcement scanner is a very efficient tool in order to get information about the diameter, spacing, and concrete cover of the in-built reinforcement. Different scan types can be used to verify existing reinforcement. Application examples can be seen in Fig. 2. The left picture shows a 'Quick scan', which IS used to determine the existence, distance, and concrete cover of reinforcement bars. The right picture shows an 'Image scan'. An 'Image scan' shows the distribution of rebars in a section and gives additional information on rebar diameter. If necessary and allowed by the customer, destructive tests on few selective locations are a suitable for validation.

An additional tool for non-destructive testing is a thermal image camera. Thermal imaging is used in areas where it is not possible to apply (non-)destructive testing or where the structure part is not easily reachable. Different heat absorption and emission of used materials show a clear separation of e.g. structural and non-structural parts.

It is a very helpful tool in order to give the structural engineer an insight about the structure. Application examples can be seen in Fig. 3. The two left pictures show the existence of a joint between two building parts in the thermal image and afterwards verified by a destructive test. The destructive was done to determine the used material for the joint.

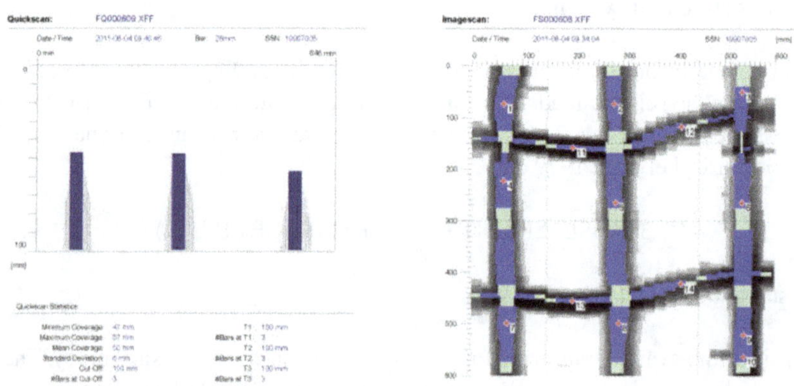

Figure 2: Use of ferroscans.

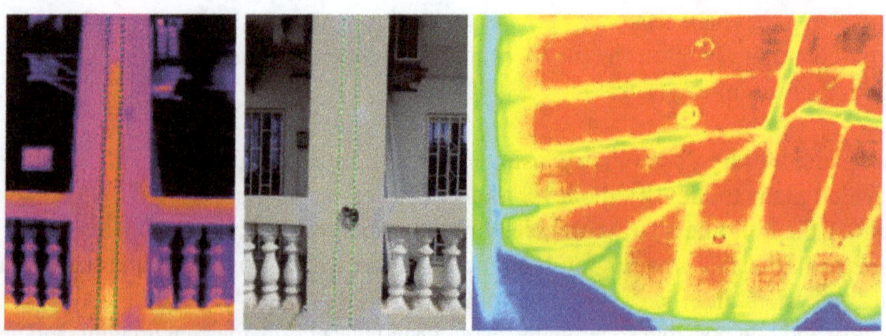

Figure 3: Use of thermal camera.

The right picture shows a load-bearing roof structure. The roof structure was covered by wooden panels from the inside but it was able to determine the metal construction by the use of thermal imaging.

3.2 Assessment

Most common methods to assess existing buildings on the base of linear and nonlinear methods are elastic linear static (lateral force) analysis, elastic linear dynamic (multi-modal response spectrum) analysis, nonlinear static (pushover) analysis and nonlinear dynamic (time history) analysis.

Selection of the method is the common decision of the structural engineer and the client. Desired level of safety, seismicity of the region, importance of the building, duration, and costs of the assessment method are the major factors effecting the selection. The client should be informed by the structural engineer about the pros and cons of each method and about the consequences. Based on the site-survey results, it could also be concluded that the building does not meet the safety requirements and no further assessment is necessary for the building.

Elastic linear dynamic analysis is conducted for building one to eight. Seismic demands were compared with the capacity of the members. Based on the results and distribution of the DCR (demand capacity ratio), the final decision of the necessity of the retrofit was made. Table 3 shows the PGA (peak ground acceleration) values used in the analysis and the type of assessment method.

For the Unreinforced Masonry Building (Building 9) linear static analyses was conducted. Building 9 has experienced 2010 Haïti earthquake and masonry walls had cracks, which reduces the strength of the walls considerably. Based on the results it was concluded that further investigation is not necessary and that the building does not meet the seismic safety criteria. A retrofit of this building is economically unfavourable.

There was no code for the seismic assessment of the existing buildings in all four countries. In agreement with the client, Obermeyer used Eurocode 8 [8], Uniform Building Code [9] and Turkish Earthquake Code [10] in order to be able to reflect the material and workmanship quality, dominant construction types, and local seismicity. Turkish Earthquake Code [10] was selected considering the construction practice and quality of the four countries. An example

Table 3: Buildings assessment overview.

		PGA(g)	Method	Retrofit
Nepal	Building 1	0,4	Linear Dynamic (DCR)	Yes
	Building 2	0,4	Linear Dynamic (DCR)	No
	Building 3	0,4	Linear Dynamic (DCR)	Yes
	Building 4	0,4	Linear Dynamic (DCR)	Yes
	Building 5	0,4	Linear Dynamic (DCR)	Yes
Djibouti	Building 6	0,1	Linear Dynamic (DCR)	No
Turkmenistan	Building 7	0,4	Linear Dynamic (DCR)	Yes
Haïti	Building 8	0,3	Linear Dynamic (DCR)	No
	Building 9	0,3	Linear Static (DCR)	Yes

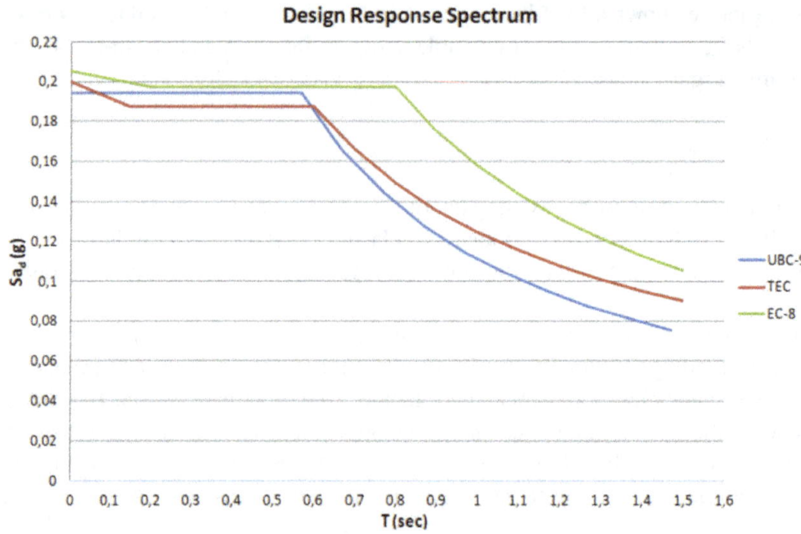

Figure 4: Design response spectrum.

response spectrum obtained from three different codes can be seen in Fig. 4. As it can be seen, constant acceleration plateau of the three different code spectra are in agreement.

4 APPLICATION EXAMPLES

4.1 Kathmandu, Nepal

Nepal with the Kathmandu valley is situated in an active seismic zone. Pandeya *et al.* [11] investigated in the recent past the microseismic activity between 1994 and 1999. Concluding, for earthquakes with magnitudes >8 the following statement will be cited here: 'We infer four 250–400 km long segments that could produce earthquakes comparable to the M = 8.4 Bihar–Nepal earthquake that struck eastern Nepal in 1934. Assuming the model of the characteristic earthquake, the recurrence interval between two such earthquakes on a given segment is between 130 and 260 years.' This is probably based on conservative model assumptions neglecting minor seismic stress releases but gives a comparable value for the hazard evaluation.

As requested by the client the Nepalese standard NBC 105:1994 is the basis for the seismic assessment. For the design of buildings the seismic level in NBC 105 is based on a return period of 50 years. Further, with the assumed lifetime of a building in Nepal therein of 30 years this corresponds to a 45 % chance of exceedance in 30 years. Compared with most international regulations for earthquake safety and design these values seem to be very unconservative.

From an engineering perspective it can not only be the aim to fulfil certain code regulations without considering background information or state of the art research and engineering knowledge. Therefore we suggested using design earthquake loads that correspond to international standards without neglecting the national specifications following from the Nepalese codes.

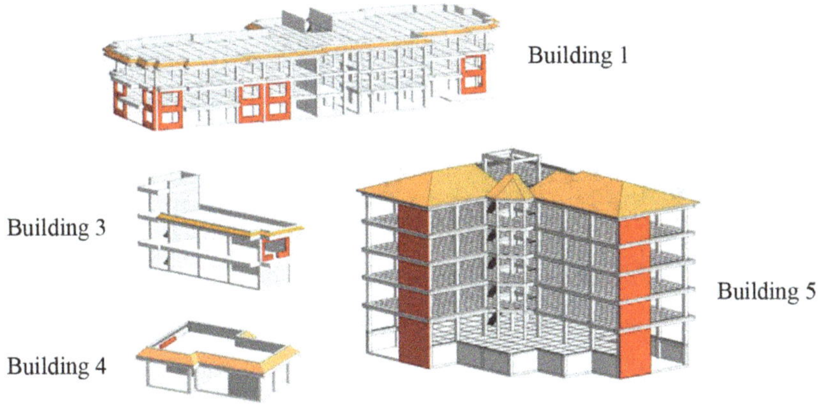

Figure 5: Retrofit design.

Parts of the structures of the buildings, which are under consideration here, already exceed the estimated lifetime of 30 years. Taking into account that the 1934 Bihar–Nepal earthquake is expected to have a return period of 130–260 years it seems reasonable to consider the 475 years return period earthquake, which refers to a 50 years building life, for structural safety calculations.

Based on the result of the seismic assessment procedure, Obermeyer planned seismic strengthening adding reinforced concrete shear walls. The proposed locations of shear walls can be seen in Fig. 5.

4.2 Aşgabat, Turkmenistan

Aşgabat is located in an active earthquake zone. In 1948 an earthquake with magnitude 7.3 has occurred near Ashgabat and destroyed nearly the whole infrastructure. The epicentre of the earthquake was located 25 kilometres southwest of Aşgabat. Peak ground acceleration could be determined as 3.8 m/s^2 for a seismic event of 10% exceedance in 50 years.

The building under consideration was built in 1993 as a hotel building. The customer will use the building as an office building. The whole reinforced concrete construction was separated into five parts using construction joints as a requirement of seismic safety. An inclined and a plane steel roof composed of steel truss members are the two parts of the steel construction.

As previously discussed, conventional parties such as local authorities, construction culture, and construction companies are important in the application of retrofitting together with quality of workmanship and availability of material.

Based on the result of the seismic assessment procedure, Obermeyer planned seismic strengthening adding reinforced concrete shear walls. The proposed locations of shear walls can be seen in Fig. 6.

However, the construction company started to use steel bracings without the agreement of Obermeyer, which would definitely not substitute RC shear walls in terms of stiffness and strength properties. Additionally, the use of steel bracings needs a proper workmanship that was not available on site. Quality of welding seams was inappropriate as well as connection to reinforced concrete parts. The only way to react to these circumstances was the addition of supplementary steel bracings.

Figure 6: Retrofit design.

Obermeyer could fortunately avoid the replacement of reinforced concrete shear walls by steel bracings in the most affected areas in the lower floors.

Figures 7 and 8 show the improper application of welding beams and bolted connections. Additionally, Fig. 8 shows the inadequate connection to existing reinforced concrete load-bearing parts.

Figure 7: Unplanned use of steel bracings.

Figure 8: Inadequate workmanship.

Figure 9: Shear walls made of reinforced concrete.

Figure 9 shows the non-consideration of the retrofit design by local HVAC and electrical installation.

5 OBSERVATIONS NEPAL 2015

Obermeyer carried out a seismic assessment of five building in Kathmandu, Nepal in 2011. Four of the assessed buildings On 25 April 2015, an earthquake occurred with a magnitude of 7.8 M having the epicentre in the Gorkha district followed by multiple aftershocks. In Kathmandu, recorded peak ground acceleration was approximately 0.10–0.15 g. Large spectral accelerations are observed due to the site amplification at the periods of 2–4 seconds [12].

Rapid damage assessments have been conducted in April and May 2015 by local contractors after the earthquake to all previously assessed and retrofitted buildings to assess the safety. Additionally, Obermeyer went to Kathmandu to observe buildings and damages related to the earthquake.

5.1.1 Building 1

Building 1 experienced only slight damage. Mostly fine hair cracks were observed in column to beam connections.

Fine cracks were observed in beam joints, joints of beam and wall, and joins of shear wall and brick masonry wall in all storeys. In first floor slab experienced hair cracks (Fig. 10).

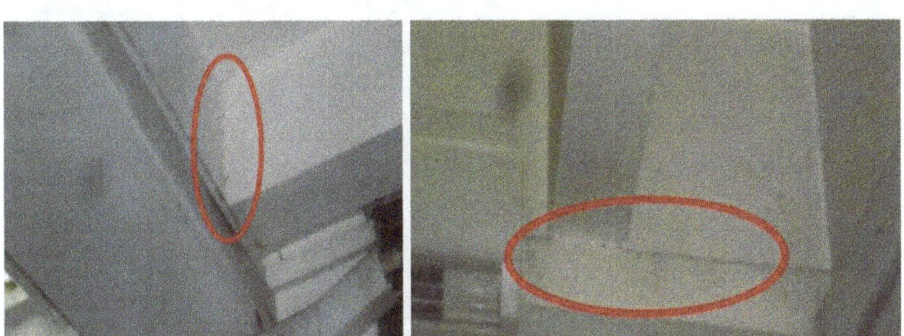

Figure 10: Results of damage assessment – building 1.

As a result of post-earthquake site survey building can be used for office occupancy. Retrofit of cracks is necessary.

5.1.2 Building 2

A retrofit of Building 2 was done prior to the assessment of Obermeyer in 2004/2005. Building 2 experienced damages but life safety criteria were still achieved.

Minor cracks on top of beam level and non-structural cracks were observed. Pounding happened between northwest staircase block and office building (Fig. 11).

Building can be used for office occupancy. Retrofit of cracks and staircase is necessary.

5.1.3 Building 3

Building 3 experienced only slight damage in form of non-structural cracks.

In joint of wall and column in ground floor separation cracks were observed together cracks in first floor walls.

Cracks are non-structural cracks. Building can be used for office occupancy. Retrofit of cracks is necessary.

5.1.4 Building 4

One storey building 4 experienced minor damage in form of cracks in non-structural walls.

Building can be used for office occupancy. Retrofit of cracks necessary by means of strengthening with injecting epoxy or grouting is necessary (Fig. 12).

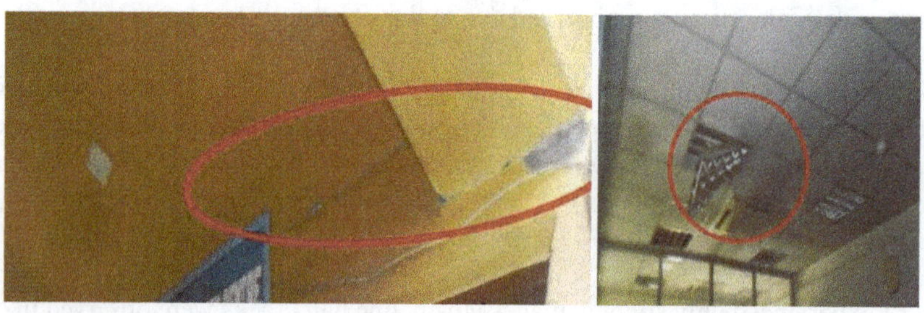

Figure 11: Results of damage assessment – building 2.

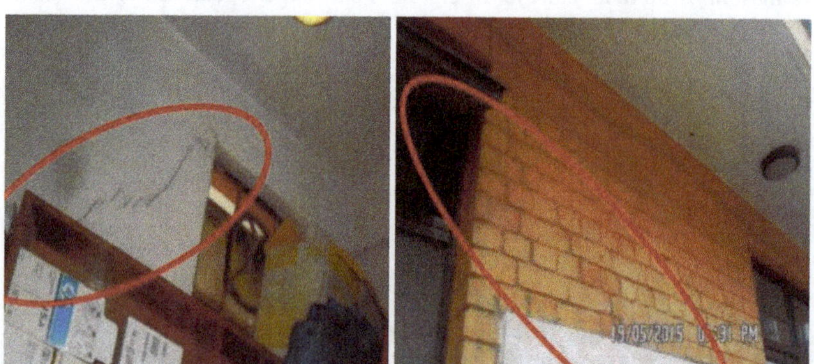

Figure 12: Results of damage assessment – building 4.

6 CONCLUSION

Seismic assessment of the existing buildings is a challenging task. Every earthquake event enhanced the knowledge of practicing engineers. Performance of buildings after a seismic event gave the engineer the opportunity to validate his design and quality of the application. Observation of damages after 2015 Nepal earthquake of five buildings previous assessed and partly retrofitted by Obermeyer is shown in this paper. Experienced slight damage by all buildings agreed to Life Safety Performance Level targeted during retrofit design.

Moreover, difficulties by the implementation of a proper retrofit design through local parties are observed in some of above mentioned countries.

As it could be concluded from this paper many factors play an important role through the whole process. Not only the type of numerical analysis and complicity of them but also the local practice and parties are very important to reach an optimum seismic assessment.

REFERENCES

[1] Holmes, T.W., *Risk assessment and retrofit of existing buildings*. Proceedings of 12. World Conference of Earthquake Engineering, Auckland, New Zealand, 2000.

[2] Martellota, L., Martens, D. & Teueffel P., *Review of seismic retrofitting strategies for residential buildings in an international context*. Proceedings of the International Association for Shell and Spatial Structures Symposium, Amsterdam, Netherlands, 2015.

[3] Nepal National Building Code NBC 105:1994, *Seismic Design of Buildings in Nepal*, 1994.

[4] Gebbeken, N., Braun, M., Hachmann, T. & Yilmaz, H., Earthquake engineering – reconnaissance and assessment of existing buildings. *International Journal of Protective Structures*, **3**(4), pp. 375–388, 2012.
https://doi.org/10.1260/2041-4196.3.4.375

[5] Règles parasismiques 1969 et annexes, *Collections UTI*, Union Technique Interprofessionnelle des Federations Nationales du Batiment et Travaux Publics (Paris), 1976.

[6] SNIP II-7-81, *Building code on construction in seismic areas*. The Ministry for Construction, 1996.

[7] ASCE 41-13, *Seismic evaluation and upgrade of existing buildings*. American Society of Civil Engineers, 2013.

[8] Eurocode 8, *Design of structures for earthquake resistance*, 2004.

[9] UBC-97, *Uniform Building Code*, 1997.

[10] Turkish Earthquake Design Code TEC 2007, *Specifications for Design of Buildings in Seismic Regions*, Ministry of Public Works, 2009.

[11] Pandeya, M.R, Tandukara, J.P., Avouac, J. & Héritier, T., Seismotectonics of the Nepal Himalaya from a local seismic network. *Journal of Asian Earth Sciences*, **17**, pp. 703–712, 1999.
https://doi.org/10.1016/s1367-9120(99)00034-6

[12] Kit Miyamoto, H. & Amir SJ Gilani., *Damage assessment and seismic retrofit of traditional and modern midrise buildings in the aftermath of 2015 Nepal earthquake.* Proceedings of the 16th World Conference on Earthquake, 2017.

APPROACHES FOR EVALUATING FAILURE PROBABILITY OF EMERGENCY POWER SUPPLY SYSTEMS IN HOSPITALS

CHI-HAO LIN, XIAOQIN LIU, CHENG-TAO YANG, YI-KUANG PAN & YU-CHIA LIAO
National Center for Research on Earthquake Engineering, Taiwan.

ABSTRACT

The unpredictable feature of earthquake is one of the major reasons for building damages and casualties occurring in earthquake events. Despite earthquake-induced structural damages to buildings, ground shaking can also impose critical effects on nonstructural components and systems in buildings. This study aimed to develop a probability-based approach for evaluating seismic damages to emergency power supply systems (EPSSs) in Taiwanese hospitals. For this purpose, four case hospitals that were somewhat damaged in the Chi-Chi earthquake (September 21, 1999) were investigated, and the component scoring system developed by the Multidisciplinary Center for Earthquake Engineering Research in the United States was referenced as the basis for developing seismic fragility curves for the components in the EPSSs of the case hospitals. Additionally, the logic tree method was employed to evaluate the failure probability of the EPSSs. The results exhibited acceptable consistency with the recorded damage status of the investigated systems, thereby verifying the feasibility of the proposed approach.

Keywords: emergency power system, failure probability, logic tree, seismic fragility

1 INTRODUCTION

First-aid hospitals are critical facilities that play an important role in receiving injured patients and performing emergency operations during severe earthquake events. Therefore, hospitals must maintain not only the safety of their building structures but also ensure the functionality of their nonstructural components and systems. These include emergency power supply systems (EPSSs), water supply systems, communication systems, and a range of medical equipment for patient care. Records of past earthquake disasters such as the 1971 San Fernando earthquake, 1994 Northridge earthquake, 1995 Hanshin earthquake, 1999 Chi-Chi earthquake indicate that many hospitals could not provide medical services because of severe nonstructural damage. In particular, EPSS is the most important system, because it provides electric power for other equipment and systems.

In this study we focused on a failure probability analysis of EPSSs. There are several approaches to assessing failure probability. Our methodology involved applying a system logic tree, assessing component fragility data and site hazards. After the Chi-Chi Earthquake in Taiwan, Chuang *et al.* investigated four severely damaged hospitals [1]. They constructed logic trees for the EPSSs for these hospitals. Five types of emergency power supply component (EPSC) are involved in an EPSS, namely a control panel (CP), a generator (G), a battery rack (BR), a diesel tank (DT) and a cooling tower (CT). The seismic fragility function of these components was studied based on the component seismic scoring system developed by the Multidisciplinary Center for Earthquake Engineering Research (MCEER) [2–3]. We utilized logic trees to estimate the failure probabilities of the EPSSs on the basis of the peak ground acceleration (PGA) of the hospital sites, and then verified the suitability of the methodology.

DOI: 10.2495/SAFE-V7-N4-568-576

2 FRAGILITY DATA OF EMERGENCY POWER SUPPLY COMPONENTS

In earthquake engineering, fragility is usually used to quantify the failure probability of structures and nonstructures. In this study, we applied PGA as the ground-motion parameter because it can be easily and accurately measured. The component fragility was assumed to be a two-parameter lognormal distribution function with median x_m and log-standard deviation β, as shown in eqn (1).

$$P(\alpha) = \Phi\left[\frac{\ln(\alpha / x_m)}{\beta}\right] \tag{1}$$

where P denotes the failure probability, $\Phi[\cdot]$ is the standardized normal distribution and α is the ground-motion parameter, e.g. PGA.

In our previous study based on MCEER, we proposed fragility data for EPSCs [3]. These fragility data are considered to provide not only high standards for installation but also to function as performance modification factors (PMFs). PMFs describe a variety of poor installation cases for nonstructural components. Tables 1–5 show the fragility data for EPSCs. The PMF descriptions reference the seismic score sheets developed by MCEER. x_m and β were originally proposed by MCEER for the high-standard installation of each component [2]. In our study, the fragility data for each PMF were identified to accurately estimate the damage probabilities of EPSCs. x_m has a markedly higher value under the high-standard installation. This indicates that the component has a high seismic capacity. If the component has an installation deficiency, x_m is significantly reduced.

Table 1: Parameters of seismic fragility data for control panels.

Control panel			x_m	β
high-standard installation			2.3	0.4
PMFs	PMF1	No anchorage	0.58	0.5
	PMF2	Poor anchorage	0.67	0.5
	PMF3	Suspect load path	0.58	0.5
	PMF4	Pounding or impact concerns	1.22	0.5
	PMF5	Inflexible concerns	0.58	0.5
	PMF6	Interaction concerns	0.46	0.5

Table 2: Parameters of seismic fragility curves for generators.

Generator			x_m	β
high-standard installation			2.0	0.4
PMFs	PMF1	No anchorage	0.78	0.5
	PMF2	Poor anchorage	0.91	0.5
	PMF3	Vibration isolator concerns	0.91	0.5
	PMF4	Rigid attachment concerns	0.58	0.5
	PMF5	Driver/generator diff. displacement	0.58	0.5
	PMF6	Interaction concerns	0.91	0.5

Table 3: Parameters of seismic fragility data for battery racks.

Battery rack			x_m	β
high-standard installation			2.5	0.4
PMFs	PMF1	No anchorage	0.67	0.5
	PMF2	Poor anchorage	0.78	0.5
	PMF3	No battery spacers	0.67	0.5
	PMF4	No longitudinal cross-bracing	0.78	0.5
	PMF5	No battery restraints	0.58	0.5
	PMF6	Interaction concerns	0.58	0.5

Table 4: Parameters of seismic fragility data for diesel tanks.

Diesel tank			x_m	β
high-standard installation			1.6	0.5
PMFs	PMF1	Tank is unanchored or the anchorage is in poor condition.	0.56	0.5
	PMF2	If anchored to a skid, the skid is unanchored.	0.89	0.5
	PMF3	Attached piping is too rigid to withstand expected displacement.	0.71	0.5
	PMF4	Legs appear to be undersized for weight of the tank, or skirt has unreinforced opening.	0.71	0.5

Table 5: Parameters of seismic fragility data for cooling towers.

Cooling tower			x_m	β
high-standard installation			NA	NA
PMFs	PMF1	Cooling tower is unanchored or the anchorage is in poor condition.	0.46	0.5
	PMF2	If anchored to a skid, the skid is unanchored.	0.91	0.5
	PMF3	Rigid attachment concerns	0.46	0.5

3 FAILURE PROBABILITY OF EMERGENCY POWER SUPPLY SYSTEMS

On September 21, 1999 at 1:47 am local time an earthquake measuring 7.3 on the Richter scale occurred in central Taiwan. Its epicenter was at 23.87°N 120.75°E in Chi-Chi Township of Nantou County and located at a depth of only 7.0 km. The earthquake originated along the Chelungpu fault line in western Taiwan and caused substantial damage. This incident was subsequently named 'the Chi-Chi earthquake.'

Chuang *et al.* investigated the four first-aid hospitals (A–D) that were damaged in Nantou County [1]. Figure 1 shows a seismic intensity map of the Chi-Chi earthquake and the location of the four hospitals. The maximum PGA was approximately 1*g*. Table 6 shows the PGA at the four hospitals. The Chelungpu fault is a reverse fault. Hospital D was located on the

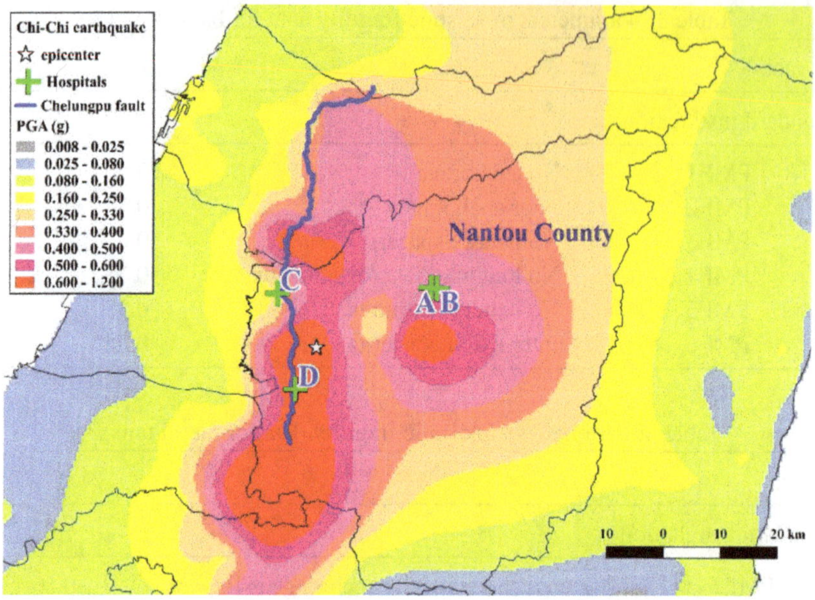

Figure 1: Seismic intensity map of the Chi-Chi earthquake and the location of the four hospitals.

Table 6: PGA values of the four hospitals in the Chi-Chi earthquake.

	A	B	C	D
PGA (g)	0.45	0.45	0.28	0.67

hanging wall and near the fault, only a short distance from the epicenter. Thus, it experienced a considerable ground motion. Although hospital C was also near the fault, it was located on the foot wall. Thus, the ground motion it experienced was not substantial. Hospitals A and B were adjacent and experienced the same ground-motion intensity.

3.1 System logic tree

Logic trees are graphical descriptions of the logical dependencies between systems and their subcomponents [4–5]. Mathematically, upper events are connected to lower events through an 'and' gate (symbol: ⌂) or an 'or' gate (symbol: ◇). Equation (2) provides the probability for the upper event through an 'and' gate. For an 'or' gate, the probability of the upper event can be derived using eqn (3).

$$P[u] = 1 - \prod_{i=1}^{n} \left(1 - P[l_i]\right) \tag{2}$$

$$P[u] = \prod_{i=1}^{n} P[l_i] \tag{3}$$

where P denotes the failure probability, u represents the upper event, l_i are the lower events for the component and $\prod$ denotes the product.

Figure 2 shows a logic tree diagram of the EPSSs in the hospitals. Only the EPSS in hospital C had no cooling tower; the other hospitals had the five major components. The floor where the component was located is indicated in parentheses in the figure. B1 refers to the basement first floor. RF denotes the roof floor. The EPSSs in hospitals A and B differ in that

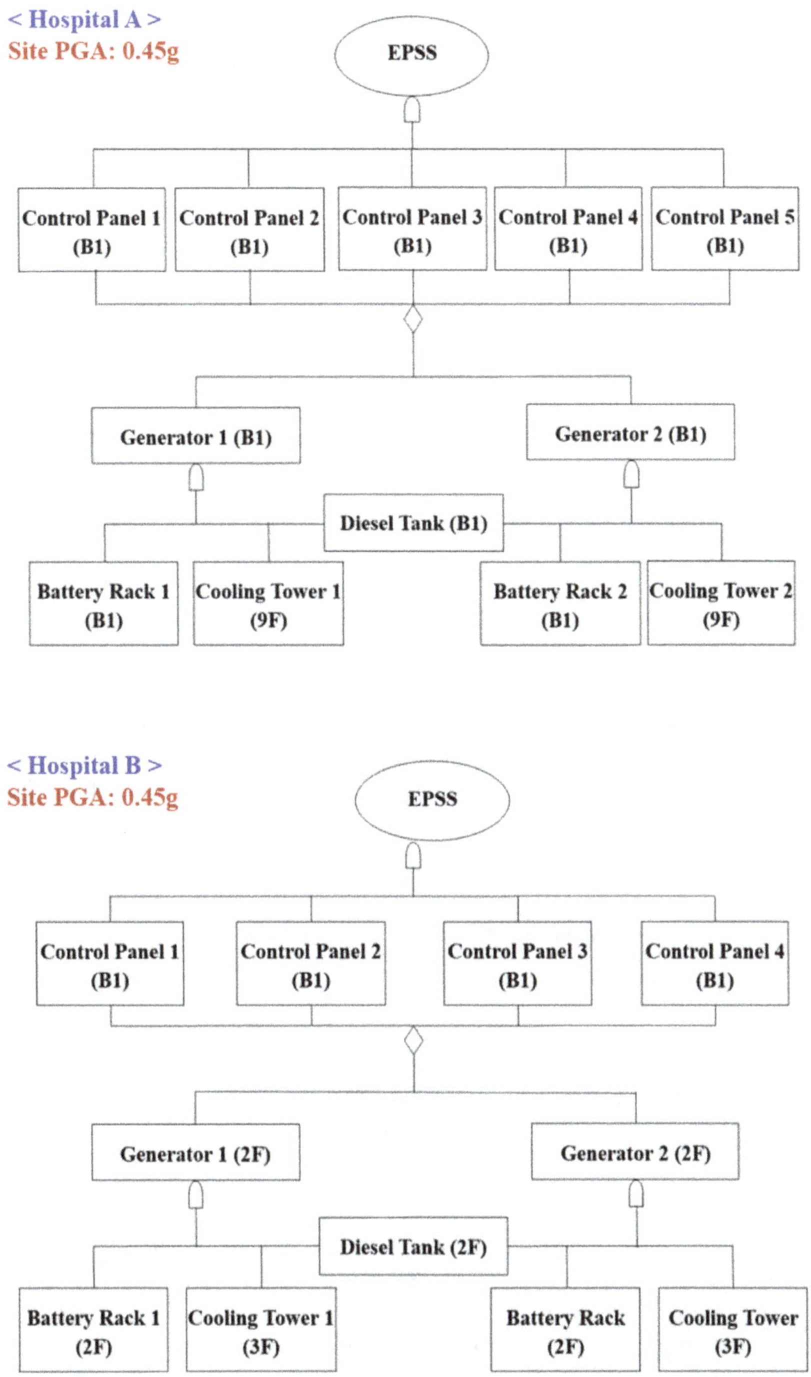

Figure 2: Logic tree diagrams of the EPSSs in Hospitals A – D.

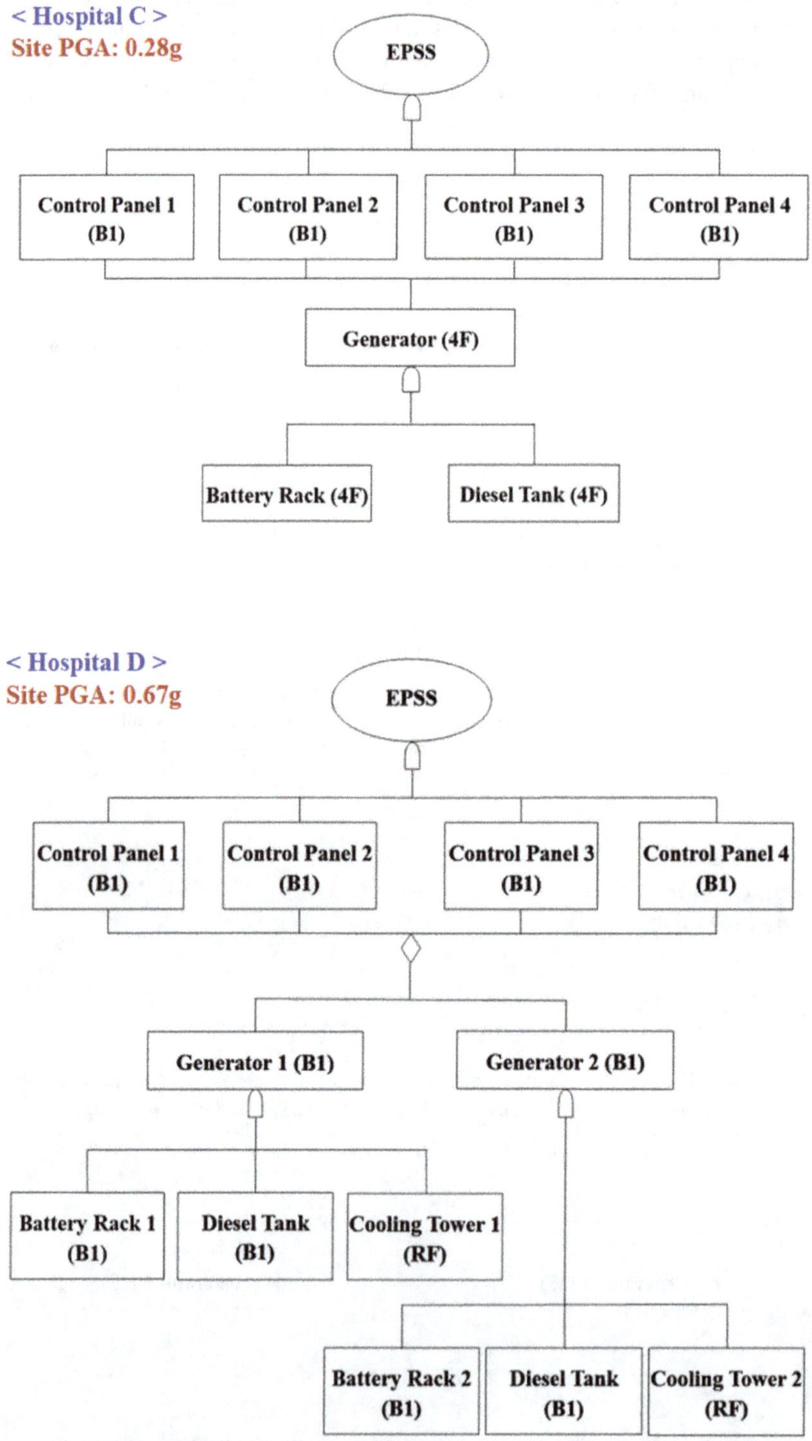

Figure 2: (*Continued*)

two generators used the same diesel tank. The failure path must be considered in the calculation of the probability. We took hospital A as the example case. The detailed process of our computation is shown in the following section.

3.2 Example case: Hospital A

The EPSS in hospital A had two generators that shared the same diesel tank. Therefore, the generator subsystems were not operated independently. Equations (2) and (3) cannot entirely solve this problem. The failure path must be considered for both subsystems. Furthermore, the peak floor acceleration (PFA) must include the acceleration amplification effect. In our study, the building floors were divided into three parts: the amplification factor of bottom part was set at 1.0, the top part was set at 2.0 and the middle part was set at 1.5.

Hospital A was a 9-story building with one basement. The cooling tower was located on the top floor (9F). In the calculation of its failure probability, the acceleration at the top floor was considered to be 2.0 times the PGA. The other components were placed in the basement. The PFA was identical to the PGA at this site. In addition, according to Chuang's report [1], each component may have experienced many seismic losses (PMFs). In our study, we selected the appropriate fragility median in accordance with the post-earthquake investigation. Table 7 shows the PMF, x_m, PFA, and failure probability for each component in hospital A.

Figure 3 shows the detailed derivation for the failure probability of the EPSS (P_{EPSS}). In general, we can obtain the failure probability of the upper event by using eqn (2) or eqn (3) according to the 'and' or 'or' gate. At Step 1, the battery rack and the cooling tower were taken as the lower events. The failure probability of Upper Event 1 (P_1) was calculated using eqn (2). At Step 2, the diesel bank was not included. The failure probability of a generator subsystem (P_2) was derived through an 'and' gate that was composed of P_1 and P_G. At Step 3, the failure probability of upper event 3 (P_3), which consisted of five control panels, was calculated. Because the two generators shared the same tank, we proposed the failure path shown in Step 4 for calculating the probability (P_4). Finally, we obtained P_{EPSS} by using P_3 and P_4, which formed an 'and' gate.

4 CONCLUSIONS

Based on the above methodology, the P_{EPSS} for Hospitals A–D were obtained as listed in Table 8, and they were then compared with the disaster survey. The results exhibit acceptable consistency with the recorded damage status. This can be explained by the following three reasons: (1) The fragility data identified in our study are appropriate for estimating the

Table 7: PMF, PFA and failure probability for each EPSC in hospital A.

Component	PMF	x_m	Floor	PFA (g)	Failure Probability
Generator	No anchorage of rubber bottom	0.78	B1	0.45	$P_G = 13.6\%$
Control Panel	Vulnerable to heavy impact	1.22	B1	0.45	$P_{CP} = 2.3\%$
Battery Rack	No battery restraints	0.58	B1	0.45	$P_{BR} = 30.6\%$
Diesel tank	Unanchored	0.56	B1	0.45	$P_{DT} = 33.1\%$
Cooling Tower	No anchorage of bottom brake	0.91	9F	0.90	$P_{CT} = 49.1\%$

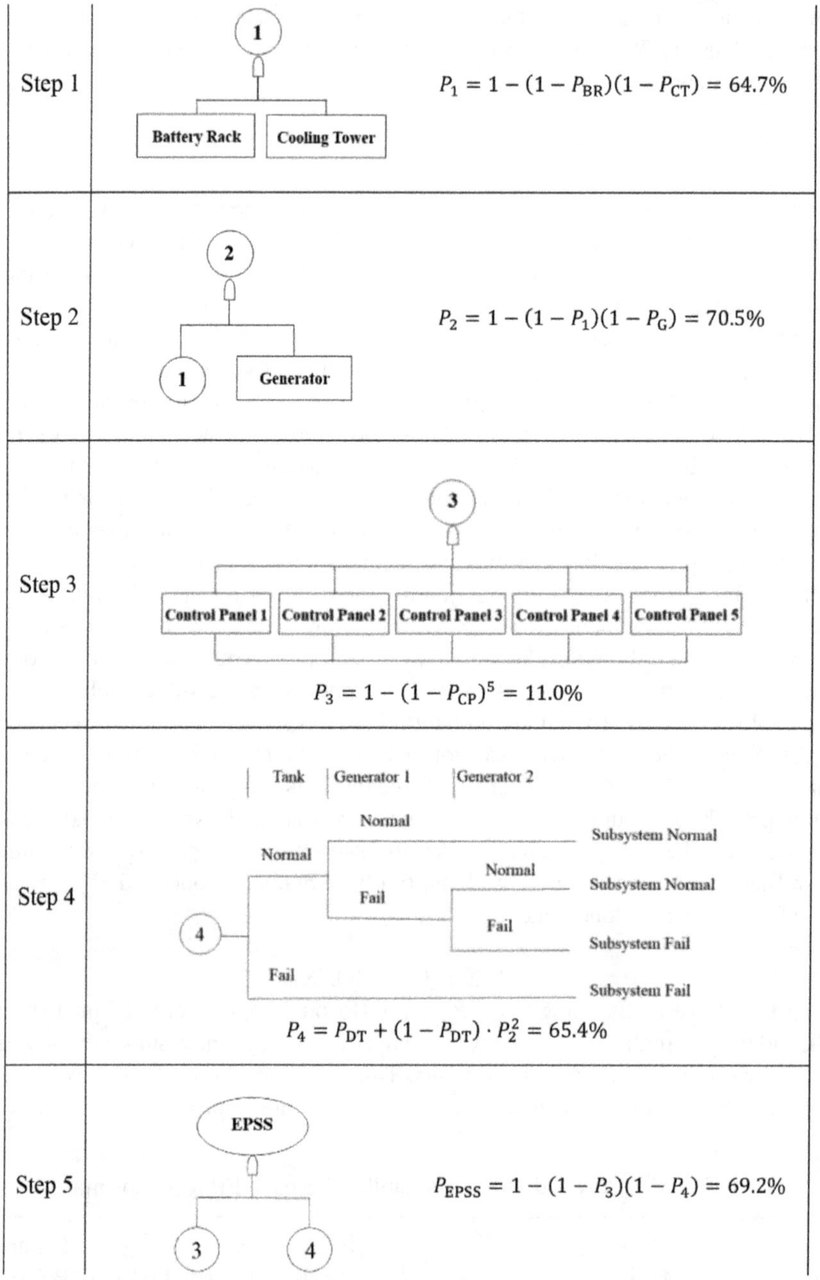

Figure 3: Derivation process of P_{EPSS} for hospital A.

Table 8: P_{EPSS} compared with the disaster survey for the Hospitals A–D.

Hospital	A	B	C	D
Disaster survey	fail	fail	normal	fail
P_{EPSS}	69.2%	83.8%	34.3%	97.5%

damage probability for each component impacted by the Chi-Chi earthquake. (2) Logic trees can fully represent the composition of each component. (3) Bottom-up probability analysis combined with failure paths can represent the seismic risk for EPSSs.

ACKNOWLEDGEMENTS

This work is supported by the Ministry of Science and Technology, Taiwan (R.O.C.) within the projects 106-2119-M-492 -003 and 105-2119-M-492 -004.

REFERENCES

[1] Chuang, C.C. & Yao, G.C., *Evaluation of seismic capacity of designated hospitals' curative equipment in Tainan region*. Department of Emergency Medicine, National Cheng Kung University Hospital, College of Medicine, National Cheng Kung University, Taiwan, 2001.

[2] Multidisciplinary Center for Earthquake Engineering Research (MCEER). *Seismic Reliability Assessment of Critical Facilities: A Handbook, Supporting Documentation, and Model Code Provisions*, Technical Report MCEER-99-0008, 1999.

[3] Lin, C.H., Liu, X.Q., Kao, L.H., Tsai, C.Y. & Yang, C.T., *Seismic fragility analysis for hospital's emergency power supply equipment*. The Thirteenth National Conference on Structural Engineering & The Third National Conference on Earthquake Engineering, Taiwan, Paper No. 1116, 2016.

[4] Porter, K. & Ramer, K., Estimating earthquake-induced failure probability and downtime of critical facilities. *Journal of Business Continuity & Emergency Planning*, **5**(4), pp. 352–364, 2012.

[5] Masri, S., Caffrey, J., Myrtle, R., Nigbor, R., Agbabian, M., Johnson, E., Petak, W., Shinozuka, M., Tasbihgoo, F., Tranquada, R. & Wellford, L., *The FEMA-USC hospital project: nonstructural mitigation in hospitals*. The Thirteenth World Conference on Earthquake Engineering, Canada, Paper No. 2480, 2004.

REVIEWING THE 2016 KUMAMOTO EARTHQUAKE DAMAGE IN MASHIKI TOWN AREA BY USING MICROTREMOR MEASUREMENTS

TARA NIDHI LOHANI[1], TAKASHI NAGAO[1] & YASUHIRO FUKUSHIMA[2]
[1]Research Centre for Urban Security and Safety, Kobe University, Japan.
[2]Eight-Japan Engineering Consultants Inc. (EJEC), Japan.

ABSTRACT

The 2016 Kumamoto Earthquake has caused vast building damages and human losses, although Japan is one of the world's well-prepared countries against earthquake disaster. The accelerometers set in Mashiki town area of Kumamoto prefecture rolled to the full scale measuring the seismic intensity as 7 in JMA scale of 7. In addition to such large shaking, there are opinions about the existence of three parallel fault lines beneath the town and the area lying on very soft volcanic soils regarding the vast damage of this particular area. In order to investigate the soil behaviour of this area, microtremor array studies were conducted starting from April 2016 to September 2016. Phase velocity obtained by spatial autocorrelation method was cross-checked with the borehole data available and was also found to be correlating well. Besides, the shear wave velocity distribution in the field was also consistent with the extent of damage observed after the earthquake.

Keywords: Earthquake, Kumamoto, Mashiki, microtremor, shear wave velocity.

1 INTRODUCTION

On April 16, 2016, Kumamoto Prefecture in west Japan was hit by a big earthquake of M7.3 and, as per Japan Meteorological Agency, JMA, Mashiki town and surrounding areas in central Kumamoto were shaken by intensity 7 in JMA seismic intensity scale. The same level of shaking was generated in Mashiki town by the foreshock 2 days before. The Mashiki town area was worst hit with large building damages and human losses. The Ministry of Land, Infrastructure and Transport (MLIT) Japan report [1] shows that three parallel fault rupture lines run in parallel beneath this small town. The right strike slip fault movement of Hinagu-Futagawa fault system that triggered the tremor, also lies in a very close proximity.

Earthquakes over magnitude 4.5 within the year 2016 from April 14th are plotted in Fig. 1. The NE-SW trend of epicentres extending from Kumamoto City to Oita City area could be visualized. Besides, the geology of the area shows that it is covered by very soft volcanic soils. In other words, large shakings, underlying fault lines and very soft layer of underlying volcanic soil have invited such vast disaster in the area. Photograph shown in Fig. 1 was taken in the field where almost 2m lateral displacement was observed.

2 MICROTREMOR FIELD SURVEY

In order to study the underground behaviour, microtremor study was done in the field [2]. Total of four visits were done in the Mashiki town area from April to September 2016 in order to assess the damage situation and conduct field measurements. The devices used in measurement had a very sensitive accelerometer of natural period 1 s.

In this research, measurements taken in July (prefix JA, JS) and September (prefix SS) 2016 (Fig. 2) are discussed. The July array measurements were done in a detailed way, with

DOI: 10.2495/SAFE-V7-N4-577-584

(a) (b)

Figure 1: a) Earthquake shakings over magnitude 5.0 within the year 2016 starting April 14th, and b) Photo shows a ground shift of about 2 m along the fault.

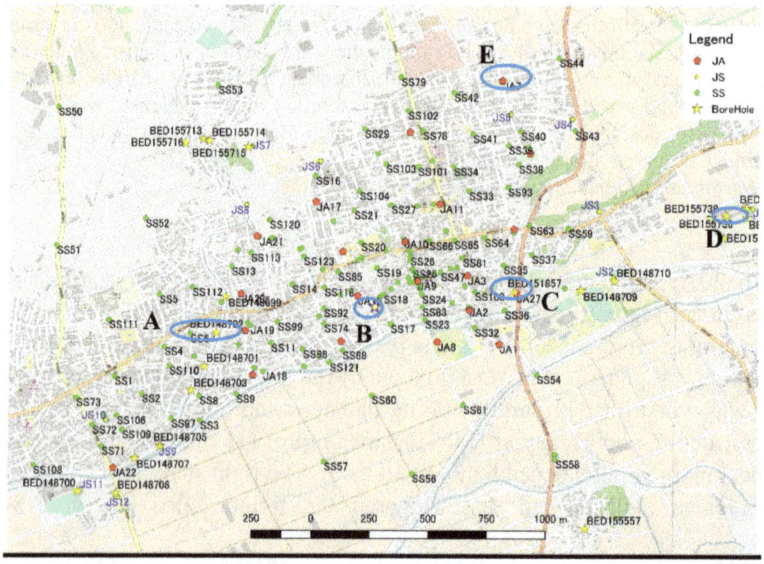

Figure 2: Microtremor measurements taken in July (JS, JA) and September (SS), in Mashiki Town, Kumamoto (A: Array, S: Single Device, measurement).

three devices set in a radial direction and one, at the centre (Fig. 3). The radial distance was varied from 4 to 32 m in general. The recording time was 6, 8, 11 and 11 min respectively. In addition, some recordings continued up to 20 min. Measuring devices were synchronized on each day by using GPS signal before performing the field measurement.

The recorded binary files consisted of the data in east-west (EW), north-south (NS) and UD (up-down) directions written at 100 Hz sampling frequency. The H/V spectra of the wave records were calculated in order to get an information of the in situ natural frequency [3]. Smoothing of both horizontal and vertical component of the recorded microtremor traces was done by using a Parzen window with a bandwidth of 0.2 Hz. By considering an influence of long period components negligible, only 40.96 s interval of the observed data at three minimum disturbed sections were selected and averaged to obtain the average H/V spectra. The horizontal

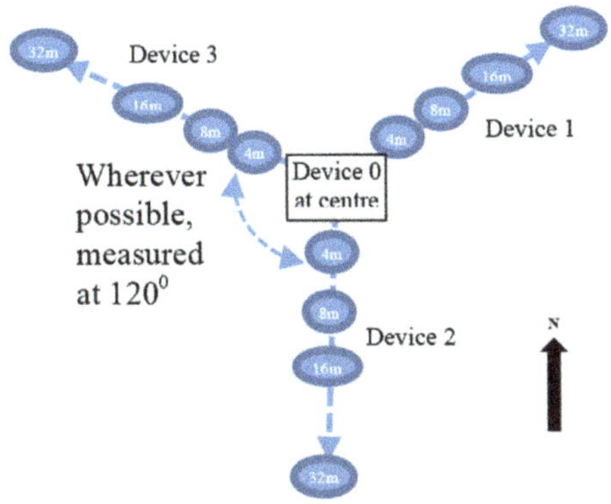

Figure 3: Array measurements done in the field with four devices, one at the centre and three other devices in radial direction.

component was obtained from the square root of the mean of the square of two orthogonal components. Figure 4 shows typical H/V plots prepared for JA7, JA9 and JA20 stations that are not so far from the locations E, B and A, in Fig. 2. The natural period of around 0.8 s could be noticed for JA9 station, where the housing damage was also found heavier during site visits.

The peak frequency (PF) distribution map (Fig. 5) was prepared by getting the peak position of H/V plots and plotting them by using inverse distance weighting method. Considering the possible non-linearity of underlying soil that might alter the natural period of the site along with an elapse period after the earthquake, only the data taken in July and September were plotted. The general trend is higher PF values (shorter natural period) in N-E region and lower PF values (longer natural period) in S-W region. Specifically, very long natural period along the Prefectural highway running EW at the centre of the map and southward could be noticed.

The phase velocity profile was obtained by considering spatial autocorrelation method [4] at all stations where array measurements were done (ref JA prefix in Fig. 5). The analysis was done by including UD component data and applying two different combinations of points in the analysis; i) between the stations at the centre and those set at the specified radial distance along the circumference, and ii) between the stations along the circumference. The average phase velocity plot was then obtained by combining all the plots for various distances together and the average trend was noted manually on this line (plotted as 'Observed in Fig. 6). Haskell

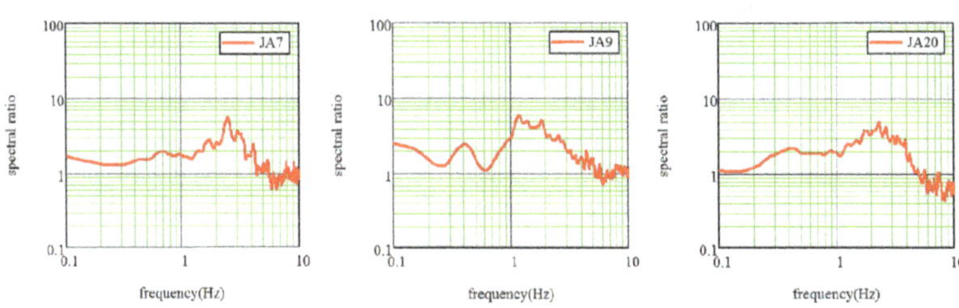

(a) (b) (c)

Figure 4: H/V plots for array stations; a) JA7, b) JA9, and c) JA20.

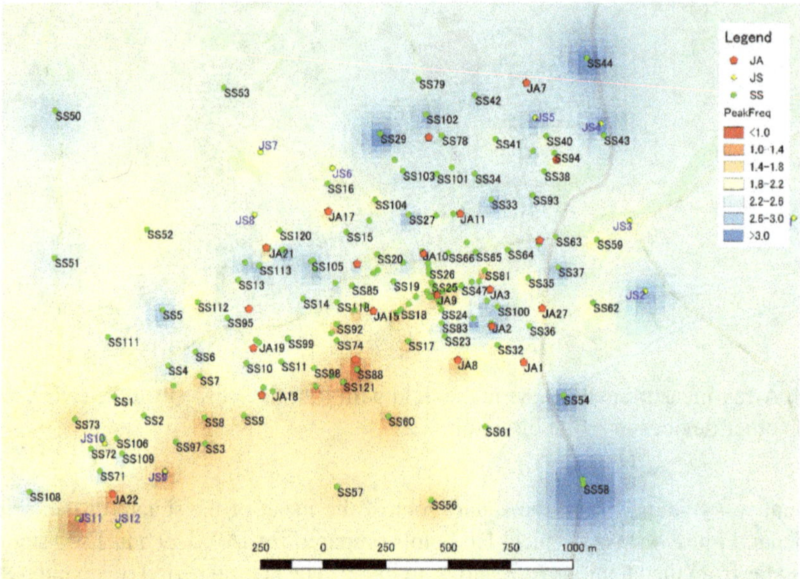

Figure 5: Peak frequency distribution map of Mashiki town area observed from H/V spectra of the data recorded in July and September 2016.

[5] method was applied to obtain the velocity profile [6] at each station. While doing this, underground soil conditions and depths were varied by observing the nearby boreholes, if exist, and phase velocity curve (plotted as 'Calculated' in Fig. 6) was obtained. For example, the JA7 point, which is very close to KMMH16, a KiK-net station [7] and borehole data is also available [8], was a very good reference to start the calculation. The phase velocity curves for three stations, JA7, JA9 and JA20 are shown in Fig. 6. Based on the velocity profile obtained in this way, an average velocity for the top 10 m, 15 m, 20 m, 25 m and 30 m was calculated. While such data are available only at array measurement stations, the distribution plot of whole area (Fig. 7) was prepared by using inverse distance weighting method, similar to that used in preparing Fig. 5. This plot will be referred again later in the following

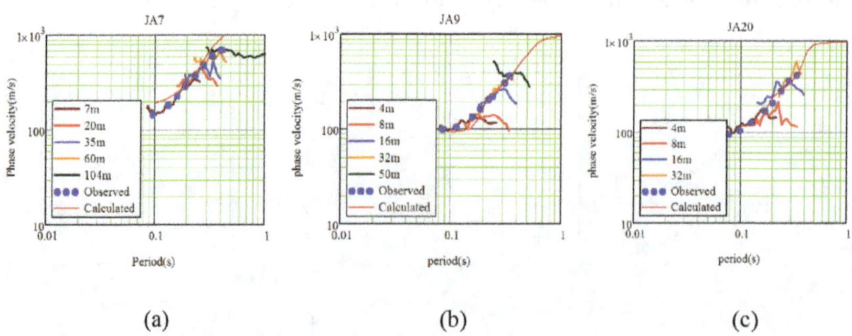

Figure 6: Calculated and observed phase velocity plotted together with spectral plots for various distances at locations; a) JA7, b) JA9, and c) JA20, respectively.

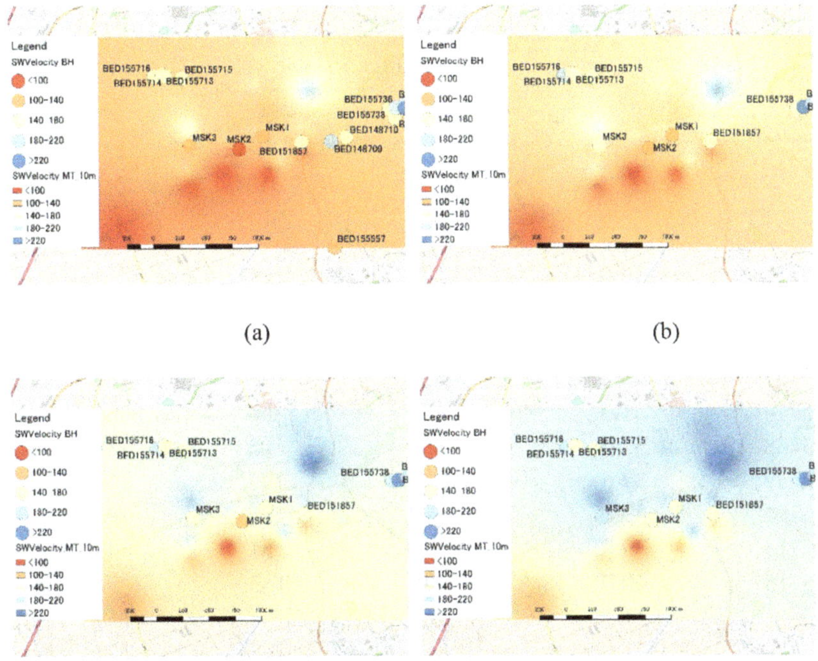

(a) (b)

(c) (d)

Figure 7: Calculated shear wave velocity from the borehole soil profiles drawn over the average shear wave velocity distribution map obtained by for the top; (a) 10 m, (b) 15 m, (c) 20 m and (d) 25 m.

paragraphs with additional data. There is an obvious and clear trend of increase in velocity with depth and localized variation of underground soil could be noticed.

3 INTERPRETATION OF BOREHOLE DATA

Boring data from Kumamoto Earthquake Reconstruction Support [9] borehole map was used to obtain the soil profile data at various borehole locations. Calculation of shear wave velocity was done from the given SPT data at various depths by using the relation proposed by Imai [10] for Quaternary and Pleistocene deposits of clayey soils (Equations 1 and 2) and Silty/ Sandy soils (Equations 3 and 4), respectively. The data also include the borehole made after earthquake (MSK1, MSK2 and MSK3) [11].

$$Vs = 102N^{0.292}. \tag{1}$$

$$Vs = 114N^{0.294}. \tag{2}$$

$$Vs = 80.6N^{0.331}. \tag{3}$$

$$Vs = 97.2N^{0.323}. \tag{4}$$

In order to evaluate the consistency of shear wave velocity (Vs) data obtained from a few of the in-situ microtremor tests, average Vs for the top 5 m, 10 m, 15 m, 20 m, 25 m and 30 m soil layer was calculated at each borehole points by using Equation (5) shown below. Some

Table 1: Average shear wave velocity from the ground surface to various depths.

	Borehole	Average Velocity, from ground to			
BH Name	Depth, m	10 m	15 m	20 m	25 m
BED148709	10.5	186			
BED148710	10.5	176			
BED155740	10.0	172			
BED155739	10.0	227			
BED155557	10.0	125			
BED155736	29.0	216	227	233	238
BED151857	25.0	153	156	165	178
BED155714	31.0	150	160	163	170
BED155738	35.0	155	165	188	203
BED155737	37.0	210	198	203	197
BED155716	39.0	168	186	188	196
BED155713	39.0	143	151	146	154
BED155715	40.0	142	143	151	159
MSK1	51.5	109	134	159	179
MSK2	71.0	92	115	137	156
MSK3	51.5	121	147	172	190

assumptions were needed at some locations where the SPT value for the surface layer was missing. The calculated average velocity data are shown in Table 1. These data points are shown as circles in Fig. 7.

$$Vs = \frac{\sum h_i}{\sum \left(h_i / Vs_i \right)} \tag{5}$$

To make an easier comparison, the colour schemes of Vs calculated from boreholes and from phase velocity are made similar. There is a sufficient match between the two and therefore, it assures the reliability of the velocity distribution map prepared.

4 DISCUSSIONS

Figure 8 shows the extracts of SPT N value and Vs values obtained from references. Except somewhat higher Vs at location D, the top 10-m soil layer of all other stations, which primarily affects the bearing and shaking behaviour of the widely distributed one- or two-storied buildings of the area, has very low SPT N value or Vs magnitude. Similar trend could be observed from the calculated profiles, such as in Fig. 7a therefore, the consistency in Vs calculation and the interpolation methods applied in this research can be assured.

The scale of building damage in Mashiki area was quite high [12] and are found to be consistent to some extent with the weaker Vs locations for the top 10-m soil layer (see Fig. 7a and 9). Peak frequency associated with the natural period of soil of such weak areas can be

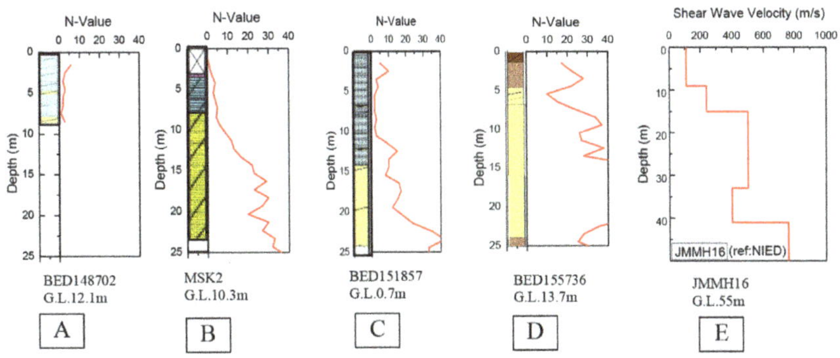

Figure 8: Borehole logs showing SPT values at locations A, B, C [9], D [11]; and Shear wave velocity profile at location E in Fig. 2 [8].

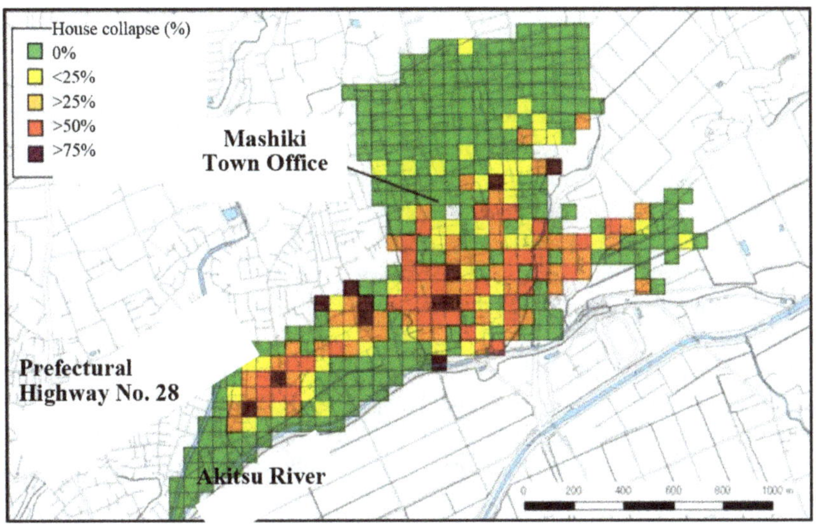

Figure 9: Building collapse situation after Kumamoto earthquake in Mashiki area (reproduced from the map on MLIT report [12]).

read from Fig. 5 as less than 1.0 second therefore, resulted one- or two-storied wooden structures of the area more vulnerable. In the vicinity of JA9 station, which is located in between B and C points (Fig. 2), the damage was the most and could be inferred from the very long natural period from in Fig. 4. However, although having lower Vs and longer natural period, the damage along the westward site of Mashiki Town along Akitsu river seems to be relatively small and the reason for such anomaly is not so clear. As the area had both newer and older buildings, the smaller scale of damage may not just be linked with the construction code followed or the year of construction. Thus, these sites need more detailed studies of underlying soil layers and building types to understand the reason of such anomaly. The general trend of higher damage at the sites where calculated velocity was small and had longer natural period to get the insight.

5 FINAL REMARKS

Shear wave velocity distribution of Mashiki Town area was obtained from this study. The Vs from spectral analysis was found to correlate well with the available borehole records. When compared with the extent of building damage, sites with calculated longer natural periods had felt more damage in the field. The accuracy of absolute Vs value and the distribution map, however, needs to be improved with more array measurements and also finding more borehole records in the field.

ACKNOWLEDGEMENT

Authors are thankful to Dr. M. Yoshimi from AIST, Japan for providing the digital data of three boring locations.

REFERENCES

[1] MLIT, Report on safety measures for reconstruction of urban areas in Mashiki Town from the Kumamoto earthquake, Ministry of Land, Infrastructure and Transport (MLIT) interim report on December, 2016, available at: (in Japanese) http://www.mlit.go.jp/report/press/toshi08_hh_000032.html. (accessed 24 April, 2017).

[2] Nagao, T., Lohani, T.N., Fukushima, Y., Ito, Y., Hokugo, A. & Oshige, J., A study on the correlation between ground vibration characteristic and damage level of structures at Mashiki town by the 2016 Kumamoto earthquake, 2016. (*Submitted to Journal of JSCE*).

[3] Kudo, K., Sawada, Y. & Horike, M., *Current studies in Japan on H/V and phase velocity dispersion of microtremors for site characterization*. 13th World Conference on Earthquake Engineering, **1144**, August, 2004.

[4] Aki, K., Space and time spectra of stationary stochastic waves, with special reference to microtremors. *Bulletin of Earthquake Research Institute* (University of Tokyo), **35**, pp. 415–456, 1957.

[5] Haskell, N.A., The dispersion of surface waves on multilayered media. *Bulletin of Seismological Society of America*, **43**(1), pp. 17–34, 1953.

[6] Nagao, T. & Konno, K., Estimation of average S-wave velocity of ground by use of microtremor array observation. *Journal of JSCE*, **696,** pp. 225–235 (in Japanese), 2002. https://doi.org/10.2208/jscej.2002.696_225

[7] Aoi, S., Obara, K., Hori, S., Kasahara, K. & Okada, Y., New strong-motion observation network: KiK-net. *Eos Transactions American Geophysical Union*, **81**, 329, 2000.

[8] KMMH16 Borehole data, available at: http://www.kyoshin.bosai.go.jp/cgi-bin/kyoshin/db/siteimage.cgi?0+/KMMH16+kik+pdf/. (accessed 24 April, 2017).

[9] Kumamoto Earthquake Reconstruction Support borehole map. available at: http://geonews.zenchiren.or.jp/2016KumamotoEQ/webgis/index.html. (accessed 24 April, 2017).

[10] Imai, T., P and S wave velocities of the ground in Japan. *Proceedings of the IXth ICSMFE*, **2**, pp. 257–260, 1977.

[11] Yoshimi, M., Hata, Y., Goto, H., Hosoya, T., Morita, S. & Tokumaru, T., *Borehole exploration in heavily damaged area of the 2016 Kumamoto earthquake, Mashiki town, Kumamoto*. Fall 2016 academic conference, Japan Active Fault Society, October, 2016.

[12] Report to analyze the causes of building damage in the Kumamoto earthquake. MLIT Building Construction Standards Committee, September 2016, available at: http://www.nilim.go.jp/lab/hbg/kumamotozisinniinnkai/20160912shiryou.htm. (accessed 24 April, 2017).

A NEXT-GENERATION OPEN-SOURCE TOOL FOR EARTHQUAKE LOSS ESTIMATION

S. MOLINA-PALACIOS[1], D. H. LANG[2], A. MESLEM[2], C. D. LINDHOLM[2] & N. AGEA-MEDINA[1]
[1] Dpto. Ciencias de la Tierra y del Medio Ambiente, Universidad de Alicante, Spain.
[2] NORSAR, Department Earthquake Hazard and Risk, Kjeller, Norway.

ABSTRACT

Earthquake loss estimation (ELE), generally also referred to as earthquake risk assessment, is a comparably young research discipline which, at first, relied on empirical observations based on a macroseismic intensity scale. Later, with the advent of methodologies and procedures that are based on theoretical simulation in estimating physical damage under earthquake loading, the analytical approach for ELE was formulated. The open-source software SELENA, which is a joint development of NORSAR (Norway) and the University of Alicante (Spain), is undergoing a constant development. One of the more recent features being included is the possibility to address topographic amplification of seismic ground motion. Additionally, SELENA has been adapted by including various methods for the analytical computation of structural damage and loss. SELENA now offers complete flexibility in the use of different types of fragility curves based on various ground motion intensity parameters (e.g. PGA, Sa, Sd), which has been suggested by many recently released guidelines (e.g. FEMA P-58, GEM-ASV, SYNER-G, HAZUS-MH). Besides, under the framework of the ongoing Horizon 2020 LIQUEFACT project, SELENA is extended in order to allow the consideration of liquefaction-induced ground displacements and respective structural damage.

In general, software tools for ELE are particularly useful in two different settings, i.e., for disaster management and (re)insurance purposes. Both sectors pose very different demands on ELE studies: while the (re)insurance sector is foremost interested in the direct and indirect economic losses caused by an earthquake to its insured physical assets, those institutions (often governmental and non-governmental organizations) in charge of disaster emergency management and response are more interested in reliable estimates on human losses and the potential short- and long-term social consequences. Being aware about these peculiar differences between software tools for disaster management and insurance applications, NORSAR/UA thereby offers two in its core similar software tools, i.e., the open-source software SELENA and the proprietary software PML (Probable Maximum Loss) which is actively used by the insurance association in Chile (South America) since 2011.
Keywords: Analytical methods, damage and loss, earthquake loss estimation, SELENA

1 INTRODUCTION

Earthquake loss estimation (ELE), generally also referred to as earthquake risk assessment, is a comparably young research discipline which somehow evolved from earthquake hazard assessment after the first studies were published in the late 1960s. Some of the first earthquake loss estimation studies were performed in the early 1970s following the 1971 San Fernando earthquake, and initially they purely relied on empirical observations based on a macroseismic intensity scale. These studies put a heavy emphasis on loss of life, injuries, and the ability to provide emergency health care. Later, with the advent of methodologies and procedures that are based on theoretical simulation in estimating physical damage under earthquake loading, the analytical approach for ELE was formulated. However, its breakthrough came first when FEMA released the first version of HAZUS in 1992 [1, 2]. HAZUS, which was then published in 2001 [3], 2002 [4] and 2003 [5], is a methodological software framework which provides a powerful technique for developing earthquake loss estimates. This framework can be used to anticipate the possible nature of an earthquake disaster and the scope of the emergency response needed to cope with an earthquake disaster, i.e., the

DOI: 10.2495/SAFE-V7-N4-585-596

development of plans for recovery and reconstruction following the event, and the mitigation of its possible consequences of earthquakes.

The following years can be described as the high period of ELE which resulted in numerous risk studies being conducted in many parts of the world, and methodologies and software tools both being developed and enhanced. Nowadays, a large variety of proprietary, free or open-source tools for ELE computation are available. Under the umbrella of the International Centre for Geohazards (ICG), NORSAR (Norway), in collaboration with the University of Alicante (Spain) started the development of an open-source seismic risk estimation tool in 2004 [6] which was later named SELENA (Seismic Loss Estimation using a Logic Tree Approach) [7, 8]. SELENA is coded in Matlab and also available as a compiled stand-alone version. In contrast to many other tools, SELENA is independent of any Geographic Information System (GIS). The user will supply built area or number of buildings disaggregated by different model building typologies, earthquake sources, empirical ground-motion prediction equations (GMPE), soil maps, topographic features and corresponding ground-motion amplification factors, capacity curves and fragility functions corresponding to each of the model building types and finally cost models for building repair or replacement. SELENA will compute, for a given intensity measure (or level of ground motion), the probability of structural damage in each one of the four damage states (slight, moderate, extensive, and complete) for the given building types. This probability is subsequently used with the built area or the number of buildings to express the results in terms of damaged area (square meters) or number of damaged buildings. Finally, using a simplified economic model, the damage is converted to economic losses in the respective input currency and human casualties, in terms of different injury types, are computed. Optionally, mean damage ratios (MDR), the amount of debris resulting from the shaking damage to buildings and the total number of uninhabitable buildings and displaced households can also be computed. The main innovation of this tool is the implementation of the computation under a logic tree scheme, allowing the consideration of epistemic uncertainties related to the different input parameters to be properly included, and the final results are provided with corresponding confidence levels.

The strength of SELENA comes not only from the use of a transparent coding, the ease of preparing input files for any region in the world but also from the availability of the NORSAR-UA consortium in preparing user-specific software versions, if required. These can be adapted to the users' requirements and applications. Since its first release, SELENA has been successfully applied to a multitude of testbeds worldwide (Fig. 1), e.g., Naples [7], Bucharest [9], the

Figure 1: Map distribution of the different countries/cities where SELENA has been applied.

Romanian-Bulgarian border region [10], Dehradun and Mussoorie (India) [11, 12], Haiti [13, 14], amongst others.

2 THE SELENA METHODOLOGY

The implemented methodology has been developed for a population of buildings and cannot be used to estimate the damage or losses associated to individual buildings unless a detailed finite element analysis that considers the peculiarities of an individual building is conducted. Further, the capacity curves and fragility functions used for different building typologies represent 'mean' curves for all buildings that belong to the respective building typology. As such, the results of SELENA are meaningful as long as they are used to estimate the damage and associated losses for a population of buildings. SELENA, like most other risk estimation software tools, considers the minimum geographical unit (geounit), i.e., the census tract, as the smallest area unit. In practice, this unit is related to building blocks or smaller city districts. The decision on the extent of each geographical unit has to be made considering different aspects such as having equal soil conditions, constant surface topography or a homogeneous level of building quality within the demarcated area. The main basis information of any ELE study consists in the building inventory database, which can sometimes be provided, e.g. by local agencies or governmental institutions. In any case, a thorough investigation of the local building stock by walk-down surveys and on-site inspections should be conducted in order to allow a representative classification of the prevalent building typologies. The building inventory database should contain a maximum of details about building materials, structural system, built area, floors of the building, height, foundations, seismic regulations used in the construction, use of the building, number of occupants, year of construction, etc. The building information is classified according to building type in each one of the geographical units which form the region under investigation. Ideally, the classification of the building type should be done following a user-defined classification scheme being most representative for the available building stock. This part of the database compilation represents the most critical phase in order to come up with accurate damage and loss estimates. Alternatively, the classification of building typologies can also be done according to existing taxonomies taken from literature, i.e., HAZUS [5], PAGER [15], GEM [16], SYNERGY [17].

2.1 Provision of seismic demand

A key point in any seismic risk assessment is the provision of the seismic ground motion (level and spectral characteristics of earthquake shaking). In order to carry out a seismic damage and loss assessment with SELENA, the user can provide seismic ground-motion estimates on three different ways: a) provision of spectral ordinates (taken out from probabilistic shaking maps) for each geographical unit (herein referred to as 'probabilistic' analysis even though the computation is still of deterministic character); b) definition of deterministic earthquake scenarios (e.g., historical or user-defined events) in combination with appropriate ground-motion prediction equations (GMPEs) in order to compute the spectral ordinates in each geographical unit (deterministic analysis); and c) provision of recorded ground-motion amplitudes at the locations of seismic (strong-motion) recording stations (analysis with real-time data).

In previous versions, spectral acceleration values had to be provided at three spectral periods, i.e., T = 0.01 s (referring to peak ground acceleration, PGA), T = 0.3 s (Sa0.3), and T = 1.0 s (Sa1.0), in order to describe the elastic design spectrum following the provisions of the International Building Code IBC-2006 [18]. Most other international seismic design codes (e.g. EN 1998, [19]) define the shape of the design spectrum such that only a single design acceleration value, generally PGA, is required to scale the amplitudes of the spectrum. Starting with SELENA v6.0, a site-specific elastic response spectrum can also be used to define the ground motion. If this option is selected, and either the probabilistic or real-time analysis is chosen, spectral acceleration values at nine different periods, i.e., T = 0.01 s (PGA), 0.10 s, 0.20 s, 0.30 s, 0.50 s, 0.75 s, 1.00 s, 1.50 s, and 2.00 s, have to be provided. In case of the deterministic analysis, SELENA will use all the spectral periods defined in the respective GMPE in order to describe the response spectrum.

2.2 Soil-dependent seismic demand – amplification of ground motion

In case of the presence of sedimentary soil materials at a site, the seismic ground motion at the ground surface is modified both in amplitude and frequency content. SELENA can use soil-dependent GMPEs or the respective amplification factors and/or corner periods, which basically describe the shape of the design spectra for the different soil classes specified by the corresponding code provisions. Currently, the procedures of IBC-2006 [18], EN 1998 [19], Indian standard IS 1893 [20], Cuban code NC 46 [21] and site-specific elastic response spectrum are incorporated. Furthermore, upcoming SELENA versions will not only include more of these international code provisions but also consider the use of natural ground motion records in order to account for record-to-record variability and hence offer more flexibility to the user.

2.3 Surface topography-dependent seismic demand – amplification of ground motion

A major shortcoming of the current generation of analytical ELE software tools is the fact that the contribution of surface topography to the ground motion characteristics is not considered by any of the existing tools; hence applying these tools in hilly regions represents a significant simplification. To address these effects in ELE, the site-specific geometry and subsoil conditions of the sites where the various building assets are located have to be properly assessed and described. In addition to a more complex definition of the seismic ground motion, buildings located at slopes are generally characterized by strong irregularities in plan, elevation and foundation design. These features lead to a complex structural behaviour, which cannot be addressed by nonlinear static-based procedures.

In a newly developed version of SELENA, the conduct of ELE for hilly regions is allowed through the implementation of a user-selectable analysis procedure addressing topographical amplification by taking into consideration the topographic amplification factors given by EN 1998 [19], Italian building code ICMS [22] or a period-dependent topographic amplification relationship recently developed [23].

2.4 Liquefaction demand and structural performance

Liquefaction demand expresses the liquefaction-induced ground deformations in horizontal (lateral spreading) and/or vertical direction (settlements) at the site of a structure or infrastructure facility. The method to be implemented for the assessment of liquefaction demand comprises three main steps:

– Evaluation of the site-specific liquefaction susceptibility that defines the tendency of the various geomaterials at a given site to undergo a severe loss of shear strength due to the pore water pressure build-up caused by earthquake ground shaking.

- Estimation of liquefaction probability, which defines the likelihood of experiencing liquefaction at a specific location given a certain amplitude and duration of ground shaking, as well as the availability of a certain ground water depth.
- Evaluation of the various modes/mechanisms of permanent ground displacement beneath a structure or infrastructure facility. These modes of liquefaction-induced ground deformations/displacements can be divided into two broad categories:
  a) Lateral spreading (ground surface deformation) in which the ground is deformed in horizontal direction. This deformation can result in considerable damage to overlying structures. Consequently, liquefaction-induced lateral spreading has been taken into consideration in the available design codes;
  b) Ground settlement deformation which is caused by the change in volumetric strain as pore water pressures dissipate after liquefaction has taken place.

The evaluation of building damage states and its overall performance to the various modes of liquefaction-induced ground deformation is essentially associated with a factor, i.e., the foundation system design (in addition to the structural system and member properties).

In general, structures or infrastructure facilities that are vulnerable to liquefaction-induced ground deformation are those with shallow foundations. In this context, the distinction will be whether the foundation is rigid or flexible.

- Structures/infrastructures on foundations that have sufficient relative stiffness, compared to soft underlying soils, and experiencing either uniform or differential movements (i.e., uniform/differential lateral spreading or uniform/differential vertical settlements), may behave as rigid bodies with no or minor damage to the structural elements.
- In case of structures/infrastructures on flexible (i.e., unrestrained) foundations, columns and walls can move independently (either under lateral spreading or vertical settlements) and thus differentially, and hence damage occurs locally in the structural elements. It is also possible to observe rigid body damage.

2.5 Damage probability computation.

Currently, the analytical method for seismic risk estimation implemented in SELENA can obtain the structural damage following three different approaches that are chosen by the user in the input files. The user can select: a) The spectral displacement-based damage estimation approach which was the existing approach in the previous SELENA versions (Fig. 2); b) The spectral acceleration-based damage estimation in which each model building type is assigned a natural period (T_1) and the structural damage is obtained by interrogating a spectral acceleration versus damage probability fragility curve using the spectral acceleration value $Sa(T_1)$; and c) The PGA-based damage estimation approach in which, once the PGA is obtained for each site, this value is used to interrogate the PGA-damage probability curves in order to obtain the corresponding damage probabilities (Fig. 3).

Additionally, in the case of the spectral displacement-based seismic risk approach, the initial performance point calculation method implemented in SELENA was the conventional capacity-spectrum method (CSM). Later, the CSM was replaced with its successor, i.e., the MADRS method, as well as the displacement coefficient method (DCM) provided in FEMA-356 [24] with the improvements proposed in FEMA-440 [25], which is referred henceforth as Improved Displacement Coefficient Method (I-DCM). Since more recently, SELENA can also obtain the performance point using the N2 Method for a bilinear elastoplastic form of the

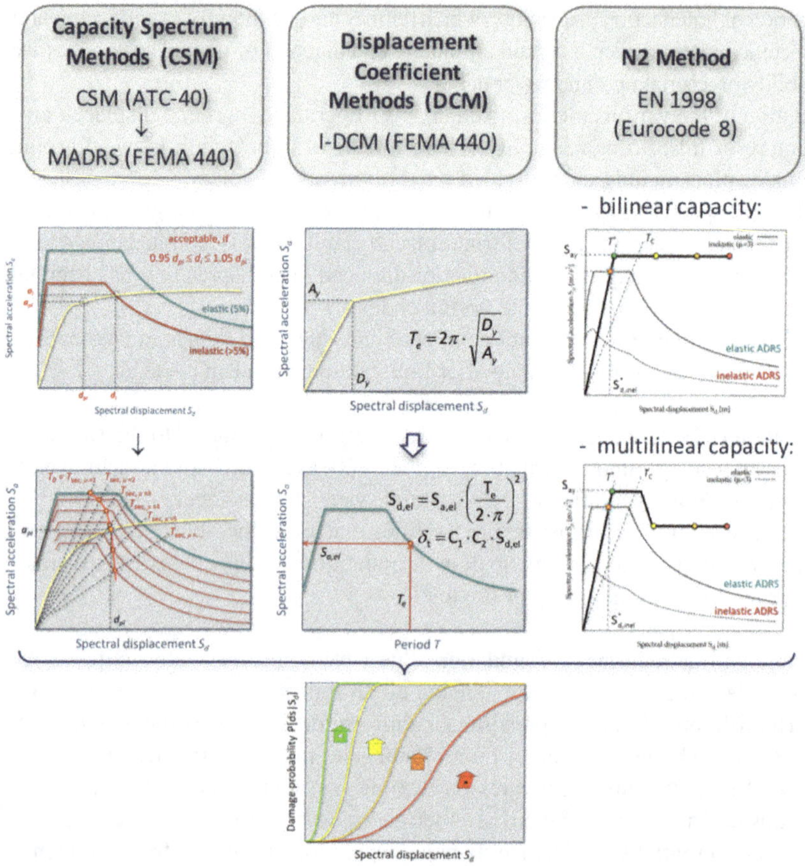

Figure 2: The spectral displacement-based approach implemented in SELENA for structural damage estimation.

capacity curve, based on earlier work of Fajfar [26] and recommended by EN1998 [19] as well as the N2 Method for a multilinear elastoplastic form of the capacity curve, based on earlier work of Dolsek and Fajfar [27].

The final damage results are given as absolute estimates (number of buildings or number of building floor area) of the respective damaged building type, so that users are able to present and further process these results using a spreadsheet program (e.g. MS Excel, OpenOffice, etc.) or any other tools or applications in any desired format (e.g., as percentage of built area [m^2] normalized by the total built area in each geographical unit or by the total built area in the studied region, i.e., summed over all geographical units).

2.6 Running SELENA

SELENA can be downloaded free of charge from the UA web page: http://personal.ua.es/es/sergio-molina/selena-rise.html or the NORSAR web page: http://www.norsar.no. The software contains all the source code (required when running the software under Matlab), an example of input files and some examples of output files (several folders with different results according to the computation method chosen) as well as the technical user manual.

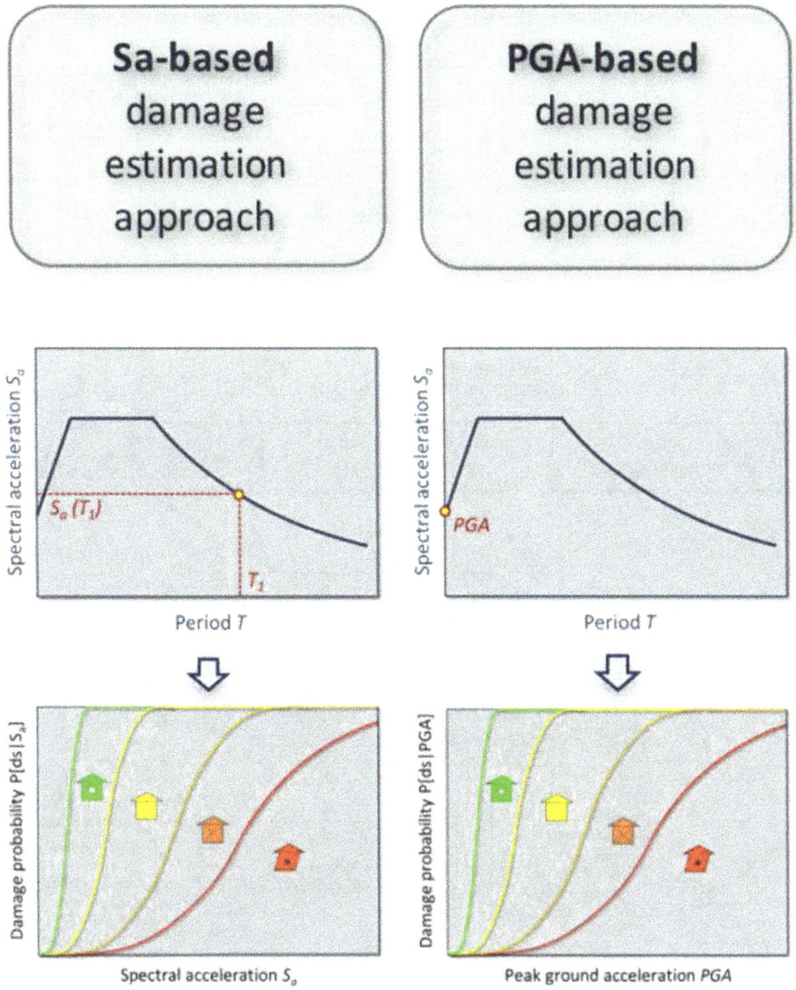

Figure 3: The spectral acceleration-based and PGA-based approach implemented in SELENA for structural damage estimation.

SELENA can be executed under Matlab or as a stand-alone executable using the command window in Windows or the Terminal under Mac OS X. Independently of the operating system, the user should prepare all the required input files into a user-defined folder. The input files need to be prepared in ASCII-format and provided as plain text files (*.txt). A detailed explanation about the formats of all the needed files is given in the user manual. Next, SELENA has to be executed within that folder according to the formats and orders given in the user-manual; i.e., selena d (deterministic earthquake scenario); selena p (probabilistic seismic hazard scenario); selena r ("real-time" ground motion scenario).

Figure 4 shows a flowchart of the current version of SELENA where all required input files and corresponding output files can be seen.

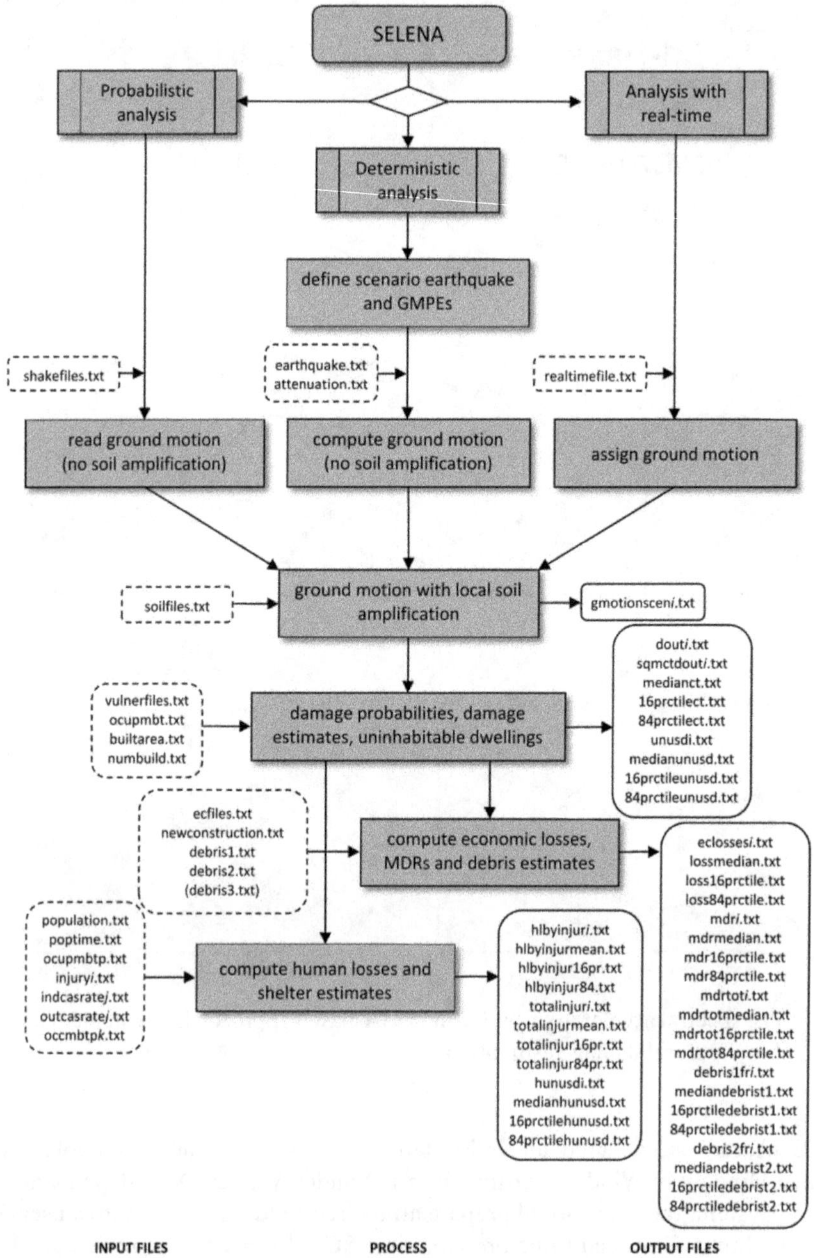

Figure 4: SELENA flowchart.

2.7 Dealing with uncertainties

Currently, SELENA computes median values as well as the 16% and 84% fractiles of the seismic risk results. This is done by means of a logic tree methodology in which the different branches of the tree can be weighted so that at the end of the computation, the risk results are

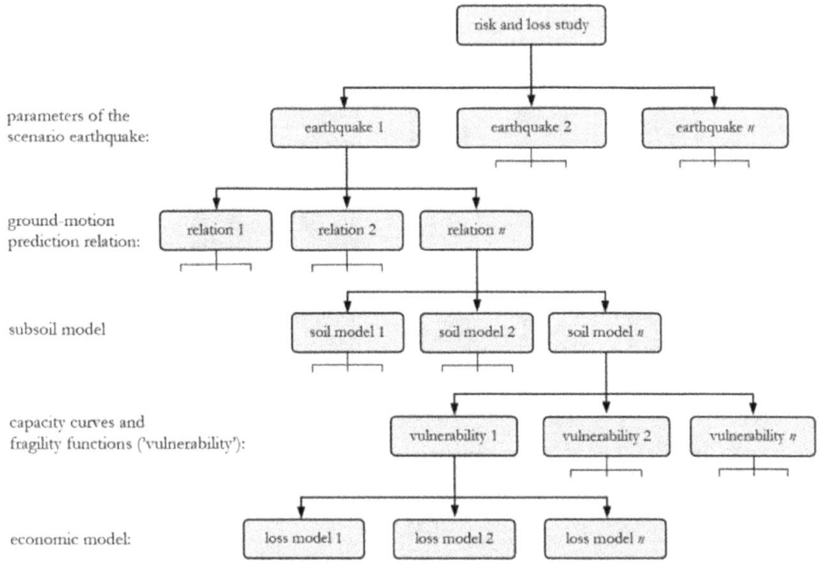

Figure 5: Logic tree structure. Each branch will be weighted in order to compute the expected values and confidence levels.

multiplied by their corresponding weights and then are fitted to a normal distribution in order to obtain the median values as well as its fractiles.

The single branches of the logic tree (Fig. 5) currently represent the epistemic uncertainties of the individual input components: i.e., earthquake source parameters, ground motion prediction equations, soil types, vulnerability curves, and economic values of the built area in case of a deterministic analysis, or ground shaking probabilistic maps, soil types, vulnerability curves, and economic values of the built area in case of a probabilistic analysis. In Fig. 6, we show an example of damage results following five branches of a logic tree and the mean structural damage for a given model building type.

3 CONCLUSIONS

The herein described Matlab-based tool SELENA can be used to provide damage results, economic and social losses for the general building stock and population of a city or country on the level of minimum geographical units (i.e., geounit or census tract). The level of resolution of the damage and loss predictions basically depends on the size of the geographical units which can be defined by the user. Both, the Matlab scripts and the ASCII input files are fully transparent allowing the user to apply own modifications and adjustments. Furthermore, it was tried to include as many comments as possible into the code such that the user can go through the scripts and easily change them if necessary.

It should be noticed that the presented tool for seismic risk and loss assessment SELENA is an ongoing development. Therefore, with any new development, SELENA will increase its capability to improve the understanding of the seismic risk in the studied area, to evaluate the effectiveness of different mitigation strategies by measuring risk and their uncertainties before and after they are implemented, to compare the seismic risk with other natural risk using annualized earthquake losses (AEL), and finally to help the insurance and re-insurance decisions by means of Probable Maximum Losses and loss exceedance curves.

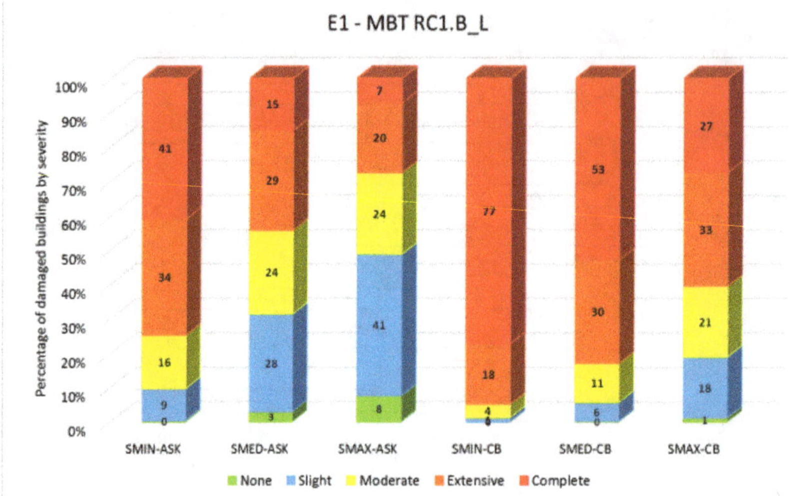

(a)

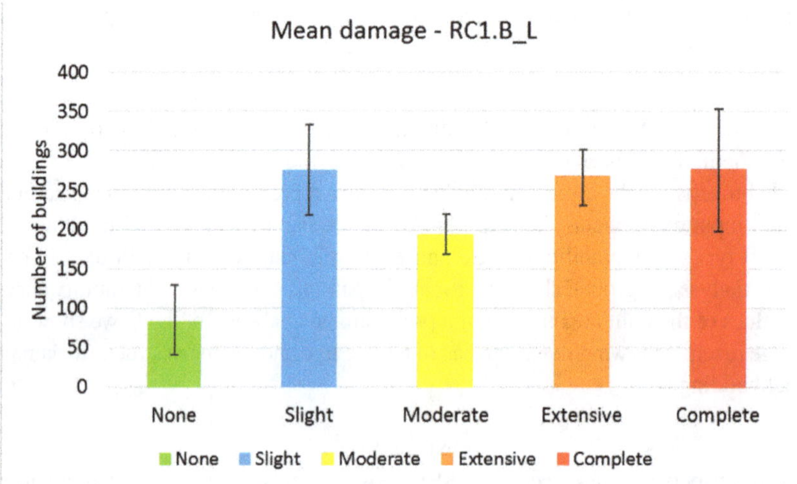

(b)

Figure 6: a) Percentage of damaged building following six branches of a given logic tree (each branch represents a combination of soil conditions and ground motion prediction equation). b) Mean value (expressed as number of damaged buildings).

ACKNOWLEDGEMENTS

The present research has been benefited from funding of NORSAR and the Univ. Alicante through research contracts (NORSAR1-14A, NORSAR1-08I), the funding of the Ministerio de Economía, Industria y Competitividad (CGL2016-77688-R) and the Generalitat Valenciana (BEST/2012/173 and AICO/2016/098). The development and implementation of the liquefaction risk assessment methodology is done under the LIQUEFACT project funded by the European Union's Horizon 2020 research and innovation programme under grant agreement (No. 700748).

REFERENCES

[1] Kircher, C.A., Nassar, A.A., Kustu, O., & Holmes, W.T., Development of building damage functions for earthquake loss estimation. *Earthquake Spectra*, **13**(4), pp. 663–682, 1997. https://doi.org/10.1193/1.1585974

[2] Whitman, R.V., Anagnos, T., Kircher, C.A., Lagorio, H.J., Lawson, R.S. & Schneider, P., Development of a national earthquake loss estimation methodology. *Earthquake Spectra*, **13**(4), pp. 643–661, 1997. https://doi.org/10.1193/1.1585973

[3] Federal Emergency Management Agency-FEMA, *HAZUS99-SR1 estimated annualized earthquake losses for the United States (FEMA 366)*, Report of the Federal Emergency Management Agency, Washington, DC, United States, 2001.

[4] Federal Emergency Management Agency-FEMA, *HAZUS99-SR2 (Service Release 2). Technical Manual*, Federal Emergency Management Agency (FEMA) and National Institute of Building Sciences (NIBS), Washington, DC, 2002.

[5] Federal Emergency Management Agency FEMA, *HAZUS-MH. Multi-hazard Loss Estimation Methodology*, Technical manual., Washington DC, USA, 2003.

[6] Molina, S. & Lindholm, C.D., A logic tree extension of the capacity spectrum method developed to estimate seismic risk in Oslo, Norway. *Journal of Earthquake Engineering*, **9**(6), pp. 877–897, 2005. https://doi.org/10.1080/13632460509350570

[7] Lang, D.H., Molina, S. & Lindholm C.D., Towards near-real-time damage estimation using a CSM-based tool for seismic risk assessment. *Journal of Earthquake Engineering*, **12**(2), pp. 199–210, 2008. https://doi.org/10.1080/13632460802014055

[8] Molina, S., Lang, D.H. & Lindholm, C.D., SELENA – An open-source tool for seismic risk and loss assessment using a logic tree computation procedure. *Computers & Geosciences*, **36**(3), pp. 257–269, 2010. https://doi.org/10.1016/j.cageo.2009.07.006

[9] Lang, D., Molina-Palacios, S., Lindholm, C. & Balan, S., Deterministic earthquake damage and loss assessment for the city of Bucharest, Romania. *Journal of Seismology*, **16**(1), pp. 67–88, 2012. https://doi.org/10.1007/s10950-011-9250-y

[10] Erduran, E., Lang, D.H, Lindholm, C., Toma-Danila, D., Balan, S.F., Ionescu, V., Aldea, A., Vacareanu, R. & Neagu, C., *Real-Time earthquake damage assessment in the Romanian-Bulgarian border region*, Presented at 15th WCEE, Lisboa, pp. 1–10, 2012.

[11] Lang, D.H., Singh, Y. & Prasad, J.S.R., Comparing empirical and analytical estimates of earthquake loss assessment studies for the city of Dehradun, India. *Earthquake Spectra*, **28**(2), pp. 595–619, 2012. https://doi.org/10.1193/1.4000004

[12] Singh, Y., Lang, D.H. & Narasimha, D.S., *Seismic risk assessment in hilly areas: case study of two cities in Indian Himalayas*. Presented at the SECED 2015 Conference: Earthquake Risk and Engineering towards a Resilient World, Cambridge UK, 2015.

[13] Molina, S., Torres, Y., Benito, B., Navarro, M. & Belizaire, D., Using the damage from 2010 Haiti earthquake for calibrating vulnerability models of typical structures in Port-au-Prince (Haiti). *Bulletin of Earthquake Engineering*, **12**(4), pp. 1459–1478, 2014. https://doi.org/10.1007/s10518-013-9563-z

[14] Torres, Y., Molina, S., Martinez-Cuevas, S., Navarro, M., Martinez-Diaz, J. J., Benito, B., Galiana-Merino, J.J. & Belizaire, D., A first approach to earthquake damage estimation in Haiti: advices to minimize the seismic risk. *Bulletin of Earthquake Engineering*, **14**(1), pp. 39–58, 2016.
https://doi.org/10.1007/s10518-015-9813-3

[15] Jaiswal, K.S. & Wald, D.J., Creating a global building inventory for earthquake loss assessment and risk management. *U.S. Geological Survey* (Open-File Report 2008–1160), available at: http://pubs.usgs.gov/of/2008/1160/. (accesed February 1, 2017).

[16] Brzev, S., Scawthorn, C., Charleson, A. & Jaiswal, K., *GEM basic building taxonomy, Version 1.0, GEM ontology and taxonomy global component project*, available at: http://www.nexus.globalquakemodel.org/gemontology-taxonomy/posts/updated-gem-basic-building-taxonomy-v1.0. (accesed February 1, 2017).

[17] Pitilakis, K. Franchin, P., Khazai, B., Wenzel, H. (eds.), *SYNER-G: Systemic seismic vulnerability and risk assessment of complex urban, utility, lifeline systems and critical facilities*, Springer-Verlag: Berlin and New York, 2014.

[18] International Code Council-ICC. *International Building Code (IBC-2006)*, United States, 2006.

[19] CEN- European Committee for Standardization, *EN 1998-1, Eurocode 8—design of structures for earthquake resistance, part 1: general rules, seismic actions and rules for buildings*, Brussels, 1998.

[20] BIS - Bureau of Indian Standards, *Indian standard—criteria for earthquake resistant design of structures, part 1—general provisions and buildings*, Technical report, ICS 91.120.25, 2002.

[21] CNBS - Cuban National Bureau of Standards, *Norma Cubana NC 46: 2013, Construcciones sismoresistentes – Requisitos básicos para el diseño y construcción, 1. Edición*, Officina Nacional de Normalización (NC), Habana, Cuba, 2013.

[22] ICMS- Dipartamento de la Proteccione Civile, *Indirizzi e criteri per la microzonacione sísmica*, ed. Grupo di lavoro "ICMS". Conferen-za delle Regioni e delle Province autonome - Dipartimento della protezione civile, Roma, 3 vol. e Dvd, 2008.

[23] Molina, S., Lang, D.H., Meslem, A. & Singh, Y., *Topographic amplification effects - towards their inclusion in ELE Studies (State-of-the-art report),* Report no. 15–001, Alicante–Kjeller–Roorkee, 2015.

[24] Federal Emergency Management Agency- FEMA. *Prestandard and commentary for the seismic rehabilitation of buildings*. Technical report, Federal Emergency Management Agency (FEMA), Washington DC, FEMA 356, 2000.

[25] Federal Emergency Management Agency- FEMA. *Improvement of nonlinear static seismic analysis procedures*, Technical report, Applied Technology Council, California, USA. FEMA-440, 2005.

[26] Fajfar, P., *Structural analysis in earthquake engineering. A breakthrough of simplified nonlinear methods*, Presented at the 12th European Conference on Earthquake Engineering, London, paper 843, 2002.

[27] Dolsek, M. & Fajfar, P., Inelastic spectra for infilled reinforced concrete frames. *Earthquake Engineering and Structural Dynamics*, **33**, pp. 1395–1416, 2004.
https://doi.org/10.1002/eqe.410

INFLUENCE OF HYDROGEOLOGICAL AND GEOTECHNICAL PARAMETERS ON THE SEISMIC BEHAVIOR OF POTABLE WATER INFRASTRUCTURES

RAMÓN EGEA PÉREZ[1], JESÚS P. CHAZARRA ZAPATA[2] & FRANCISCO J. LÓPEZ PEÑALVER[3]
[1]EMUASA, Municipal Water and Sanitation Company, Murcia, Spain.
[2]Engineering Department, University Miguel Hernandez, Orihuela (Alicante), Spain.
[3]Superior Polytechnic School, Alicante University, Spain.

ABSTRACT

This study analyzes the influence of the urban hydraulic infrastructure behavior, the soil typology and the characterization of the hydrogeological zonal risks, proposing constructive dispositions and procedures, which will increase the reliability of the pipes consisting of a fragile nature, guaranteeing the service continuity even during seismic events with intensity higher than VII (IMM). That would greatly reduce the degree of vulnerability of the urban hydraulic infrastructure, especially in the case of Southeast Spain, which presents a high risk of seismic intensity. Although the effects on the vulnerability of pipelines, of the displacement of the ground, such as fault movements, settlements and lateral expansions, have been widely studied and analyzed, there is no comprehensive methodology, incorporating other factors related to the Management of the infrastructure, and therefore a general application method is proposed that will evaluate in addition to the effects according to the soil typology and the hydrogeological characteristics zonal (hydrological hazard, seismic hazard, and specific hydrogeological factors), the typology of the infrastructure linear hydraulic, based on operational analysis, that allows to correlate the behavior of the soils and the possible affectation in the potable water pipes, thus serving as planning element and preventive tool for the mitigation of damages.
Keywords: ground motion, hazard, hydrological, liquefaction, management, risk, seismic, vulnerability

1 INTRODUCTION

As it has been observed in different earthquakes, saturated, weakly cohesive granular soils are more susceptible to an increase in the interstitial pressure of the water, causing very loose sands and loss of initial resistance to cutting. The increase in interstitial pressure in the pores generates the amplitude of the oscillatory filtration, according to the size, shape and gradation of the particles, the confinement pressure acting on the soil, the texture of the soil and the rate of over-consolidation of this.

According to Japanese experiences, the damage in linear underground hydraulic infrastructures is much greater in the earthquakes associated with the liquefaction of the soil than in those that are not associated with that effect, the soils being more susceptible, those of flat topography. Therefore, large deformations in the ground are not easily perceived, which makes it difficult to determine the type of soil deformation, which has caused damage in a pipe following a seismic movement.

In general, there are two forms of seismic deformation in the soil, linked to the liquefaction and that could cause damage to buried pipes:

- The dynamic deformation or alternating tension and compression.
- Static strain or tension in one direction.

In addition, in buried pipes, ascending hydrostatic pressures (saturated soils whose behavior can be assimilated to a static fluid) can be produced by the favor buoyancy of the same,

DOI: 10.2495/SAFE-V7-N4-597-611

causing ruptures or superficiality of these, as well as an increase of the pressure of water contained in the pores of the soil that favors behavior [1].

Impacts on pipeline vulnerability, land displacement, such as fault movements and lateral expansions, although widely studied, have not yet been proposed for a general application method that assesses the effects of soil-based conditions in a statistical analysis, which allows to correlate the characteristics of behavior of the soils and the susceptibility to the damages of the pipes according to their nature, and constructive disposition.

According to the existing statistics based on different earthquakes occurring in Japan, New Zealand, Mexico, Chile, Italy and other countries with significant seismic activity, and as indicated, it has been verified by the following:

- Damage is concentrated in areas where irregular terrain conditions occur; being the non-uniformity of the displacement generated as a product, in turn, of the variability in the nature of the terrain, the cause of the greater deformations of the soil.
- The degree of vulnerability depends to a large extent on the type of piping system and its material nature. Depending on the material used, less flexibility of the pipe or its joints, greater effect on the vulnerability of linear infrastructure is generated.
- The vulnerability also depends on the intensity of the seismic movement, which is represented by the maximum acceleration [2].

It is for this reason that liquefaction – enhanced by the characteristics of the original subsurface – is considered the main cause of damage in the piping system during a seismic movement. As one of the specific damages, the lifting and/or displacement of the pipes, is due to the following occurrences or factors:

- The existence of a liquefied layer of land exerting a pressure or push on the top layer of soil.
- The location of an impermeable layer, which precludes or limits the dissipation of the water high pressure in the pores of the land.
- The generation of an amplification of the seismic movement, increasing the degree of liquefaction of the land located on the lower layer of the subsoil.

The impacts of seismic movement on hydraulic infrastructures, and more specifically on pipes and fittings, may vary to a greater or lesser degree, depending on the following factors:

Geology, geotechnics and characterization of local underground hydrology:

- Nature and characteristics of the land.
- Variability and power of strata constituting of the land.
- Existence of groundwater levels and their seasonal variability.
- Geology and geotechnics of the area.
- Contamination of sources and areas of surface runoff.

Typology and characteristics of linear hydraulic infrastructure:

- Type of pipe.
- Nature of the pipe material.
- Type of installation.
- Buried or surface pipeline.
- Possible interaction with other linear infrastructures (roads, railways, oil pipelines, power lines and telecommunications).

Next, in the organizational chart of Fig. 1, the process of the seismic action on the infrastructure, related to the zonal geology is described.

In the case of Southeastern Spain, and especially the City of Murcia, whose seismic hazard is moderate, presents a characterization of soils, Fig. 2, especially susceptible to liquefaction and subsidence phenomena, increased by other adverse climatological factors, such as the aridity of the territory and periodic droughts, heavy rains [3].

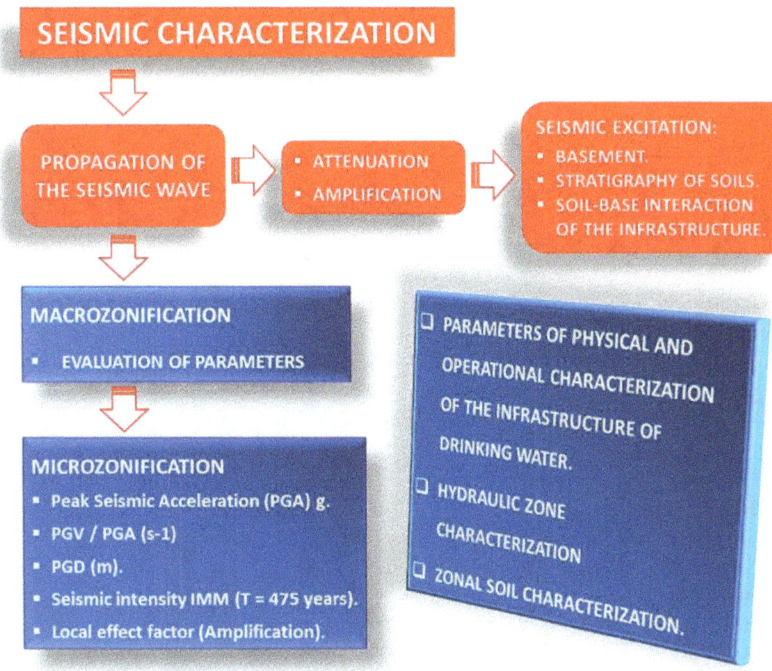

Figure 1: Organization chart of the characterization of the seismic risk and its geological component in relation to the Infrastructure of the drinking water network.

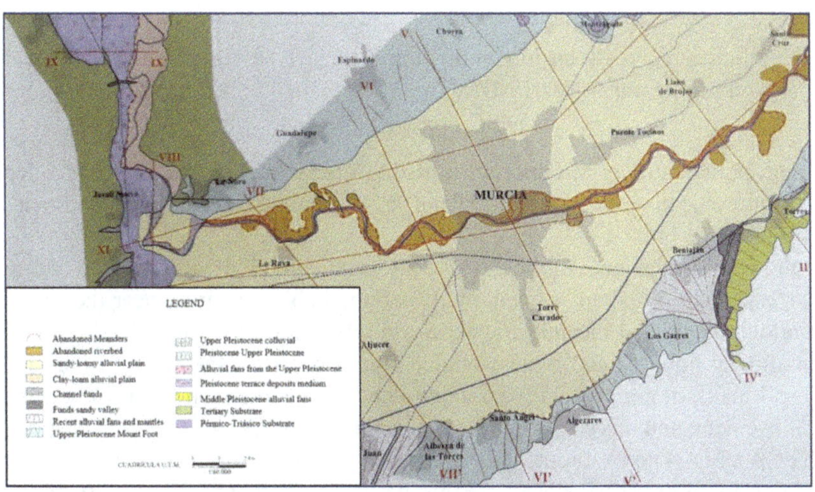

Figure 2: Geological map of quaternary surface formations metropolitan area of the city of Murcia. (Source: IGN).

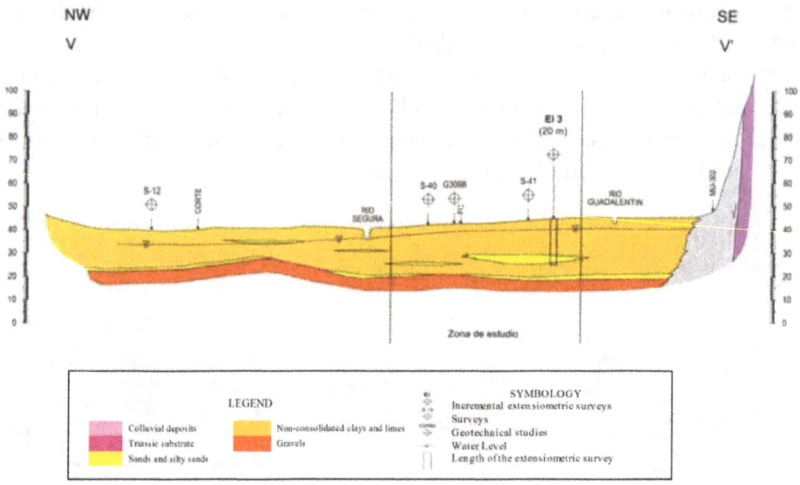

Figure 3: Geological profile representative of the urban nucleus of the city of Murcia (VV ‘). Scale 1:1250. (Source: IGN)

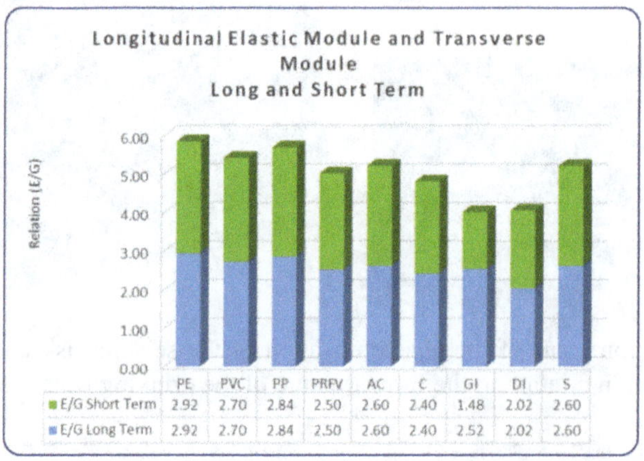

Figure 4: Relationship between longitudinal and transverse elastic modules, and drinking water piping materials. (EMUASA).

The metropolitan area of Murcia City is also located in an alluvial plain characterized by fillers and anthropic material in the superficial strata, cemented on colluvial lands and surface aquifers, Fig. 3.

The following are the areas of the city of Murcia that are more vulnerable to hydrogeological risks, especially flood and seismicity. It is remarkable, the areas near the river Segura, where the old abandoned meanders appear, and where the most superficial layers of land, settle on alluvial floodplains. In these circled areas, the characterization of the type V and type VI soil makes them especially susceptible to the liquefaction phenomenon in case of seismic movement, and this contributes to the fact that the expected maximum seismic acceleration (PGA), Fig. 4 and Table 1, in these zones can reach values of 0.24 g for an estimated seismic intensity of grade VII (IMM). In this, case, the effects of that intensity are enhanced by the ‘site effect’ by a factor, Fa = 2.5 (Borcherdt, 1994a, 1994b), [4].

Table 1: Amplification factors for different soil types and base acceleration values, PGA. (Source: RISMUR-Update 2015), [2].

PGA	Site class FEMA (NEHRP-2009)	Mapped MCE geometric mean peak ground acceleration, PGA (g)					Site class RISMUR 2015
		PGA ≤ 0.1	PGA ≤ 0.2	PGA ≤ 0.3	PGA ≤ 0.4	PGA ≤ 0.5	
0	A	0.80	0.80	0.80	0.80	0.80	I Null
0.1	B	1.00	1.00	1.00	1.00	1.00	IIa Very low
0.2	C	1.20	1.20	1.10	1.00	1.00	IIb Low
0.3	C	1.20	1.20	1.10	1.00	1.00	IIIa Half
0.4	D	1.60	1.40	1.20	1.10	1.00	IIIb Half
0.5	D	1.60	1.40	1.20	1.10	1.00	IV High
	E	2.50	1.70	1.20	0.90	0.90	V High
	F	Zonal specific study					VI Very High

The marked areas correspond to locations that are particularly vulnerable to hydrogeological phenomena, with a potential risk of liquefaction and differential settlements, Figs 5 and 6, so that existing drinking water supply networks must be subject to strict controls, the incorporation of resilient elements and the planning of actions that guarantee the continuity of operation, according to the elastic modulus of the pipe material, Fig. 7.

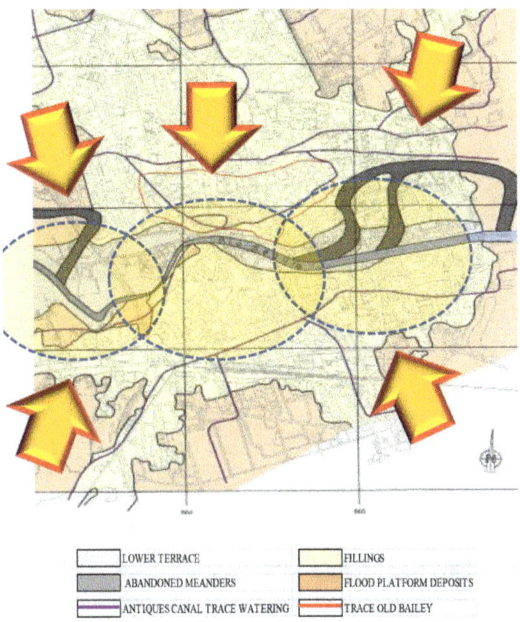

Figure 5: Geological Map of the City of Murcia at scale 1 /12,500 (Source: IGN)

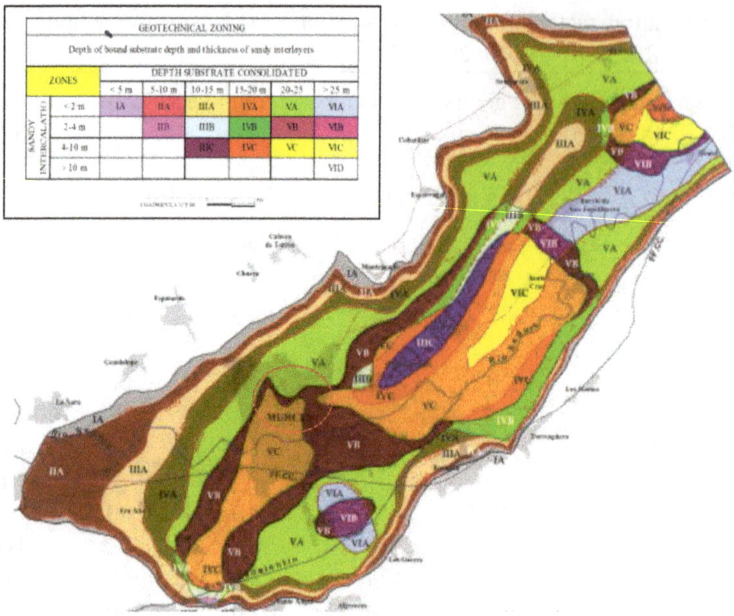

Figure 6: Map of geotechnical zoning of the middle Vega of the Segura River. (Source: IGN)

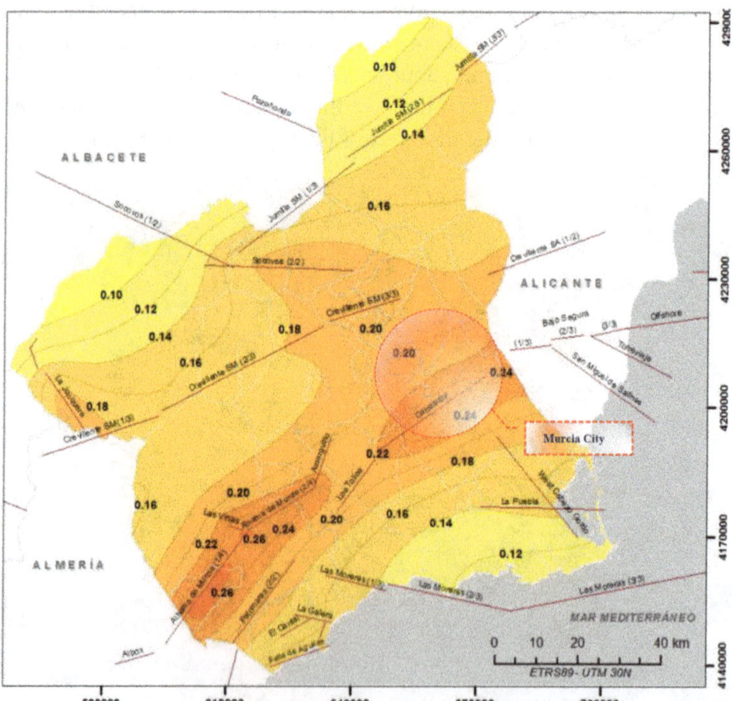

Figure 7: Map of seismic hazard in rock of the region of Murcia for periods of return of 475 years. Average maximum accelerations of the horizontal components of the PGA movement (g), [11].

2 METHODOLOGY

The analysis of the geotechnical parameter and the selection of the most representative ones in their interaction with the effects generated by seismic and hydrological events, is essential to provide the linear hydraulic infrastructure (supply and sanitation networks) with a response capacity or of sufficient resilience to guarantee the operation of the System and to counteract the intensity of the natural event that occurred [5–9].

The parameters to be considered in the risk assessment of the linear hydraulic infrastructure are:

- Tectonic movement:
  - Characterization of the active fault.
  - Seismic return period.
  - Directionality of the fault.
- Soil movement: PGV, PGA
- Liquefaction: the susceptibility of the ground to the displacement can be up to 10 times greater than in normal conditions. If the terrain topography is flat, there is no lateral displacement.
- Side propagation (PGD).
- Landslide.
- Differential settling.
- Shaking.

To estimate the damage to the components of the linear infrastructure, it must be characterized based on:

- Material of the pipe.
- Type of retention joint or joint.
- Length of the pipe.
- Medium conditions and definition of the elements of protection against corrosion.

As a measure of damage mitigation, it is necessary to characterize the soils, and the seismic or hydrological potential of the area, if applicable. For this, the degree of affectation of the linear hydraulic infrastructure and the evolution of the soil-infrastructure interaction are evaluated and defined, due to the occurrence of effects generated by an earthquake.

The basic parameters, which define the linear water infrastructure, as well as other local geological and geotechnical parameters that must be characterized, are summarized in the following tables (Table 2).

Different phases of proposed geological a geotechnical parameter influence value methodology is described below in schematic form. Some of those were analyzed and defined earlier (Fig. 8).

3 RESULTS AND DISCUSSION

The analysis of the proposed characterization model leads to the establishment of preventive measures and planning of actions to be carried out, as well as to provide operational management of the potable water distribution system with fundamental information to limit the impact of a seismic event, and increase the reliability of the Hydraulic Infrastructure, Fig. 9.

Table 2: The parameters that define the linear infrastructure of drinking water, in relation to the geological and local geotechnical characterization.

Physical characterization Hydraulic sector or sub-sector
Hydraulic sector designation: Location (UTM): Designation pressure zone: Nature (material) and length networks (km): Ductile cast iron-DF (km): Polyethylene-PEHD (km): Steel-S (km): Gray-GF casting (km): Asbestos cement-AC (km): Sr (relative rigidity) Tons/m^2: Average age of the network by type of material (years): Number of sectorization valves (units) Max. Pipe (mm): No counters zone (uds):
Operational characterization of the Hydraulic sector or subsector
Average supply pressure (m.c.a.): Number of breaks/km net (hydraulic basin or sub-basin): Degree of criticality (I-IV) from lowest to highest: Strategic supply (Yes/No) Network class: Tertiary network (Yes/No) Secondary network (Yes/No) Primary Network (Yes/No) Transport network (Yes/No) Existence of alternative supply (Yes/No): Supply flow (m^3/h): Q average (l/s) Qmax. (L/s) Water quality: Dosage chlorination (NaClO) mgr/1-ppm: Turbidity index (NTU) Type of maintenance: I - VI (lowest to highest grade) Seismic zone definition: Peak seismic acceleration (PGA) g: PGV/PGA (s-1): PGD (m): Seismic intensity IMM (T = 475 years): Local effect factor (amplification):

Table 2: (*Continued*)

Hydrological definition
Flood zone (Yes/No):
No. of flood episodes/year:
Average height of water (m)
Maximum terrain slope (%):
Coef. Runoff (C):
Geological and geotechnical characterization
Soil typology (RISMUR-2015 classification) (I-VI)
Permissible voltage or load bearing capacity (N / mm^2):
N30 (SPT):
Phreatic level depth (m):
Complementary characterization
Artistic heritage area (Yes/No)
Number of inhabitants (Inhab.)
Location of strategic centers (YES/No):
Existence health centers (Yes /No):
Existence of other critical infrastructures (Yes/No):

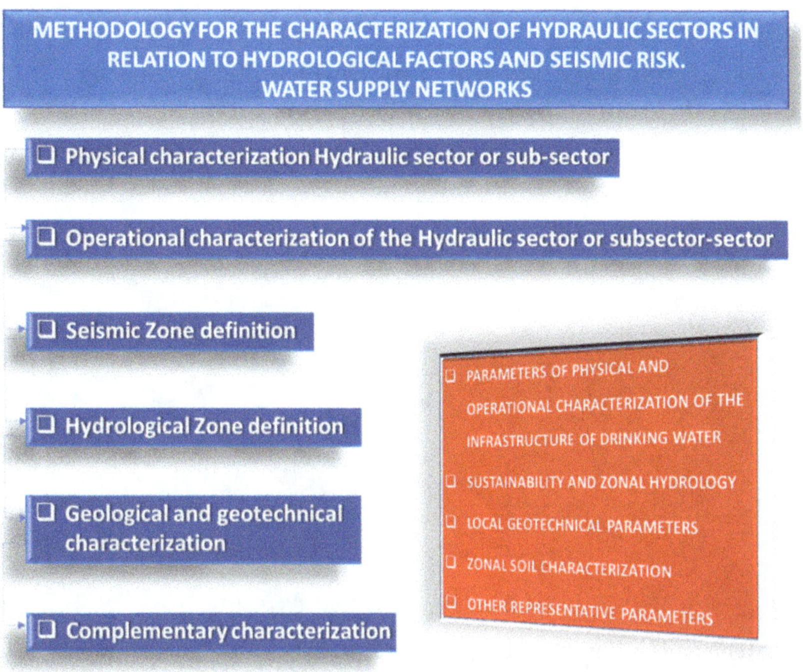

Figure 8: Methodology for the characterization of hydraulic sectors in relation to hydrogeological factors and seismic risk in water supply networks. (EMUASA).

Figure 9: Hydraulic subzones in various neighborhoods of the City of Murcia. (EMUASA).

Likewise, it is a tool that allows structuring and channeling to the essential information necessary for real-time decision making. Next, we analyze various hydraulic sectors of the metropolitan area of the city of Murcia, where some main water supply pipelines are reflected, Fig. 10.

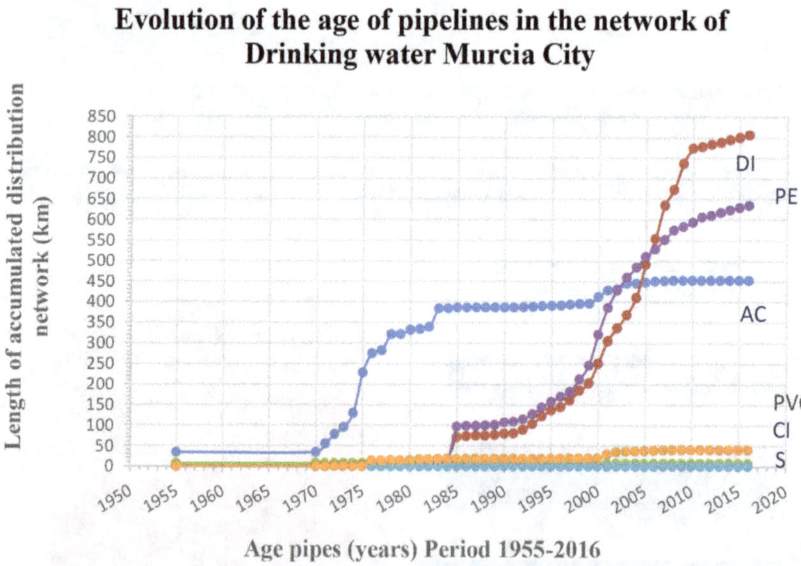

Figure 10: Temporal evolution and typology of material in the potable water distribution network of the City of Murcia (1955–2016) – DI (Ductile Cast Iron), PE (Polyethylene), AC (Asbeste cement), PVC (polyvinyl chloride), CI (Cast Iron), and S (Steel). EMUASA

As already indicated, in non-cohesive plots located in strata higher than the water table or in the transition zone, differential settlements can be generated with displacement of the connections or even rupture of the pipe in extreme conditions, being the heterogeneity and the different stratification of the terrain, a susceptibility factor, which can influence the behavior of linear hydraulic infrastructure [7], especially during seismic and/or hydrologic events, increasing damages and disrupting the system's integral operation. Next, there are representations of thicknesses of fills, limes and clays of the metropolitan area of the City of Murcia (B° de San Miguel, Sta. Eulalia, San Pedro, Infante J. Manuel, Vistabella, and others), Figs 11 and 12.

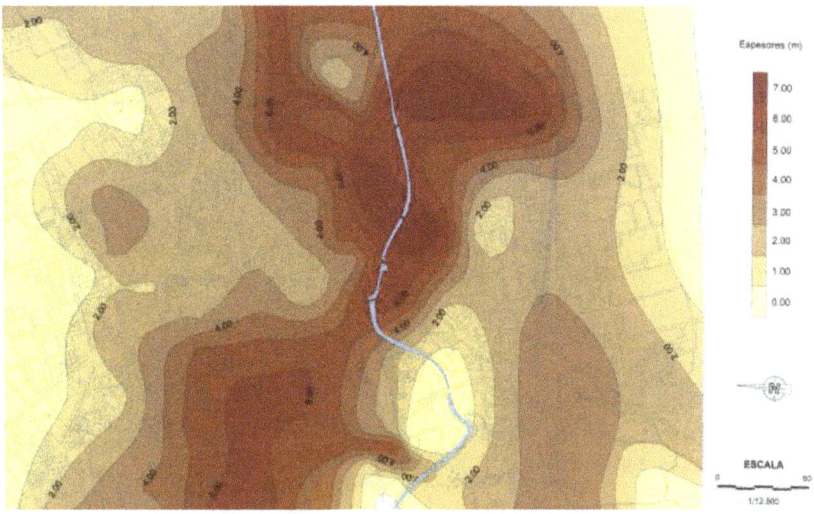

Figure 11: Thickness of fillings in the metropolitan area of Murcia. The Land Subsidence in the City and Metropolitan Area of Murcia IGME 2010.

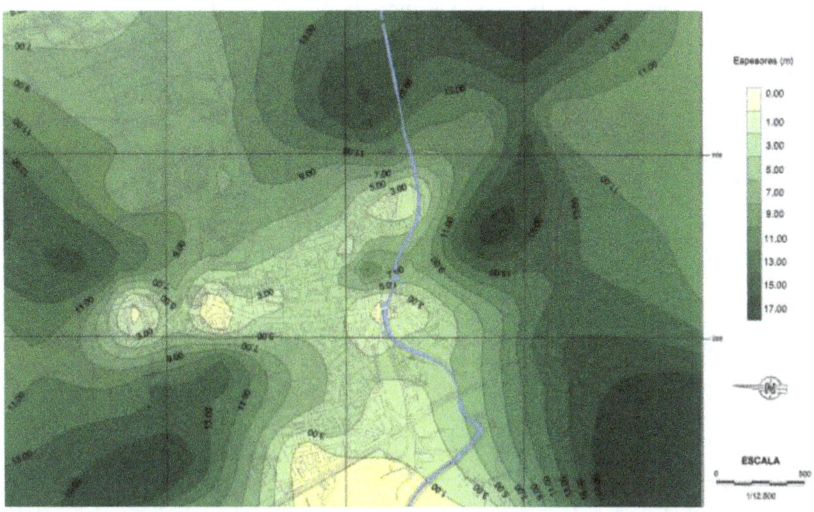

Figure 12: Thickness of clays and limes in the metropolitan area of Murcia. The land Subsidence in the City and Metropolitan Area of Murcia IGME 2010.

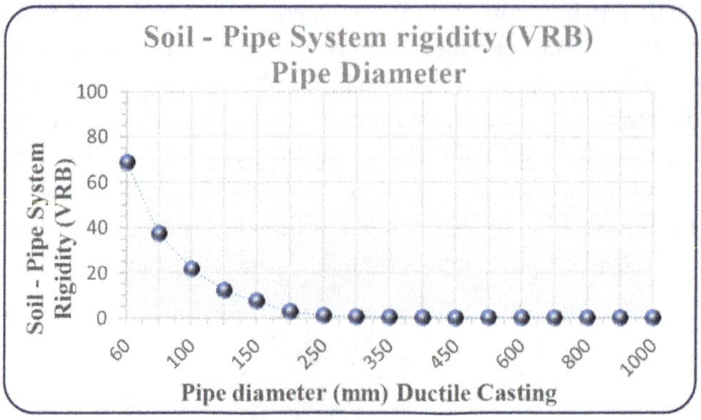

Figure 13: Representative graph of the Terrain structural system - Ductile cast iron pipe and pipe diameter (EMUASA).

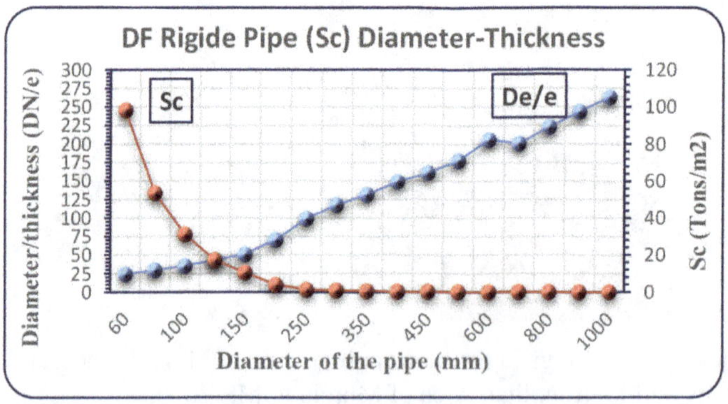

Figure 14: Representative graph of the evolution of the long-term circumferential stiffness, external diameter and installed DF pipe thickness (EMUASA).

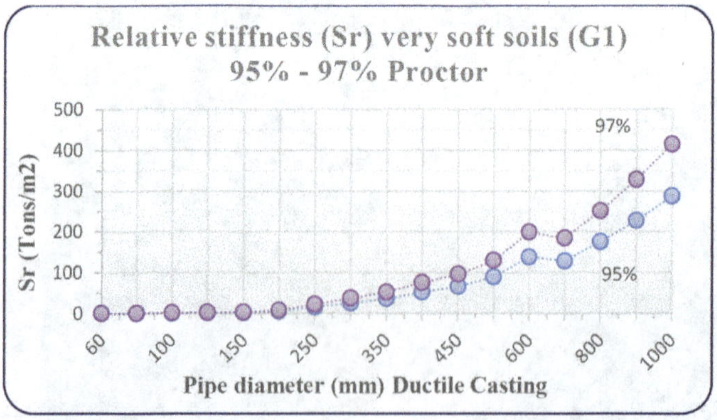

Figure 15: Relative stiffness (**Sr**) in very soft soils according to pipe diameter and degree of compaction of the filling. Ductile Casting (EMUASA).

Table 3: Classification of drinking water pipes according to the level of service and the period of return of occurrence of an earthquake. (FEMA Seismic Guidelines for Water pipelines 2005).

Pipe function class	Importance	Tipology	Probability of exceedance 50 years	Return period T (years)
I	Very low to None	-	100%	Undefined
II	Ordinary, normal	-	10%	475
III	Critical	-	5%	975
IV	Essential		2%	2475

Table 4: Calculated values of PGD, based on FEMA Seismic Guidelines for Water pipelines 2005 recommendations, according to failure movement and service level of the pipeline.

Pipe function	Design movement (PGD)	Peak ground displacement (PGD) (m)
I	-	-
II	1AD	0.631
III	1.5AD	0.946
IV	2.3AD	1.451

Following are some graphic representations showing the behavior of drinking water pipes according to their physical characteristics (material nature, diameter, ..), and circumferential stiffness, as well as the nature and degree of soil compaction (Figs. 13, 14 and 15).

According to studies included in the FEMA Seismic Guidelines for Water pipelines 2005 [8], in relation to drinking water pipes, pipes are classified into 4 classes according to the degree of importance and the period of return T of occurrence of an earthquake, as indicated in the Table 3.

By means of these recommendations, the parameter PGD (Peak Ground Displacement) was obtained, as well as the maximum slip (AM) generated for an earthquake of magnitude 6.8 (maximum probable), according to the characteristics of the Carrascoy Fault located at 6 Km of the metropolitan area of Murcia, as indicated in the Table 4 [9].

The value of these parameters has been calculated based on the equations proposed by (Wells & Coppersmith, 1994) [10].

4 CONCLUSIONS

The analysis of the geological and geotechnical parameters of the area of implementation of the exposed methodology is essential for the development of a comprehensive, preventive and resilient planning for the Sustainability of the linear infrastructure of urban supply.

This research proposes the structured planning of the different phases that define the integration of the characterization of the potable water supply system and its interaction with the hydrogeological zonal risks.

The different milestones that define the methodology that is proposed are characterized by hydraulic sectors or subsectors and are described below:

- Characterization of hydraulic sectors and subsectors.
- Identification of pressure zones.
- Physical characterization of the Linear Infrastructure according to the main components listed.
- Operational characterization of the Hydraulic Sector or Subsector according to the described parameters.
- Incorporation of parameters that characterize water quality.
- Location of 'sensitive' subscribers, and/or strategic.
- Definition of seismic parameters representative of the area.
- Definition of hydrological and hydrogeological parameters representative of the area.
- Characterization of the most representative parameters that define the geology and geotechnics of the area.
- Additional information, relevant to characterize the hydraulic sector.

Through the characterization of the linear Infrastructure of drinking water supply linked to the geological and geotechnical definition of the particularized area of analysis, it allows its correlation with the probable degree of affectation of the same, in the case of a seismic event, as well as the risks associated with it.

ACKNOWLEDGEMENTS

The authors wish to thank EMUASA their cooperation in conducting this study, without which this study would not have been possible.

NOTATIONS

PGA = Peak ground acceleration (g o m/s^2)
PGV = Peak ground velocity (cm/s o m/s)
PGD = Peak ground displacement (cm o m)

REFERENCES

[1] Tsige, M. & García Flórez, I., Proposed geotechnical-amplification classification of geological materials in Murcia. *Geogaceta*, **40**, pp. 39–42, 2006.

[2] Akkar, S. & Bommer, J.J., Empirical equations for the prediction of PGA, PGV and spectral accelerations in Europe, the Mediterranean Region and the Middle East. *Seismological Research Letters*, **81**(2), pp. 195–206, 2010. https://doi.org/10.1785/gssrl.81.2.195

[3] Manual para Enfrentar Inundaciones Ocasionadas por Rotura de Matrices de Agua Potable o Colectores de Aguas Servidas. Access, 15 Junio 2016, available at: http://www.siss.gob.cl/577/w3-printer-8208.html.

[4] Borcherdt, R.D., Estimates of site-dependent response spectra for design (Methodology and Justification). *Earthquake Spectra*, **10**, pp. 617–653, 1994. https://doi.org/10.1193/1.1585791

[5] SRMLIFE. *Development of a global methodology for the vulnerability assessment and risk management of lifelines, infrastructures and critical facilities. Application to the metropolitan area of Thessaloniki*. Research Project, General Secretariat for Research and Technology, Greece, 2003–2007.

[6] ALA. *Development of guidelines to define natural hazards performance objectives for water systems, Volume I,* prepared by ASCE, FEMA and NIBS, 2002.

[7] ALA *Wastewater system performance assessment guideline*, Part I and II. FEMA and NIBS, 2004.
[8] ALA. *Seismic guidelines for water pipelines*, prepared by ASCE, FEMA and NIBS, 2005.
[9] Alexys Herleym Rodriguez Avellaneda, Ingeniero Civil, especialista en Telecomunicaciones. Tesis "Análisis y evaluación de riesgo sísmico en líneas vitales. Caso de estudio Bogotá D.C.". Universidad Nacional de Colombia, 2011.
[10] Wells, D.L. & Coppersmith K.J., New empirical relationships among magnitude, rupture length, rupture width, rupture area and surface displacement. *Bulletin of the Seismological Society of America*, **84**, pp. 974–1002, 1994.
[11] RISMUR II Seismic Risk Plan of Murcia, 2015.

IMPROVEMENT OF AN INTEGRATED MANAGEMENT SYSTEM RESULTING IN HIGHER INDUSTRIAL SAFETY EFFICIENCY

M. BEREZYUK[1], A. RUMYANTSEVA[1] & G. CHEBOTAREVA[2]
[1]Department of Environmental Economics, Ural Federal University, Russia.
[2]Academic department of banking and investment management, Ural Federal University, Russia.

ABSTRACT
LukOIL PermNefteproduct llc management considers industrial and labor safety system as an essential part of efficient production management. Improvement of company's personnel training and further training system in the field of industrial safety, produce quality, workplace health and safety, environmental management as part of integrated management system functioning is the necessary condition of industrial safety management. The educational electronic resource, developed by authors, is the Web-site accessible to all company's employees that consists of guidance and information-reference materials.
Economy effect of the resource usage is cost- and learning time saving, increase of company stuff's competence level.
Keywords: educational electronic resource, industrial safety, integrated management system, labor safety, personnel training.

1 INTRODUCTION

Relevance of industrial and labor safety issues is undeniable both for the world community and for the Russian Federation.

According to official statistics, 2.34 million people die because of industrial accidents and professional diseases every year. According to International Labor Organization, global economy loses up to 4% of gross domestic product due to professional injuries and diseases, deaths, worktime fund loss, payments of guarantees and compensations. Nowadays, International Labor Organization resolutely took the course to globalization of tendencies of labor safety [1].

About 2000 manmade disasters happen in our country every year, people die. Usually any accident on the hazardous facility is preceded by numerous gross violations.

The research object is the integrated management system (IMS) of LukOIL PermNefteproduct llc, which is one of the largest interregional oil products supply company of LukOIL Group. It is well-known that specifics of large international energy companies' operations are their international market orientation and accordant international standard requirements in labor and environment safety they have to meet. In addition, the Russian Federation energy industry is in top three fields in terms of volume of negative effect to the environment [2].

The decision on IMS implementation in the organization nowadays is more and more widespread and gives a number of undeniable advantages – it helps to promote company on national and international markets, increases trust from investors, insurance and credit organizations. The implementation of IMS happens to be less time-consuming and more economically profitable than the implementation of several parallel systems.

In order to efficiently meet legal and regulatory requirements, LukOIL PermNefteproduct llc created, implemented and supported the integrated management system – The Health, Safety and Environment Management System (further on – HSE Management System) [3].

DOI: 10.2495/SAFE-V7-N4-612-626

Research objective is improvement of LukOIL PermNefteproduct llc's integrated management system by creation of a new educational electronic resource.

The Company management considers industrial and labor safety system as an essential part of efficient company production management and undertakes obligations in management of production risks that influence employees' life and health, inventory and property.

It is worth noting that industrial safety level directly depends on qualification of employees working at hazardous facility. Low level of service staff qualification is often the reason for emergency situations. That's why it is essential for organizations that work at hazardous facilities to pay special attention to personnel training. These enterprises must systematically provide employees with training on rules of industrial safety.

At the same time, the integral part of any management system is enterprise's requirement to define needs in personnel training, to provide training and to assess its results, to raise level of knowledge and expertise in industrial, labor and environmental safety.

In order to achieve the given objective we solved further issues:

- analyzed International Organization for Standardization statistics data on ISO certification in the Russian Federation and in the world, analyzed IMS structure varieties at facilities;
- studied the dynamics of employees' industrial injuries and professional diseases rate in the Russian Federation, volumes of funding for preventive measures aimed at industrial injuries and professional diseases reduction in 2008–2015;
- run a general audit check and analysis of LukOIL PermNefteproduct llc's IMS;
- defined basic barriers and complications in the functioning process of LukOIL PermNefteproduct llc integrated management system;
- developed an educational electronic resource for Company's employees;
- created a platform for its allocation.

2 DYNAMICS OF FACTORS THAT CHARACTERIZE THE LEVEL OF INDUSTRIAL AND LABOR SAFETY IN THE RUSSIAN FEDERATION

The status of labor conditions and safety directly depends on specifics of country's economic development. Nowadays we see harsh competition between organizations, issues of efficient spending become the priority, and in this context labour conditions and safety are downplayed.

Among factors that characterize the level of industrial and labor safety only industrial injuries rate has positive dynamics given at the diagram (Fig. 1) [1, 4].

Frequency rate of industrial injuries is also falling down (number of accident victims per 1000 workers): 2000 – 5.1; 2005–- 3.1; 2010 – 2.2; 2014 – 1.4; 2015 – 1.3. Number of deaths due to industrial accidents fell too. 4404 people died in 2000, 3091 people in 2005, 2550 people on 2008, 1456 people in 2014, 1288 people in 2015. Rate of fatal injuries frequency fell: 0.138 in 1995, 0.149 in 2000, 0.124 in 2005, 0.094 in 2010, 0.067 in 2014, and 0.062 in 2015.

Nevertheless, industrial injuries rate is still high and 4–5 times exceeds rates in economically developed countries. Other factors that describe labor conditions and safety don't have such stable lowering tendency.

From 2008 to 2014, proportion of employees working with hazardous labor conditions rose from 26.2% to 39.7% and to 39.1% in 2015. Within this framework, the fraction of workers with at least one set type of compensations is high. According to Rosstat data, this fraction amounted 41.4% of workers headcount in 2005, 46.7% in 2014 and 44.3% in 2015 [4].

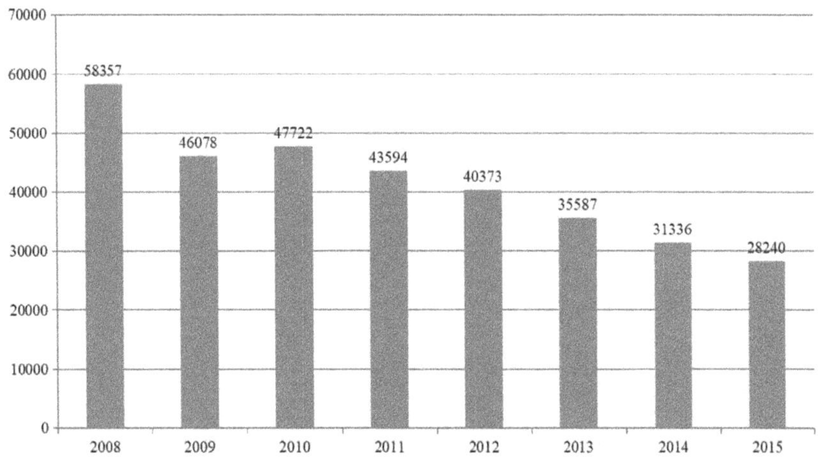

Figure 1: Dynamics of industrial injuries followed by loss of function and death in 2008–2015.

Professional diseases rate doesn't fall (Fig. 2), and chronic diseases (intoxication) prevail in the structure of diagnosed diseases: 98.8% in 2005 and 93% in 2014. Number of professional diseases per 10 thousand workers rose: 1.81 in 2000, 2.1 in 2005 and 2.3 in 2015 [5].

At the same time, volumes of funding for preventive measures aimed at industrial injuries and professional diseases reduction rose (according to the RF Social Security Fund data) from 1.2 bln rbls in 2003 to 9.5 bln rbls in 2015 (Fig. 3) [4].

Many efforts were made that focused on labor conditions and safety improvement at the facilities over the last years of economic reforms. But, unfortunately, number of labor safety violations and industrial injuries as a result, are still rather high in comparison to developed countries.

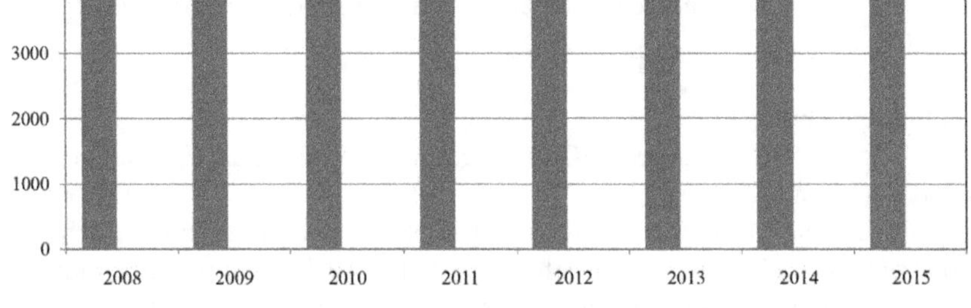

Figure 2: Dynamics of workers' professional diseases in 2008–2015.

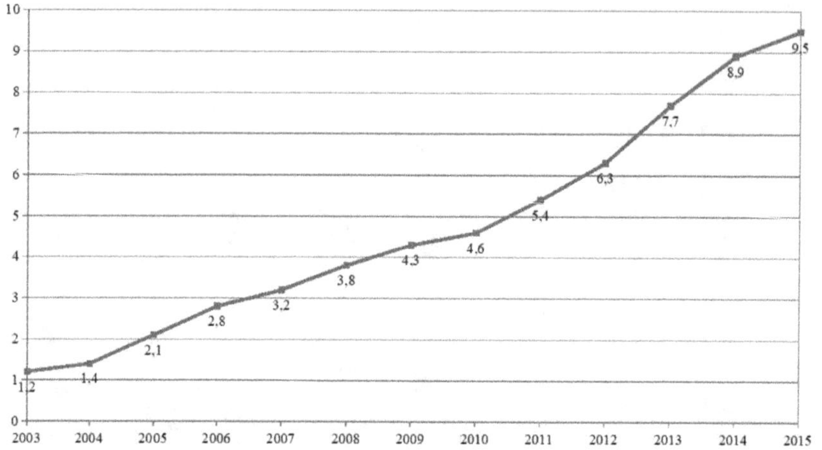

Figure 3: Funding for preventive measures on industrial injuries and professional diseases reduction, billion rubles.

High industrial injuries rate causes losses for businesses and for the government due to loss of working time: according to experts, losses from industrial injuries and compensations for workers in 2015 comprised 107.9 mln resource days, and losses from common industrial injuries – 11.5%. Underproduction of output and services within this framework turns into loss of gross domestic product – about 0.6%. Altogether, economic losses from the state of labor conditions in the Russian Federation comprised 1.29 trn rbls (1.8% of GDP) [1, 4].

Thus, for achievement of higher standards, improvement of labor conditions, increase of industrial and environmental safety level, it is necessary to use the vast experience of economically developed countries with modern legislative systems that set standards and requirements for labor safety issues.

3 DYNAMICS OF QUANTITY OF ISSUED ISO CERTIFICATES

Integrated management system is a combination of two or more management systems functioning as a whole. The most widespread components of organizations' integrated management systems in the Russian Federation are:

- quality management system based on ISO 9001:2015 standard (Russia-adapted analogue of the standard is GOST P ISO 9001:2015) [6];
- environmental management system based on ISO 14001:2015 standard (Russia-adapted analogue of the standard is GOST P ISO 14001-2016) [7];
- professional health and safety management system OHSAS 18001:2007 (GOST 12.0.230-2007) [8];

and other management systems.

Dynamics of quantity of issued ISO certificates in the Russian Federation is given at Table 1 and Figs 4 and 5 [9].

All the biggest energy companies in Russia implemented and certified EMS; explosive growth of implementations was seen before 2010. Thereat, the employees' motivation for EMS implementation in these companies is practically zero. The ownership of code certificate does not guarantee improvement in environment quality [10].

Table 1: The rate of change in the number of issued ISO international standard certificates in the Russian Federation.

Standard/year	2010	2011	2012	2013	2014	2015
ISO 9001	62265	13308	12488	11764	11301	9084
ISO 14001	1953	1093	1090	1272	1263	1156
ISO 50001	–	1	8	25	81	118
ISO 27001	72	31	27	48	50	55
ISO 22000	191	143	171	279	482	733

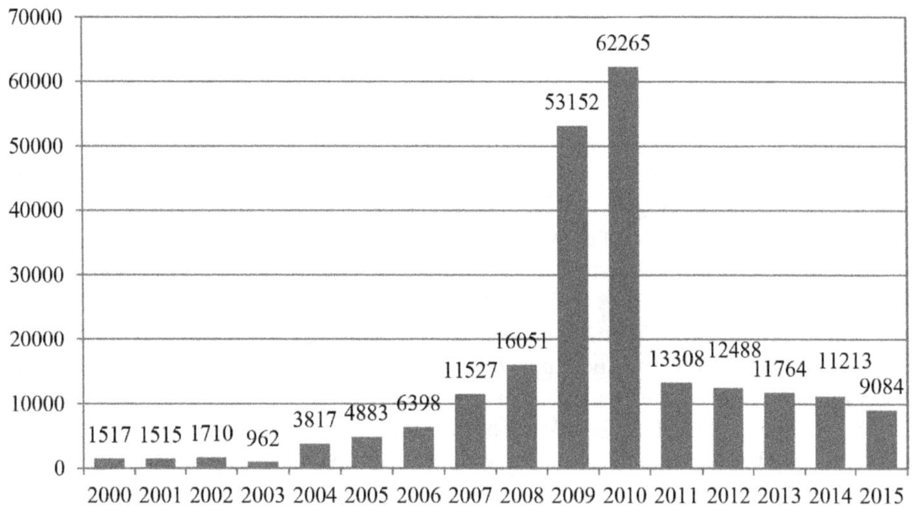

Figure 4: Dynamics of quantity of issued ISO 9001 certificates in the Russian Federation.

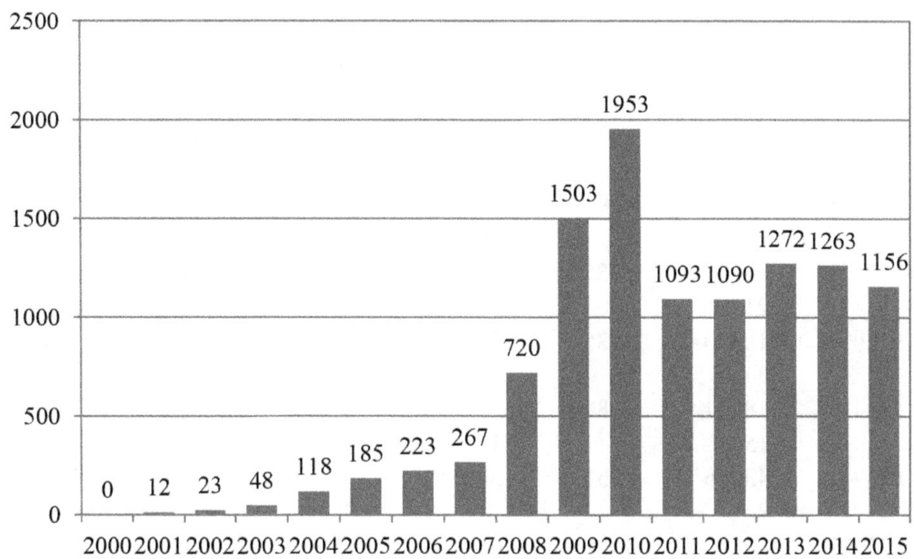

Figure 5: Dynamics of quantity of issued ISO 14001 certificates in the Russian Federation.

Data analysis of State pollution monitoring in the Russian Federation shows that the pollution load of natural habitats, both in separate factors and in full assessment, is practically unchanged over the last years [11].

Comparison of ISO 14001 certificates quantity per 100 000 people of country's population with Environmental Performance Index also doesn't give substantive statistical connection [12].

Data analysis shows that the situation in the Russian Federation corresponds with international trends mostly. There is a drop in ISO 9001 and ISO 14001 standards implementation, but trends of ISO 50001, ISO 27001 and ISO 22000 standards implementation are steady. As it was told before, in order to efficiently meet legal and regulatory requirements LukOIL PermNefteproduct llc created, implemented and supported the integrated management system – Health, Safety and Environment Management System (further on – HSE Management System) [3].

IMS for Storage and Realization of Oil Products activities functions in compliance with requirements of Russian legislation, ISO 14001:2015 'Environmental Management Systems. The Requirements and Usage Guidelines' and OHSAS 18001:2007 'Professional Health And Safety Management System – The Requirements'. Collective results of International Standardization Organization are given in Tables 2, and 3 [9].

A total of 1,036,321 certificates were issued to ISO 9001 (including 4190 issued to the 2015 version published in September 2015) a slight decrease of 0.2% on last year. A total of 319,324 certificates were issued to ISO 14001 (including 947 issued to the 2015 version published in September 2015) up 8% on last year. A total of 11,985 certificates for ISO 50001 for energy management were issued (up 77% on last year) and 3133 certificates issued to ISO 22301 for business continuity management (up 78% on last year). The strong growth seen for these two standards is likely to be related to the fact that they are both relatively new to the market. ISO and IEC's standard for information security, ISO/IEC 27001 saw a 20% increase

Table 2: Total number of issued ISO international standard compliance certificates, % change by year.

Standards	2010	2011	2012	2013	2014	2015
ISO 9001	1076525	1009845	1017279	1022877	1036321	1076525
%	1	-6	1	1	1	-0,2
ISO 14001	239880	243393	260852	273861	296736	319324
%	8	1	7	5	8	8
ISO 50001	–	459	2236	4826	6765	11985
%	–	–	387	116	40	77
ISO 27001	15626	17355	19620	21604	23005	27536
%	21	12	13	10	6	20
ISO 22000	18580	19351	23278	24215	27690	32061
%	34	8	20	4	14	16
ISO 13485	18834	19849	22317	25655	26280	26255
%	15	5	12	15	2	-0,2
ISO 16949	43946	47512	50071	53723	57950	62944
%	7	8	5	7	8	9
ISO 22301	–	–	–	–	1757	3133

Table 3: Countries with the most ISO international standard certified companies in 2015.

	International standards/Amount of the certificates							
Country	ISO 9001	ISO 14001	ISO 50001	ISO 27001	ISO 22000	ISO 13485	ISO 16949	ISO 22301
China	292559	114303	262	2469	9949	1961	25498	3
Italy	132870	22350	470	1013	818	2635	1345	17
Japan	47101	26069	44	8240	1089	1064	1482	200
Spain	32730	13310	390	676	590	379	959	51
Germany	52995	8224	5931	994	256	2508	3473	27
United Kingdom	40161	17824	1464	2790	121	1651	629	345
Russia	9084	1156	118	55	733	125	329	0
USA	33103	6067	53	1247	210	5231	4345	40
France	27844	6847	500	227	669	1400	1012	8
Total	1033936	319324	11985	27536	32061	26255	62944	3133

to 27,536 certificates worldwide, while ISO 22000 for food management and ISO/TS 16949 for the automotive sector were up 16% and 9%, respectively.

This year we added a new management system standard, ISO/IEC 20000-1:2011 Information technology – Service management – Part 1 Service management system requirements, after recommendations from international accreditation and certification experts that we consult annually. This brings the total of standards surveyed to 9. A total of 1,519,952 certificates were issued worldwide in 2015, compared to 1,476,504 the previous year, an increase of 3%.

There is not much information on OHSAS 18001 influence in the world. In particular, according to information, published at British Standards Institute (BSI) web-site, OHSAS 18001 was used in 16,000 organizations in 80 countries in 2005. 54,000 OHSAS 18001 certificates in 116 countries were used in 2009. There is no summarized data on the Russian Federation shared.

There are also translation problems, even in the terms area. There is no 'labor safety' term in the world practice, where 'occupational safety and health' or 'occupational health and safety' terms are used, that might be translated in Russian verbatim as 'production (industrial) safety and health'. There is no such word as 'labor' in English terminology. 'Safety' in Russian means 'security', 'occupational' – 'professional'. In literal translation in Russian - 'professional security'. This term is close in its meaning with 'industrial safety', because it means emergence of highly dangerous situations and relates not only to production, but to all the economy branches [13].

Regarding problems of OHSAS 18001 implementation, it is worth to note the existence of different interpretations of Standard requirements by different independent assessment officers – certification authorities that perform independent audit. Some difficulties still appear while building of an integrated management system based on OHSAS 18001, ISO 9001 and ISO 14001, but these three standards are mostly compatible. These problems probably will be solved when OHSAS will become an ISO standard, because now all the management system standards of International Standardization Organization have unified form and, as much its specifics allow, unified contents.

As is well-known, new ISO 9001:2015 and ISO 14001:2015 standards were approved in 2015. The structure and terminology of ISO 14001:2015 significantly differ from the previous version. These changes were introduced for better integrated implementation of management systems' standards in the same organization and to implicate uniform procedure management consistent with quality, environmental, labor safety requirements [14–16].

Although, the implementation of ISO 14001 standard's new version in the nearest time might become a serious challenge for Russian businesses due to specifics of national environmental legislation, poor development of environmental infrastructure and technologies, weariness of basic environmental funds; all of these require preliminary preparations at the strategic management level. Particular complications of adaptation to new basic requirements involve the Russian certified organizations' need to implement such environmental management instruments as assessment and management of product/service's life cycle, environmental performance indicators, multistakeholder dialogue, organization's context analysis, managers' leadership, records management, process and risk-oriented approaches to management [17].

Additionally, publication of new ISO 45001 standard is expected in summer of 2017; it will be the first ISO standard for labor safety management systems. ISO/DIS 45001 consists of requirements for basic parties of interest and is built on widespread OHSAS 18001 specifications. OHSAS implementation in Russia broadly corresponds to the world practice. And identity of GOST P 54934-2012 and OHSAS 18001:2007 allows companies to take part in international economic activities [18].

Therefore, the necessity and importance of personnel training improvement, generation of personnel knowledge and new competencies are particularly acute at the stage of standards update and necessity of organizations' certification according to the new standards.

4 RESEARCH METHODOLOGY

In order to assess efficiency of The Health, Safety and Environment Management System at LukOIL PermNefteproduct llc, the HSE Management System audit is performed every year. The questionnaire survey of Company's facilities was proposed as the audit method. 12 facilities, based in Yekaterinburg and the Sverdlovsk region (15% of all facilities), were audited (with questionnaire) in 2014.

Based on audit results, authors concluded that Company's personnel don't have a full comprehensive idea of environmental culture and Company's Policy, basic documents that concern providing industrial and environmental safety. Workers and specialists don't understand the importance of their specialization and the ability to make adjustments to the existing system, they are not aware of environmental aspects [19].

Moreover, only 57 (out of 574) employees of the Sverdlovsk regional department completed training in industrial and environmental safety in 2014, what comprises only 9.9% of department personnel.

All of the aforesaid defines importance and necessity of improvement of such integrated management system element as personnel training. In order to improve the situation in the Company, authors developed an educational electronic resource for all levels of personnel to acquire required professional competences within existing IMS, with special focus on environmental management system.

There is a vast classification of education types, forms and methods; each of them with a variety of advantages and disadvantages.

The traditional education is classroom activities where teacher's work is aimed mostly at delivery of knowledge and methods of operation, given to the students in prepared form, and intended for reproducing learning. The teacher is the only proactive side of educational process. Bases of this education were laid almost four centuries ago by Y.A. Komenskiy (The Great Didactics). Such education form is still the most widespread.

Remote learning is an educational technology carried out mostly with the help of IT networks and with mediated (at a distance) interaction between students and teachers (Federal law № 273-FZ 'On Education in the Russian Federation' from 29.12.2012). One of the types of remote learning is relatively new type – e-learning.

E-learning is transfer of knowledge and management of education process with the help of new information and communication technologies. Interactive electronic tools of information delivery are used in the process of e-learning, usually the Internet and companies' corporate networks, but other ways are possible, too, for example, CDs.

New information technologies implementation into the process of education is directed to activation of learning process, learning speed increase, rise of time for learner's individual and independent study. E-learning usage in education has lots of advantages and it is a widespread trend in world educational community. Let's take a closer look at advantages and disadvantages of e-learning.

E-learning advantages are:

- no temporal constraints for learning;
- learning flexibility, broad reach and mobility;
- high quality of training sessions (based on the usage of multimedia tools);
- usage of multimedia tools: texts, sounds, pictures, videos, animation;
- availability for all ranks of learners;
- broad geography of training;
- teacher-student interaction;
- physical presence of teacher and students in the classroom is not necessary;
- reduction of travel time and costs for students;
- no classroom rent expenses;
- ability to teach a large amount of students at the same time;
- ability to register students' activities and academic progress;
- ability to create different education models.

E-learning disadvantages are:

- it requires skills of Internet usage, which is not available for everyone and, inversely, requires other expertise;
- overdependence on technologies;
- lack of teacher-student interaction;
- nonexistence of face-to-face communication;
- inadequate attention to mass communications;
- lack of practical activities;
- low level of participation in educational process.

The basic principle of integrated management system functioning is participation of all employees and departments in the work of the system. So, the IMS policy must be reported to every employee, but often it is too complicated for understanding in the written form and requires thorough studying; therefore, the company employees must have a special set of professional objective competences.

All the standards comprised in IMS have relevant clauses on employees' awareness, necessity of providing within-the-company accessibility to any information that might be relevant for providing personnel safety. IMS implementation allows carrying out unified policy within the company, using resources optimally, creating common personnel training system.

All the institutions provide obligatory briefings and trainings on labor and industrial safety. Top managers and specialists that are responsible for management system creation, support and improvement take obligatory training courses at certified consulting centers. The training is provided in order to create among company's employees common understanding of terminology and standard requirements taken as the base for integration. Training of employees, responsible for creation and implementation of IMS must be taken before the start of these operations and procedures.

5 DEVELOPMENT OF EDUCATIONAL ELECTRONIC RESOURCE

As a part of the research, in order to improve the element of training, awareness and competence of LukOIL-Permnefteproduct employees, it was decided to develop an educational electronic resource for Company's employees and to create a platform for its allocation. This software solution includes further sections – basics of industrial safety; operations with chemicals and petrochemicals; organization of safe gas hazardous works; operations with hazardous waste; introduction to environmental management; further training in environment protection, etc.

Training may be provided by scaled programs according to the level of employee's involvement in IMS operations. Several blocks may be allocated: for top managers, for experts (auditors) in internal audit (inspections), for engineers and for workers. Providing consultations, seminars and trainings with the help of outside experts would be too costly because of large number of company's employees, which is not profitable for the company. Therefore it is more rational and efficient to create EER (educational electronic resource), which can solve the assigned issues and improve LukOIL Permnefteproduct's integrated management system with minimal costs.

Educational electronic resource is an automated learning system which includes methodic and informational materials on issues of LukOIL Permnefteproduct's integrated management system functioning [3, 20, 21].

EER is an assembly of Web pages with shared theme, design, interconnected with hyperlinks system, located in the Internet; in other words, EER is the Web site. It has hypertext structure which allows trainee define optimal way to study material with comfortable tempo. This software solution is designed for acquaintance with Health, Safety and Environment Management System (HSE Management System).

Educational electronic resource contains 5 structural elements [19]:

1. ***The homepage section*** gives the information on the resource, its target audience, course objectives and proposed outcome, Fig. 6;
2. ***The 'PJSC LukOIL Policy' section*** – a fundamental document of integrated management system which includes intentions and principles of the Company on issues of industrial, labor and environmental safety. The section gives abstracts from the Policy, users may also read the full version of the document by clicking hyperlink, Fig. 7;
3. ***The 'Health, Safety and Environment Management System' section*** gives the description of HSE Management System activities, Fig. 8;
   The 'Industrial safety' subsection gives definitions of 'industrial safety', 'accident', sets types of industrial safety activities, liabilities of the company working at hazardous facility.

Figure 6: The homepage of educational electronic resource.

Figure 7: The 'PJSC LukOIL Policy' section of educational electronic resource.

Figure 8: The 'Health, Safety and Environment Management System' section of educational electronic resource.

The 'Labor Protection' subsection gives definition of 'labor safety', sets the components of labor safety system, employer's liabilities in providing safe labor conditions, workers' rights and liabilities in labor safety.

The 'Environmental Protection' subsection gives definition of 'environmental safety', sets company's strategic goals in provision of environmental safety, gives company's environmental policy and information on significant environmental issues.

4. ***The 'Education' section*** consists of 2 training courses on HSE Management System for different levels of qualification: for top-managers and specialists and for other company employees.

   Targets, contents, volume and evaluation criteria of training courses differ.

   When employee passes the final test, he/she fills in his/her personal data: last name, first name, position held.

   The program forms the list of 20 questions for the test; answering time shouldn't exceed 40 minutes. All the questions are numbered and have a description according to the question type (the only correct answer, several correct answers, match the answers, etc.). If the examinee has complications with choosing the right answer, he/she may switch to the next question. All the unanswered questions can be found in 'The List' menu.

   When the test if over, statistics data appears on the screen, it shows the score of correct answers, required score for a positive mark, time spent on the test and test results.

   The report on test results is emailed to the trainee and the trainer, who makes sure that all the company's employees have passed the course.

5. ***The 'Help' section*** gives recommendations in case of difficulties' occurrence while browsing the web-site or passing the training course. Trainer's contacts are also shown at this page, so as the feedback form.

   Educational electronic resource is available for all the Internet users, which is one of the requirements of international standards of employees' information support.

6 ANALYSIS OF THE RESEARCH RESULTS

There are other forms of education at the company within the existing environment management system. Specific accidents are thoroughly studied and sorted out, corresponding protocols are drawn up, than the management draws up a conclusion and develop new instructions, if unavailable. Beyond that, of course, the company provides opportunities for information exchange and studies the experience of accidents at other holding subsidiaries. After that, the management draws up appropriate protocols and orders adapted to our company.

In order to optimize the personnel training process at LukOIL Permnefteproduct llc, two forms of training were proposed: vocational (with day release) and remote (e-learning without day release).

Top managers and specialists that are responsible for management system creation, support and improvement take 72-hours-or-more training courses at certified consulting centers, with exclusive requirements for acquired knowledge assessment. Other company employees (specialists that don't work with IMS, workers) take training with the help of educational electronic resource during the first month of placement or assumption of the office.

Therefore, the integration of two types of training – day release seminars for top and mid-level managers, responsible for IMS functioning; and e-learning for other employees, - allows boosting company's management system performance and efficiency, and cutting costs for personnel training.

The training course passing allowed company's employees to understand:

a) Importance of compliance with company policy on industrial, labor and environmental safety, IMS procedures and requirements;
b) Significant environmental aspects, dangers and professional risks and related actual and potential exposure to environment and human health;
c) His/her liabilities and responsibility for reaching compliance with integrated management system;
d) Possible after-effect of breaking set procedures.

7 CONCLUSIONS

The manufacturing enterprise that created its own integrated management system must perform regular checks of all the system components along with routine check of specific production processes and provide employees with appropriate knowledge of industrial safety and environmental management.

The software solution is designed for acquaintance with company's Industrial, Labor and Environmental Safety Management System. It allows employees to build up clear understanding of management system functioning, to get acquainted with requirements of environmental management system, ISO 14001:2015 instructions, occupational health and safety management system requirements, OHSAS 180016:2007 instructions.

At the moment, it is too early to make conclusions about actual results of improvement of any indicators of industrial safety due to EER usage, because the usage time elapsed does not give objective information. This aspect will definitely be assessed and analyzed in our upcoming works.

Personnel training on recommended practices of safe production process operations, including environmental safety, with the help of EER-based educational programs on industrial, labor and environmental safety, allow reaching these main results for Company's operations:

- training time reduction;
- training cost reduction;
- more accurate and precise performance of duties;
- decrease of employees' industrial injuries and health damage;
- decrease of compensations;
- increase of employees' working efficiency.

The analysis of internal audit checks shows that elaborate personnel training allows decreasing amount of violations detected amid increased number of checks, which further proves the only thing – increase of efficiency of the management system and overall company management.

Audits help to control whether the system was correctly implemented at the enterprise and will be supported in a proper manner, whether the enterprise follows requirements and norms of management system international standards, how efficiently this system supports company's goals and objectives realization and performance improvement.

In the long run, it would allow defining the state of the company's management system, its principles implementation in everyday operations and efficiency and operability of given formed system.

REFERENCES

[1] Rumyantseva, A. & Samarskaya, N., Labor safety issues: foreign experience. *Labor Safety in the Industry*, **6**, pp. 16–21, 2016.

[2] Berezyuk, M. & Rumyantseva, A., A concept of transition to the best available technology as a basis for sustainable development of power industry. *International Journal of Sustainable Development and Planning,* **10**, pp. 635–649, 2015. https://doi.org/10.2495/sdp-v10-n5-635-649

[3] Zhomayko, O. & Berezyuk, M., Optimization of company's personnel training process within integrated management system improvement. *Proceedings of the 9th International Conference, Environmental safety management system,* pp. 196–200, 2015.

[4] The report of the Russian Federation Ministry of Labor and Social Protection 'On realization of state policy on labor conditions and safety in the Russian Federation in 2014' [Digital source]. available at: http://www.rosmintrud.ru/docs/mintrud/salary/24.

[5] State report 'On the Condition of Population Sanitary and Epidemiological Welfare in the Russian Federation in 2015' [Digital source], available at: http://24.rospotrebnadzor.ru/documents/ros/147604.

[6] National standard of the Russian federation, [in Russian], Quality management systems. Requirements. [Digital source]. available at: http://docs.cntd.ru/document/1200124394.

[7] National standard of the Russian federation, [in Russian], Environmental management systems. Requirements with guidance foruse. [Digital source]. available at: http://www.pqm-online.com/assets/files/pubs/translations/std/iso-14001-2015-(rus).pdf.

[8] National standard of the Russian federation, [in Russian], Occupational safety standards system. Occupational safety and health management systems. General requirements. [Digital source]. available at: http://docs.cntd.ru/document/gost-12-0-230-2007-ssbt.

[9] The ISO Survey. [Digital source], available at: http://www.iso.org/iso/home/standards/management-standards.

[10] Itkin, B., Environmental management systems certification and quality of environment. *Ecology of Production*, **6**, pp. 32–41, 2015.

[11] State report 'On the State and Environmental Protection of the Russian Federation in 2015' [Digital source]. available at: http://www.mnr.gov.ru/regulatory/detail.php?ID=286341

[12] 2014 EPI Results. Yale University, 2015 [Digital source]. available at: http://epi.yale.edu/epi/country-rankings

[13] Azarov, V., Alimov, N., Gotlib, Y., Mashkova, A. & Spasova, E., Analysis of GOST 12.0.230-2007 terminology from the perspective of current system of labor safety management. *Internet Herald of VSASU. Polythematic series*, **1**(10), pp. 1–7, 2010.

[14] Khoroshavin, A., The new version of ISO 14001 standard. *Ecology of Production*, **3**, pp. 40–43, 2015.

[15] Itkin, B., About the new version of ISO 14001 standard. *Ecology of Production*, **3**, pp. 46–53, 2016.

[16] Golubeva, S., Strelnikov, Y. & Cokornova, T., Environmental management system: new version of the standard. *Ecology of Production*, **9**, pp. 40–46, 2016.

[17] Slavinskiy, D., Khoroshavin, A. & Smirnova, M., Analysis of new international requirements for environmental management systems within situation in Russia. *NRU ITMO Scientific Magazine. Economics and Environmental Management Series*, **4**, pp. 335–341, 2015.

[18] GOST P 54934-2012/OHSAS 18001:2007 Labor Safety Management Systems. Requirements. [Digital source]. available at: http://gostrf.com/normadata/1/4293789/4293789191.pdf

[19] Berezyuk, M., Rumyantseva, A. & Chebotareva, G., Improvement of the integrated management system of energy companies by the implementation of electronic educational resources. *WIT Transactions on Ecology and the Environment*, **250**, pp. 11–20, 2016.

[20] Jomayko, O. & Berezyuk, M., Process of company's personnel training with functioning integrated management system. *Western Siberia Academic Magazine*, **57**(2), pp. 89–90, 2015.

[21] Kuznetsova, A. & Berezyuk, M., Environmental management system as economic and standard regulator of companies operations. *Western Siberia Academic Magazine*, **57**(2), pp. 90–91, 2015.

SHAKING TABLE TEST OF ADJACENT BUILDING MODELS CONSIDERING POUNDING

K. FUJII & Y. SAKAI
Chiba Institute of Technology, Japan.

ABSTRACT
Seismic pounding may cause severe structural damage to buildings, such as partial or total collapse, and/or significant damage to non-structural elements. This may be caused by the difference in the dynamic properties of each building, and also an insufficient gap between each building. In this study, a shaking table test of building models was carried out to investigate the seismic pounding of (1) low-rise buildings and (2) a low-rise to a mid-rise building. The structural specimens considered in this study were two single-storey models with a different horizontal stiffness, and one two-storey model. The test parameters were a) pairings of building models, b) the size of gaps, and c) ground acceleration records. Based on the test results, the increment of kinetic energy during the collision was evaluated. The main findings from the test results are as follows: (1) In the case of the pounding of low-rise buildings, the peak displacement of the stiffer building increases, while that of the more flexible building decreases; (2) In the case of the pounding of a low-rise building to a mid-rise building, the peak inter-storey drift of the low-rise building increases. In the mid-rise building, the peak inter-storey drift of the upper storey increases, while that of the lower storey decreases; (3) The sum of the increment kinetic energy during the collision was larger as the gap between buildings was smaller. A significant loss of kinetic energy was seen in buildings whose maximum kinetic energy is larger. (4) The building model of smaller kinetic energy may gain more energy owing to collisions, and the unfavourable effect of seismic pounding to the response is predominant.
Keywords: kinetic energy, pounding of adjacent buildings, shaking table test.

1 INTRODUCTION

In the downtown area of major cities, there are many buildings whose height and/or structure are different within a limited space. When the gap between adjacent buildings is insufficient, seismic pounding may occur as a result of the difference in the dynamic properties of each building. This may cause severe structural damage to buildings, such as partial or total collapse, and/or significant damage to non-structural elements. Figure 1 shows the damaged buildings possibly caused by seismic pounding during the 2016 Kumamoto Earthquake. Such damaged buildings have been observed in previous major earthquakes, such as the 1985 Mexico Earthquake [1], the 1989 Loma Prieta Earthquake [2, 3], and the 2011 Christchurch Earthquake [4].

The seismic pounding of building structures has been analytically studied by many researchers, such as Kasai [2, 3, 5], Anagonostopoulos [6, 7], Wada *et al.* [8], and Jankowski [9, 10]. In Wada *et al.* [8] and Jankowski [10], it has been found that structural pounding during an earthquake has a significant influence on the behaviour of the lighter buildings. Experimental studies of seismic pounding have also been carried out by Papadrakakis and Mouzakis [11], Chau *et al.* [12], and Jankowski [13]. The study by Papadrakakis and Mouzakis [11] showed that the seismic input energy of a stiffer structure increases as a result of pounding.

During seismic pounding, the kinetic energy of each structure is drastically changed during the collision period; one structure may lose while the other may gain kinetic energy. Energy dissipation also occurs as a result of the non-elastic collisions. Therefore, from the viewpoint of the authors, it is important to focus on the change in the kinetic energy of each structure during the collision to understand the phenomenon of seismic pounding.

DOI: 10.2495/CMEM-V6-N5-857-867

Figure 1: Damaged buildings possibly caused by the seismic pounding during the 2016 Kumamoto Earthquake. (a) Overview of the damaged buildings, and (b) a damaged non-structural element.

In this study, a shaking table test of building models was carried out to investigate the seismic pounding of (1) low-rise buildings, and (2) a low-rise building to a mid-rise building. Based on the test results, the increment of kinetic energy during collisions is evaluated and discussed.

2 OUTLINE OF SHAKING TABLE TEST

2.1 Building models

The test building models considered in this study are shown in Fig. 2. The two single-storey models shown in Fig. 2a and b are referred to as Models 1A and 1B, respectively, and the single two-storey building model shown in Fig. 2c is referred to as Model 2. The floor mass of both Models 1A and 1B are 2.51 kg, while the floor mass of Model 2 at levels 1 and 2 are 3.24 kg and 2.51 kg, respectively. The columns of each building model are aluminium plates of thicknesses 1.5 mm for Models 1A and 2, and 2.0 mm for Model 1B. To adjust the structural damping of each model, viscoelastic material is installed on each storey, with a rigid support column free to execute rocking movements.

The natural period and the damping ratio of single-storey models are obtained from the free vibration test. The natural period and the damping ratio of Model 1A are 0.195 s and 0.096, respectively, while those of Model 1B are 0.144 s and 0.035, respectively. The natural period and the damping ratio of each mode of Model 2 are obtained from the elastic vibration test without pounding. The natural period is 0.400 s for the first mode and 0.141 s for the second mode. The damping ratio, based on the half-power method, is 0.055 for the first mode and 0.041 for the second mode.

2.2 Test parameters

Figure 3 shows the arrangement of the pounding test. Because of size limitations during the shaking table test, the two building models were set in parallel. On the level 1 of each model, the pounding arm element was installed to facilitate collisions of two building models during the test. Therefore, the pounding in this test was asymmetric pounding, which caused a

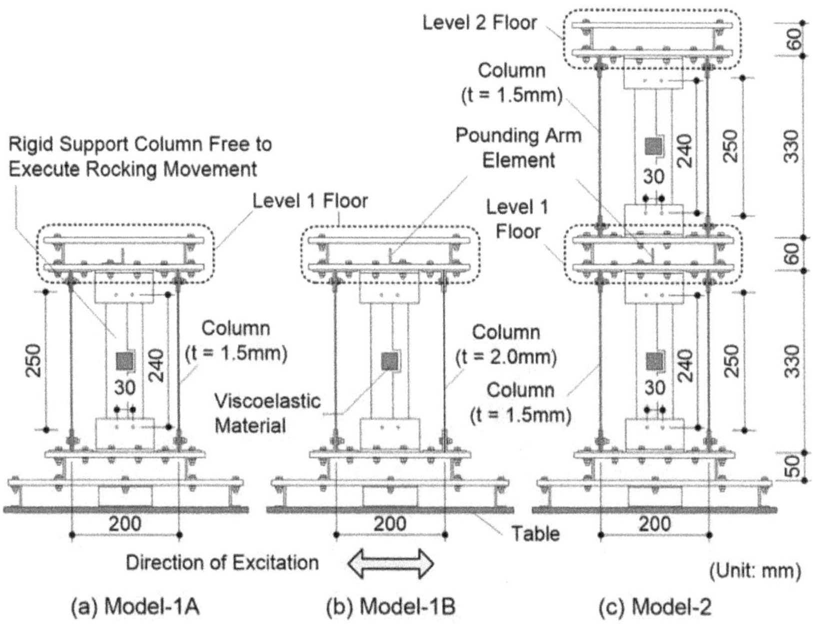

Figure 2: Building models.

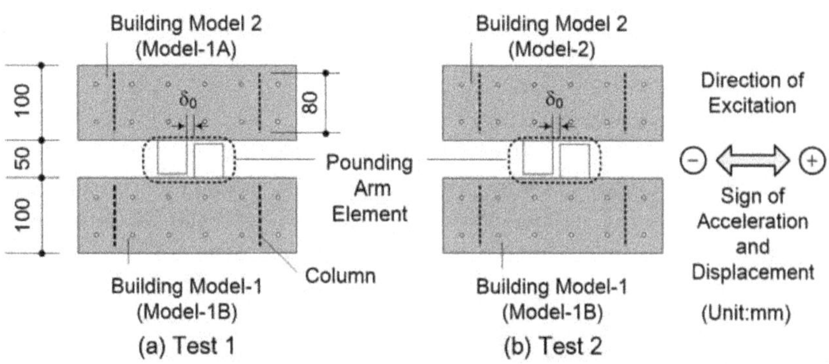

Figure 3: Arrangement of pounding test.

torsional response in both models. To minimise the torsional response, the width of the columns in each storey was determined as 80 mm and the distance between two columns as 200 mm to provide a large storey torsional stiffness.

In the tests, the following parameters were studied: a) the pairing of the two building models, and b) the size of the gap δ_0.

a) Pairing of the two building models. In Test 1, the seismic pounding of the two single-storey building models (Models 1A and 1B) shown in Fig. 3a were investigated. In Test 2, the seismic pounding of a single-storey building model (Model 1B) and a two-storey building model (Model 2) were investigated, as shown in Fig. 3b.
b) The size of the gap δ_0. The gap between the two building models was considered as shown in Fig. 3. The size of the gap δ_0 was set to 0, 5 mm, 10 mm, and infinite (without pounding).

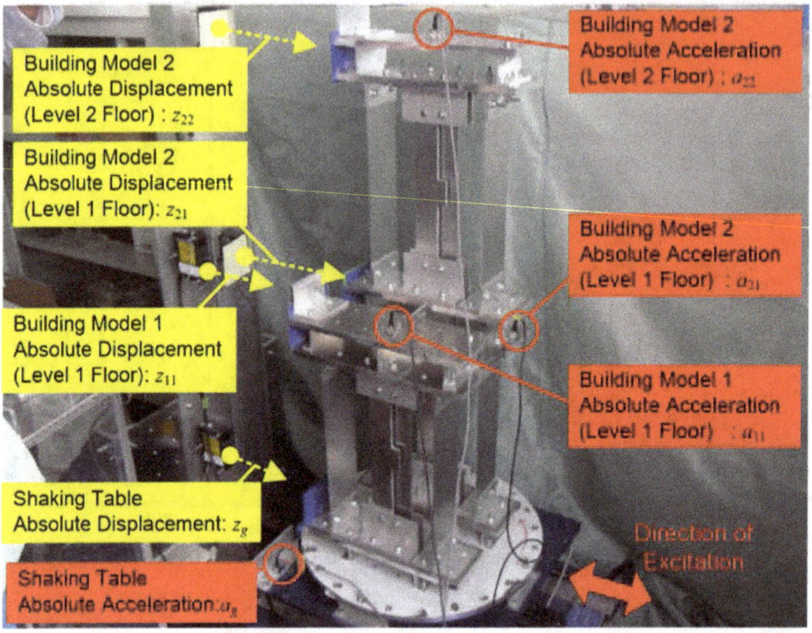

Figure 4: Experimental set-up.

2.3 Experimental setup and seismic input

Figure 4 shows the experimental set-up, where the absolute accelerations of all floors (absolute acceleration at level 1 of building Model 1, a_{11}, and at level 1 and 2 of building Model 2, a_{21} and a_{22}, respectively) and at the shaking table, a_g, are shown as measured by the accelerometers. The absolute displacement of all floors (absolute displacement at level 1 of building Model 1, z_{11}, and at levels 1 and 2 of building Model 2, z_{21} and z_{22}, respectively) and at the shaking table, z_g, were measured by laser sensors. The sampling interval is 0.001 s. The relative displacement of all floors, y_{11}, y_{21}, and y_{22} are calculated from

$$y_{11}(t) = z_{11}(t) - z_g(t), y_{21}(t) = z_{21}(t) - z_g(t), y_{22}(t) = z_{22}(t) - z_g(t). \quad (1)$$

Figure 5 shows the pseudo-velocity spectrum and input energy spectrum proposed by Akiyama [14]. As shown in this figure, the El Centro 1940 North-South (NS) ground motion record (ELC), the Tohoku University 1978 NS ground motion record (TOH), and the Japan Meteorological Agency Kobe 1995 NS ground motion record (JKB) were used for the shaking table test. In this test, the peak acceleration was scaled to 40% of the acceleration due to gravity, and the time interval of three ground motions were scaled by half for the following reasons: i) to simulate the seismic pounding of both low-rise buildings (Test 1) and a low-rise to a mid-rise building (Test 2); ii) to adjust the maximum movement of the shaking table within the allowable limit; and iii) to adjust the response of the building model within the elastic range.

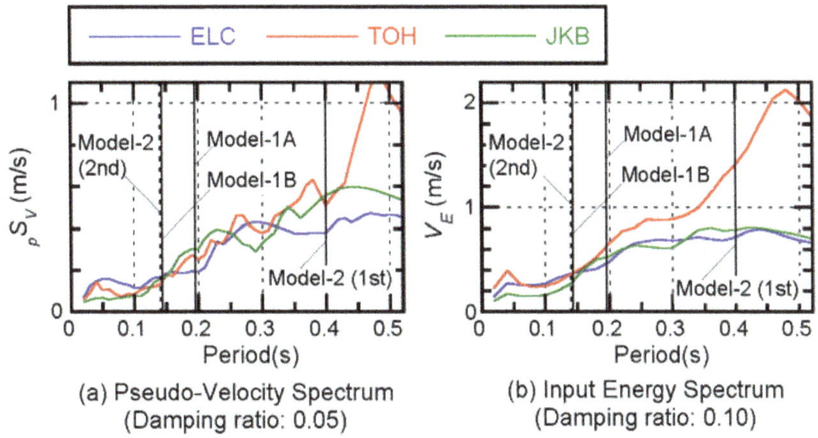

Figure 5: Response spectrum of seismic input.

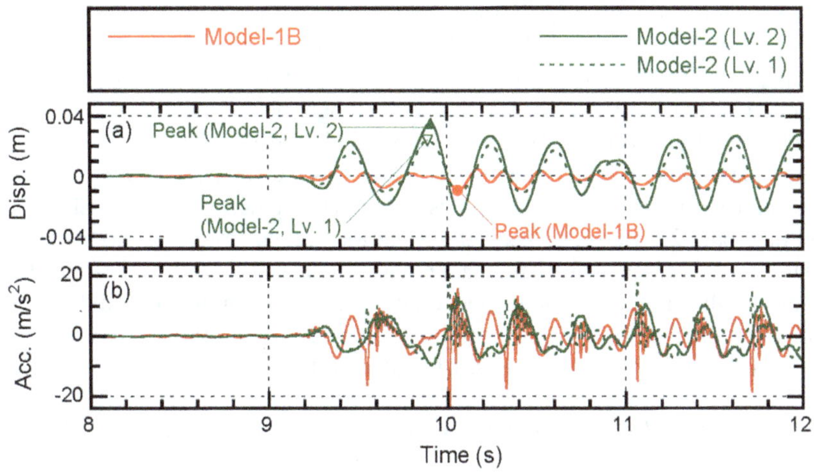

Figure 6: Time-history of Test 2 (seismic input: JKB, $\delta_0 = 0$). (a) Relative displacement at the floor, and (b) absolute acceleration at the floor.

3 TEST RESULTS

3.1 Time-history of seismic pounding

Figure 6 shows the time-history of Test 2 (seismic input: JKB, $\delta_0 = 0$), where the relative displacement (Model 1B: y_{11}, Model 2: y_{21} and y_{22}) and the absolute acceleration (Model 1B: a_{11}, Model 2: a_{21} and a_{22}) are displayed.

Sharp spikes in the acceleration response are evident when the relative displacement of Model 1B is almost equal to that of level 1 in Model 2. A sharp spike in the acceleration response is considered here as a 'collision'.

3.2 Normalised inter-storey drift

Figures 7 and 8 show the normalised inter-storey drift for each case, defined as the ratio of the peak inter-storey drift with pounding, divided by that without pounding (δ_0 = infinite).

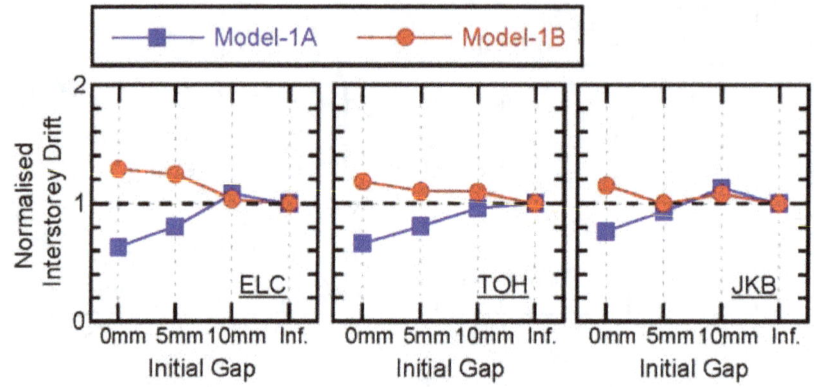

Figure 7: Normalised inter-storey drift (Test 1).

It is observed in Test 1 from Fig. 7 that the normalised drift of Model 1A is smaller than 1 in most cases; the normalised inter-storey drift is smaller as the gap δ_0 is smaller. On the contrary, the normalised drift of Model 1B is larger than 1 in most cases. This observation indicates that, in the case of the pounding of low-rise buildings, the peak displacement of the shorter period building increases, while that of the longer period building decreases.

In Test 2, it is observed from Fig. 8 that the normalised drift of Model 1B is larger than 1 in all cases; the normalised inter-storey drift is larger as the gap δ_0 is smaller. This tendency is more significant when the seismic input is either TOH or JKB rather than ELC. However, the trend of Model 2 depends on each storey: in the second storey, the normalised drift is larger than 1 in most cases, while it is smaller than 1 in the first storey in all cases. This observation indicates that, in the case of the pounding of a low-rise building to a mid-rise building, the peak storey drift of the low-rise building increases. In the mid-rise building, the peak storey drift of the upper storey increases, while that of the lower storey decreases.

4 CHANGE IN THE KINETIC ENERGY DURING THE COLLISION

The discussion that follows focuses on the change in the kinetic energy of each structure during the collision: the increment of kinetic energy during each collision is evaluated.

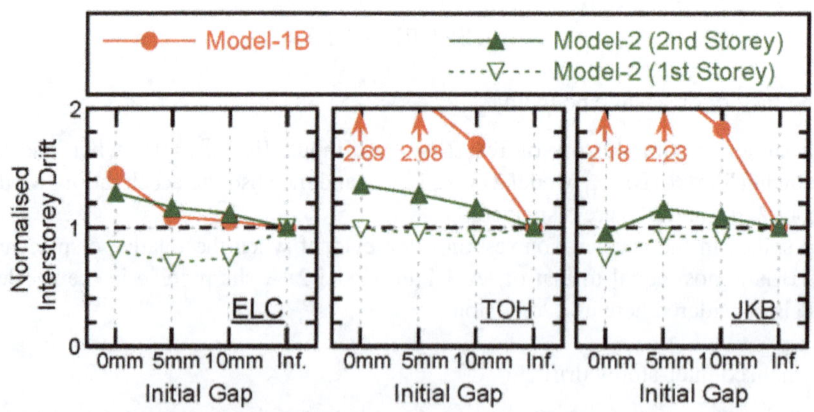

Figure 8: Normalised inter-storey drift (Test 2).

In this study, the kinetic energy of building models 1 and 2 at time t, $W_{K1}(t)$ and $W_{K2}(t)$, respectively, are calculated from

$$W_{K1}(t) = \frac{1}{2} m_{11} \{\dot{y}_{11}(t)\}^2 \tag{2}$$

and

$$W_{K2}(t) = \begin{cases} \frac{1}{2} m_{21} \{\dot{y}_{21}(t)\}^2 & \text{for Test 1} \\ \frac{1}{2} \sum_{j=1}^{2} m_{2j} \{\dot{y}_{2j}(t)\}^2 & \text{for Test 2} \end{cases}. \tag{3}$$

Here, m_{11} is the floor mass of building Model 1, while m_{21} and m_{22} are the floor masses of building Model 2 at levels 1 and 2, respectively.

4.1 Definition of collision time

The relative displacement between the building model $\delta(t)$, is defined as

$$\delta(t) = y_{21}(t) - y_{11}(t). \tag{4}$$

Figure 9 shows the definition of the collision time in this study. The collision time of the ith pounding, $\Delta_i t_c$, is defined as

$$\Delta_i t_c = {}_i t_{c1} - {}_i t_{c0}. \tag{5}$$

Here, ${}_i t_{c0}$ and ${}_i t_{c1}$ are the beginning and ending time of the collision during the ith pounding, respectively, and are defined as:

$$\text{beginning time of collision } {}_i t_{c0}\text{: } \delta(t) = \delta_0 \text{ and } \dot{\delta}(t) \geq 0 \tag{6}$$

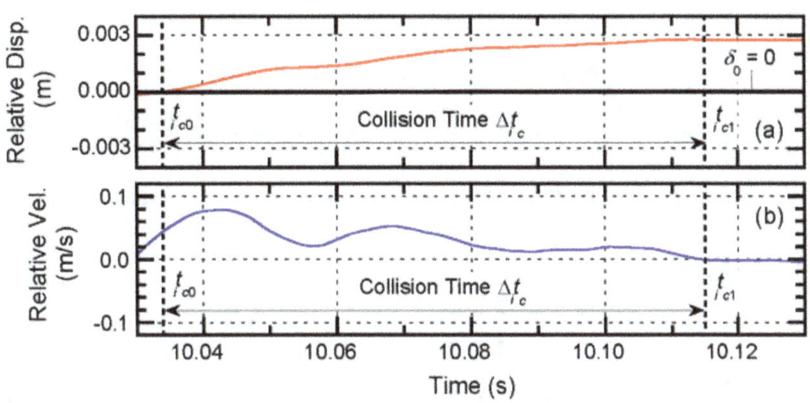

Figure 9: Definition of collision time in the case of an initial gap $\delta_0 = 0$. (a) Relative displacement between the building model $\delta(t)$, and (b) relative velocity between building model $\dot{\delta}(t)$. Note that the beginning time of collision t_{c0} depends on δ_0, while the ending time t_{c1} is independent of δ_0.

and

$$\text{ending time of collision } {}_it_{c1}\text{: } \delta(t) \geq \delta_0 \text{ and } \dot{\delta}(t) = 0. \qquad (7)$$

Note that this definition of the collision time is consistent with 'the approach period of the collision' in [9]. The increment of kinetic energy during a collision of building Models 1 and 2 of the ith pounding, $\Delta_i W_{K1}$ and $\Delta_i W_{K2}$, respectively, are then calculated from

$$\Delta_i W_{K1} = W_{K1}\left({}_it_{c1}\right) - W_{K1}\left({}_it_{c0}\right), \Delta_i W_{K2} = W_{K2}\left({}_it_{c1}\right) - W_{K2}\left({}_it_{c0}\right). \qquad (7)$$

4.2 Sum of the increment of kinetic energy and the variation of maximum kinetic energy

Figures 10 and 11 show the sum of the increment of kinetic energy of building models, $\Sigma(\Delta_i W_{K1})$ and $\Sigma(\Delta_i W_{K2})$, for Tests 1 and 2, respectively.

In Test 1 (Fig. 10), the sum of the increment of kinetic energy of both Models 1A and 1B are negative except Model 1B for the case of a TOH seismic input, and the initial gap δ_0 is 0. In general, both building models lose kinetic energy as a result of the collisions. These trends are more significant in the case of a smaller initial gap δ_0. In Test 2 (Fig. 11), the sum of the increment of kinetic energy of Model 1B is positive, while that of Model 2 is negative in all cases. These trends are more significant in the case of a smaller initial gap δ_0.

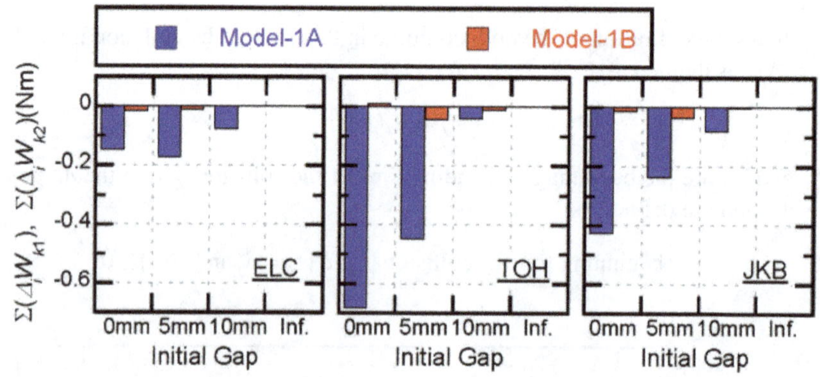

Figure 10: Sum of the increment of kinetic energy (Test 1).

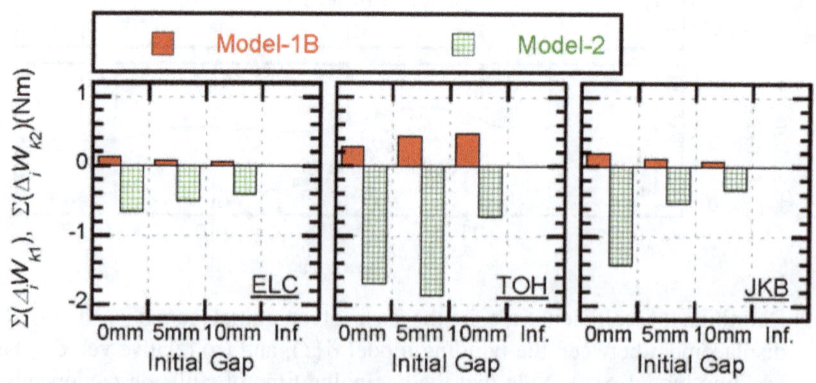

Figure 11: Sum of the increment of kinetic energy (Test 2).

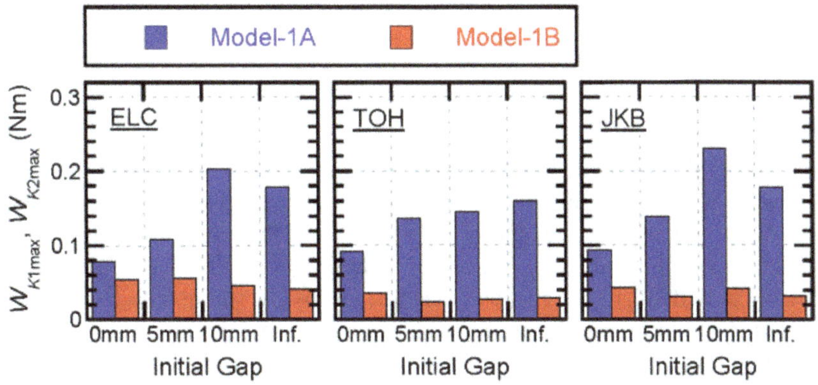

Figure 12: Maximum kinetic energy (Test 1).

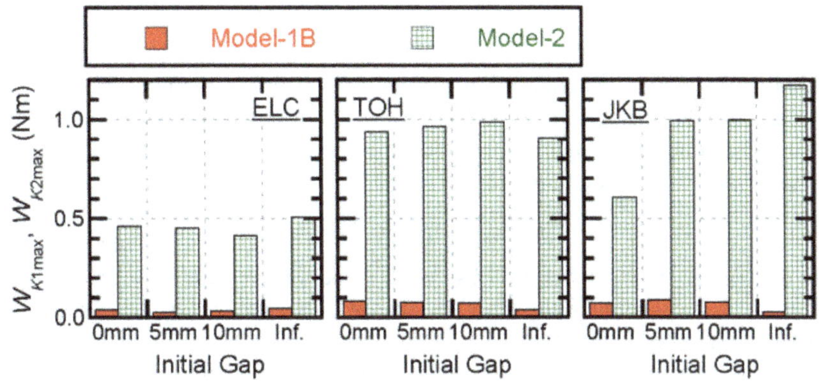

Figure 13: Maximum kinetic energy (Test 2).

Figures 12 and 13 show the maximum kinetic energy of the building models, $W_{K1\max}$ and $W_{K2\max}$, for tests 1 and 2, respectively.

In Test 1 (Fig. 12), the maximum kinetic energy of Model 1A is larger than that of Model 1B. This figure also shows that, in general, the maximum kinetic energy of Model 1A decreases while that of Model 1B increases due to the collision. In Test 2 shown in Fig. 13, the maximum kinetic energy of Model 2 is larger than that of Model 1B. The trend of the variation of the maximum kinetic energy due to the collisions of each model depends on the seismic input and initial gap δ_0.

From the discussions on the change of the kinetic energy of each structure during the collision (shown in Figs 10 and 11), and the results of the tests shown in the previous section, it is found that the building model with more kinetic energy loses more energy due to collisions, and the unfavourable effect of seismic pounding to the response is less significant. On the contrary, the building model with lesser kinetic energy may gain more energy from the collisions, and the unfavourable effect of seismic pounding to the response is predominant.

5 CONCLUSIONS

In this study, shaking table tests of building models were carried out to investigate the seismic pounding of (1) low-rise buildings, and (2) a low-rise to a mid-rise building. Based on the test results, the increment of kinetic energy during the collision is evaluated and discussed. Although the test cases shown here are limited, the main findings of this study are as follows:

1. In the case of the pounding of low-rise buildings, the peak displacement of the stiffer building increases, while that of the more flexible building decreases.
2. In the case of the pounding of a low-rise building to a mid-rise building, the peak inter-storey drift of the low-rise building increases. In the mid-rise building, the peak inter-storey drift of the upper storey increases, while that of the lower storey decreases.
3. The sum of the increment of kinetic energy during the collision was larger as the gap between the buildings was smaller. A significant loss in kinetic energy was evident in those structures for which the maximum kinetic energy is larger.
4. The building model with more kinetic energy loses more energy from the collisions, and the unfavourable effect of seismic pounding to the response is less significant. On the contrary, the building model with lesser kinetic energy may gain more energy from the collisions, and the unfavourable effect of seismic pounding to the response is predominant.

It should be noted that the maximum kinetic energy of buildings without seismic pounding can be easily estimated from the response spectrum. Therefore, buildings more susceptible to the dangers of seismic pounding may be detected from the response spectrum. This point will be taken up in the next phase of this study.

REFERENCES

[1] Bertero, V.V., Implications of observed pounding of buildings on seismic code regulations. *Proceeding of the Eleventh World Conference on Earthquake Engineering*, Elsevier: paper no. 2102, 1996.

[2] Kasai, K., Jeng, V., Patel, P.C. & Munshi, J.A., Seismic pounding effects – survey and analysis. *Proceeding of the Tenth World Conferenec on Earthquake Engineering*, Balkerma: pp. 3893–3898, 1992.

[3] Kasai, K., Jagiasi, A.R. & Maison, B.F., Survey and analysis of building pounding during 1989 Loma Prieta Earthquake. *Proceeding of the Eleventh World Conference on Earthquake Engineering*, Elsevier: paper no. 2103, 1996.

[4] Cole, G.L., Dhakel, R.P. & Turner, F.M., Building pounding damage observed in the 2011 Christchurch earthquake. *Earthquake Engineerings and Structural Dynamics*, **41**, pp. 893–913, 2012.
https://doi.org/10.1002/eqe.1164

[5] Maison, B.F. & Kasai, K., Dynamics of pounding when two buildings collide. *Earthquake Engineerings and Structural Dynamics*, **21**, pp. 771–786, 1992.
https://doi.org/10.1002/eqe.4290210903

[6] Anagnostopoulos, S.A., Pounding of buildings in series during earthquakes. *Earthquake Engineerings and Structural Dynamics*, **16**, pp. 443–456, 1988.
https://doi.org/10.1002/eqe.4290160311

[7] Anagnostopoulos, S.A. & Spiliopoulos, K.V., An investigation of earthquake induced pounding between adjacent buildings. *Earthquake Engineerings and Structural Dynamics*, **21**, pp. 289–302, 1992.
https://doi.org/10.1002/eqe.4290210402

[8] Wada, A., Shinozaki, Y. & Nakamura, N., Collapse of building with expansion joints through collision caused by earthquake motion. *Proceeding of the Eighth World Conference on Earthquake Engineering*, **4**, pp. 855–862, 1984.

[9] Jankowski, R., Non-linear viscoelastic modelling of earthquake-induced structural pounding. *Earthquake Engineerings and Structural Dynamics*, **34**, pp. 595–611, 2005. https://doi.org/10.1002/eqe.434

[10] Jankowski, R., Earthquake-induced pounding between equal height buildings with substantially different dynamic properties. *Engineering Structures*, **30**, pp. 2818–2829, 2008.
https://doi.org/10.1016/j.engstruct.2008.03.006

[11] Papadrakakis, M. & Mouzakis, H.P., Earthquake simulator testing of pounding between adjacent buildings. *Earthquake Engineerings and Structural Dynamics*, **24**, pp. 811–834, 1995.
https://doi.org/10.1002/eqe.4290240604

[12] Chau, K.T., Wei, X.X., Guo, X. & Shen, C.Y., Experimental and theoretical simulations of seismic poundings between two adjacent structures. *Earthquake Engineerings and Structural Dynamics*, **32**, pp. 537–554, 2003.
https://doi.org/10.1002/eqe.231

[13] Jankowski, R., Experimental study on earthquake-induced pounding between structural elements made of different building materials. *Earthquake Engineerings and Structural Dynamics*, **39**, pp. 343–354, 2010.

[14] Akiyama, H., *Earthquake-resistant limit-state design for buildings*, University of Tokyo Press: Tokyo, 1985.

EFFECTS OF MODELLING PARAMETERS ON THE SEISMIC ANALYSIS OF BRIDGES

YULING GAO & LAN LIN
Department of Building, Civil and Environmental Engineering, Concordia University, Canada.

ABSTRACT
The objective of this study is to investigate the effects of the uncertainties of major modelling parameters on the bridge seismic response. The parameters examined include the superstructure mass, concrete compressive strength, plastic hinge length and damping. For the purpose of the evaluation, an existing 3-span reinforced concrete highway bridge located in Montreal, Canada, was selected for the analysis. A three-dimensional model of the bridge was developed using SAP2000. Nonlinear time-history analyses were conducted in order to assess the effects of the uncertainty of each of modelling parameters mentioned above. Fifteen records obtained from earthquakes around the world were used as seismic excitations in the time-history analysis. The deck displacement and the column curvature ductility were chosen to investigate the effect of the uncertainty of the modelling parameters on the seismic response of the bridge.
Keywords: bridge, finite element, modelling, nonlinear, records, responses, seismic, uncertainty.

1 INTRODUCTION

Earthquakes are one of the main natural hazards that have caused devastations to bridges around the world. Given the observations from past earthquakes, substantial analytical and experimental research work related to bridges has been undertaken in Canada and other countries. The analytical research is focussed primarily on the prediction of the seismic performance of existing bridges. It includes bridge-specific investigations that are mainly conducted using deterministic approach, and investigations of bridge portfolios that are based on probabilistic approach. In both cases, nonlinear time-history analyses are extensively used. To conduct analysis on a given bridge, analytical (i.e. computational) model of the bridge is required. It is known that the seismic response predictions depend greatly on the accuracy of the input of the parameters considered in the bridge model.

A number of parametric studies were conducted in the past to evaluate the effects of modelling parameters on the seismic response of bridges in North America. Choi *et al.* [1] conducted a study on the effects of the material properties on four typical bridges including two multi-span simply supported bridges (one is steel, the other is pre-stressed concrete), and two multi-span continuous bridges. They concluded that fixed bearings in multi-span simply supported steel girder-type bridges were more vulnerable to earthquakes than any other components. Padgett *et al.* [2] performed seismic analysis on a multi-span simply supported steel girder bridge. The purpose of their study was to investigate the effects of restrainer cables; effective stiffness of elastomeric bearings; yield strength of the steel jackets; and the reinforcing of the shear keys on the bridge behaviour. They reported that the bearing stiffness, rotational stiffness of the foundation, and the gap width had significant effects on the bridge response. Some studies focussed on a specific modelling parameter on structure response, e.g. number of spans [3]; incident angle of seismic motions [4]; modelling techniques for abutments [5]; etc.

Our study is mainly focussed on the parameters that would affect the dynamic characteristics and nonlinear behaviour of the bridge models, namely, superstructure mass, concrete

DOI: 10.2495/CMEM-V6-N5-868-879

compressive strength, plastic hinge length, damping, yield strength of the reinforcing steel, yield displacement of the bearing and post-yield stiffness of the bearing. However, only the results for the first four modelling parameters are presented in this paper due to the page restrain.

2 BRIDGE DESCRIPTION

A three-span continuous reinforced concrete bridge (Fig. 1) located in Montreal, Canada was selected for the study. The bridge was built in 1982. The total length is 45.5 m, which consists of two end spans of 15 m each and a middle span of 15.5 m. The overall deck width is 16.85 m. It is a straight bridge without skew angle. The bridge is considered to be representative of typical highway bridges in Québec according to Tavares [6].

As shown in Fig. 1, the superstructure of the bridge consists of a slab supported by seven 900 C.P.C.I girders (Canadian Precast Prestressed Concrete). The thickness of the slab is 225 mm, and the center-to-center spacing of girders is 2.4 m. The concrete deck is reinforced with 15M bars on the top and bottom. There are 6 straight tendons and 8 parabolic tendons on the bottom of each girder across each span. The cross section of the cap beam is 1 m wide, and 1.2 m deep. The multi-column bents consist of three rectangular columns. The center-to-center spacing between columns is 5.3 m. The dimensions of each column are 1 m (in the transverse direction) × 0.75 m (in the longitudinal direction). The average height of the column is 3.8 m. Each column is reinforced with 16–30M bars providing longitudinal reinforcement ratio of 1.49%. Column ties with 15M at a spacing of 300 mm were used as transverse reinforcement. The transverse reinforcement ratio is about 0.24%. Strip footing is used for the column bents. Seat-type abutments are on a pile foundation, as illustrated in Fig. 1. Expansion bearings are used at abutments and the pier 2, while fixed bearings are used at the pier 1. All the concrete members were designed using the compressive strength $f'_c = 30$ MPa except 35 MPa was used for the girders, while the yield strength of the reinforcing steel bars was assumed to be 400 MPa.

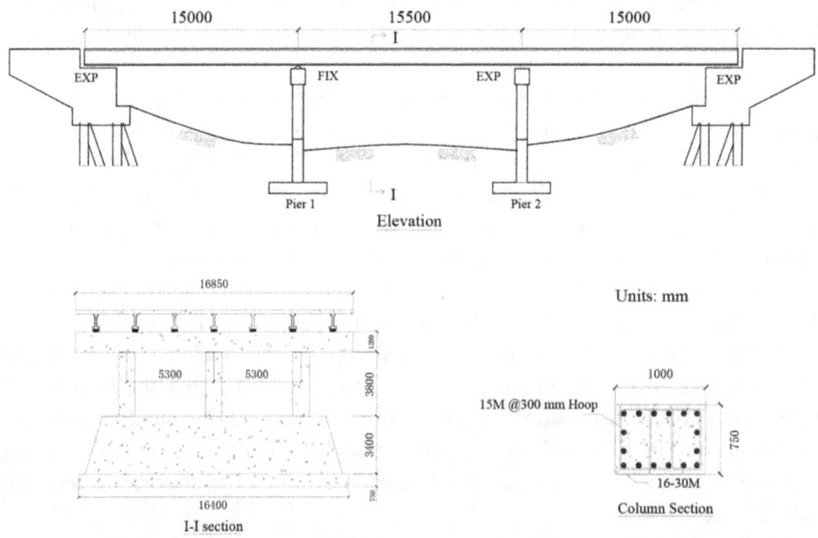

Figure 1: Configuration of the examined bridge.

3 BRIDGE MODELLING

In this study, a three-dimensional finite element model of the bridge was developed using the structural analysis software SAP2000 [7]. The advantage of SAP2000 is that a number of elements (e.g. link element) are available that can be used to model the nonlinearity of different components of a bridge system during seismic excitations including bearings and plastic hinges. It is necessary to mention that another program OpenSees [8] can also be used for the nonlinear analysis of bridges. However, a study conducted by Aviram *et al.* [5] showed that these two programs provided very similar results. SAP2000 was selected in this study due to its simplicity in modelling and less time consuming in the analysis compared with OpenSees.

3.1 Superstructure

The superstructure of the bridge was modeled as one line of elastic beam elements (Fig. 2), which is referred to as spine model in SAP2000. Each span is discretized into 10 equal segments in order to achieve higher accuracy of results though only four elements per span are required for modeling the superstructure in accordance with ATC 32 [9]. The section properties, such as the cross-sectional area, moment inertia, shear area and torsional constant of the superstructure, were determined based on the geometry presented in Fig. 1. In this study, a factor of 0.75 was applied to reduce the moment of inertia of the deck section, and no reduction factor was applied for the girder's sections according to Caltrans Seismic Design Criteria [10]. Furthermore, reduction of the torsional moment of inertia was not considered since the bridge is regular [10].

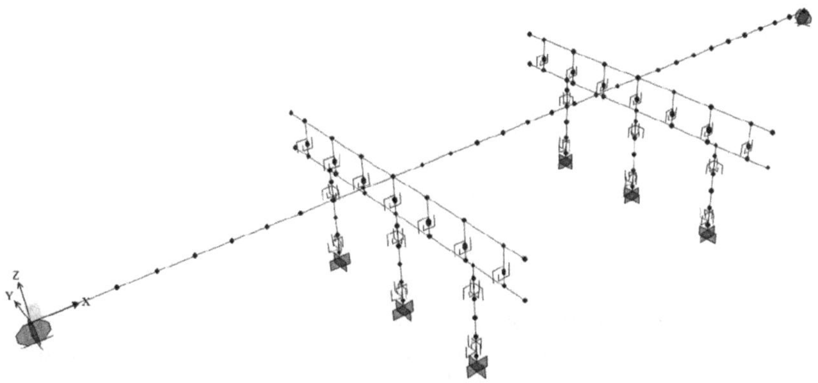

Figure 2: Finite element model of the bridge.

3.2 Bearing

Figure 3 presents an idealized bilinear curve representing the behaviour of the expansion bearing in the longitudinal direction, as recommended by Kelly [11] and DesRoches *et al.* [12]. Parameters K_1 and K_2 denote the initial shear stiffness (i.e. elastic stiffness) and post-yield stiffness, respectively. In this study, K_1 was calculated using eqn (1), while K_2 was taken as one-third of K_1, as suggested by DesRoches *et al.* [12]. Parameters D_y and D_u represent the yield displacement and maximum displacement of the bearing, respectively. Following the recommendations made by DesRoches *et al.* [12], D_u was assumed to be equal to the height of the elastomer, while D_y was taken as 10% of D_u.

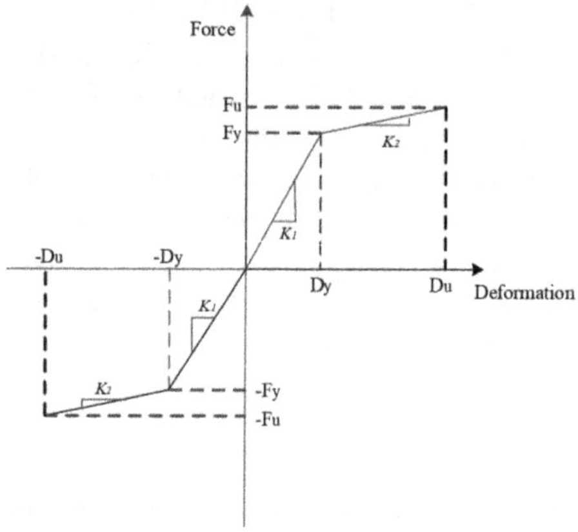

Figure 3: Bilinear behaviour of elastomeric bearings in the longitudinal direction.

$$K_1 = \frac{GA}{H_r} \tag{1}$$

- G = shear modulus, which is taken as 0.80 MPa based on Caltrans [10].
- A = plan area of the elastomeric pad.
- H_r = total thickness of the rubber.

3.3 Column and cap beam

Plastic hinges are expected to form on the bottom and/or the top of columns during larger earthquake events. In this study, the length of the plastic hinge was determined according to the seismic provisions of CHBDC [13]. For the purpose of defining the link element in SAP2000, which was used to simulate the behaviour of column in the plastic hinge region, moment–curvature relationships for the end sections of the column were developed using fibre analyses of the cross sections. The concrete stress–strain relationship including the effect of confinement was determined based on the model proposed by Mander *et al.* [14]. Nominal values for the material strengths were used in the fibre analysis, i.e. concrete and reinforcement resistance factors $\Phi_c = \Phi_s = 1$. In addition, the axial force used to develop the moment–curvature relationship includes the force resulting from the dead load only. Based on the computed moment–curvature relationship, Multi-linear kinematic plasticity model defined in SAP2000 was selected to model the column in the plastic hinge regions. The column was assumed to behave elastically outside the plastic hinge regions. In total, each column was divided into five equal segments.

Rigid bars were used to connect the nodes between superstructure and bearings, bearings and columns, bearings and cap beams, and cap beams and columns. The bottom of the columns was assumed to be fully fixed.

3.4 Abutment

Model of the abutment has a significant impact on the nonlinear response of bridges during earthquake events [15, 16]. Recently Aviram *et al.* [5] conducted a comprehensive investigation of the effect of abutment modelling on the seismic response of bridges. In their study, three abutment models were examined, which were roller model, simplified model, and spring model. It was found that the dominant periods of the examined bridges based on the three models were relatively close except that the first mode period using the spring model was significantly shorter than that using the other two models. Given this, the roller model was selected for this study due to its relatively low modelling complexity.

4 SEISMC EXCITATIONS

Since the lack of strong ground motion records in eastern Canada, Naumoski *et al.* [17, 18] compiled a number of records obtained from earthquakes around the world, which were considered representative of the characteristics of the ground motions in eastern Canada. These records were grouped into three ensembles, namely, high A/V, intermediate A/V, and low A/V according to the ratio of the peak ground acceleration A (in g) to peak ground velocity V (in m/s). As reported in Naumoski *et al.* [17, 18], ground motions in eastern Canada were typically characterized by high frequency content and high A/V ratios. Therefore, the high A/V ensemble was selected for this study. It consists of 15 pairs of horizontal and vertical records. It is necessary to mention that the vertical records were not considered in the analysis because the vibration of the bridge model was not governed by the vertical mode. The magnitudes of the selected records were between 5.3 and 6.9, and distances from 4 km to 26 km. These ranges cover the dominant magnitude and distance ranges that have the largest contribution to the seismic hazard in Montreal [19].

Figure 4 presents the 5% damped response spectra of the selected records scaled to the spectral acceleration at the fundamental period $T_1 = 0.99$ s of the bridge model. For the purpose of comparison, the design spectrum for Montreal for the probability of exceedance of 10% in 50 years for soil profile type I (i.e. rock or stiff soil) is also shown in the figure. It can be seen in Fig. 4 that the mean spectrum of the selected records exceeds the design spectrum for the periods shorter than 1.0 s, and it is slightly below the design spectrum for the periods longer than 1.0 s.

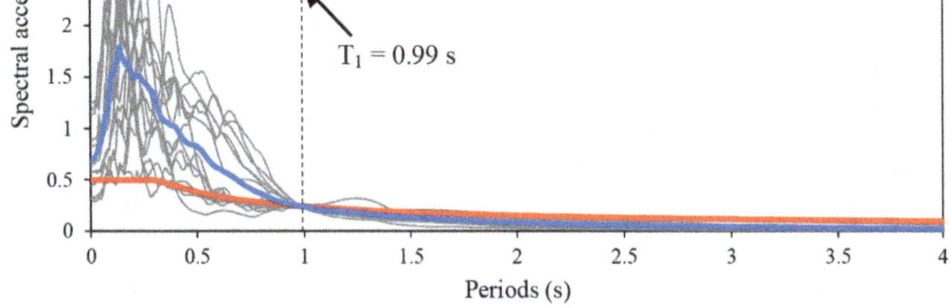

Figure 4: Scaled response spectra of the records, 5% damping.

5 ANALYSIS RESULTS

Nonlinear time-history analyses were conducted by subjecting the bridge model to the selected records scaled at a series of excitation levels from 0.2Sa(0.99s) (i.e. 0.048 g) to 5.0Sa(0.99s) (i.e. 1.20 g). These intensity levels were selected in order to cover a wide range of the bridge response from elastic to inelastic. The response parameters considered in the study are the deck displacement and the column curvature ductility that represent the global deformation and local deformation, respectively. Table 1 provides the nominal value of each modelling parameter under investigation along with its variation.

For each excitation motion, the maximum value of each response parameter was computed. Given a large number of analyses conducted in the study, it is not practical to consider the response from a single time-history analysis. Therefore, the responses resulting from the set of excitations were statistically analysed to compute the mean (M) value. The use of the mean values is considered appropriate because these values are more 'stable' (i.e. they have smaller variations) than the mean plus one standard deviation (M+SD) values.

5.1 Effects of the superstructure mass

Figure 5a shows the results of the deck displacement corresponding to the different superstructure mass from 70% to 130% of its nominal value, at 11 intensity levels. Each point in the figure represents the maximum mean displacement at a given excitation level. The main observation of the results is that the deck displacement increases with the increasing of the superstructure mass. It should be noted that the change of the superstructure mass by ±30%, with respect to its nominal value, shifts the fundamental period T_1 of the bridge slightly. However, this shifting has negligible effect on the design spectral acceleration. As a result, the inertia force is directly proportional to the superstructure mass. It can be seen in Fig. 5a that the superstructure mass does not affect the deck displacement for the excitations lower than about 0.50 g, which is equivalent to 2.0Sa(0.99s). Beyond this level, the effect becomes noticeable. Furthermore, it is found that change of the superstructure mass by 30% causes the change of the displacement by 10% at the highest intensity level 5.0Sa(0.99s).

The results for the column curvature ductility at pier 1 (Fig. 1) are presented in Fig. 5b. It can be seen in the figure that the column behaves elastically (ductility = 1.0) for the excitations up to the design earthquake level at Sa(0.99s) = 0.24 g. Furthermore, the ductility is proportional to the ratio of the intensity level to the design spectral acceleration at Sa(0.99s). For example, the ductility of 3 approximately corresponds to the ground motions of three times Sa(0.99s) which is about 0.72 g. It is necessary to mention that the force reduction factor used for the design of the bridge substructure is 3 according to CHBDC [13]. The

Table 1: List of nominal values and variations of uncertainties considered.

Modelling parameters	Nominal value	Variation
Superstructure mass	100% of the mass based on the geometry	70% ~ 130%
Concrete compressive strength	30 MPa	20 MPa ~ 40 MPa
Plastic hinge length	1.0 m	1.0 m ~ 1.5 m
Damping	5%	2% ~ 8%

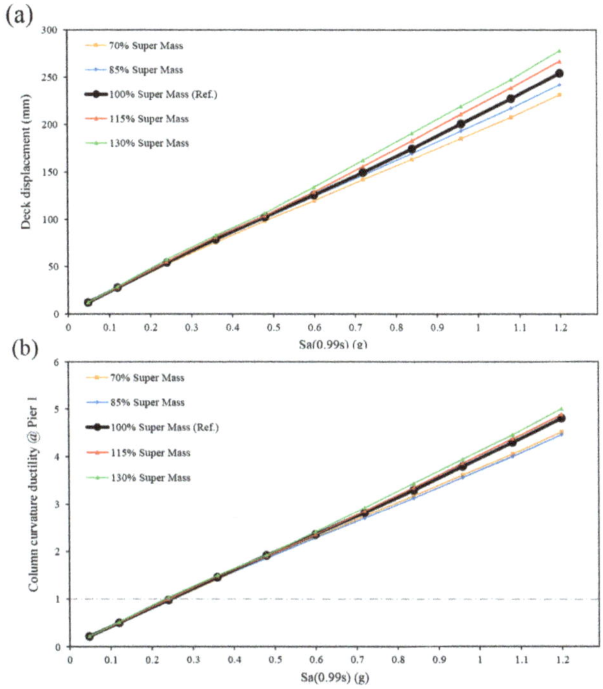

Figure 5: Responses due to change of the superstructure mass, (a) deck displacement; (b) column curvature ductility.

results in the figure clearly show that the columns on the bridge have satisfactory behaviour, i.e. they could survive in an earthquake event about 3 times the design earthquake. The acceptable performance of the substructure is mainly because the bridge was retrofitted in 1990s to improve its seismic resistance by replacing the original bearings with elastomeric bearings. Otherwise it would have sustained at least major damage under the design earthquake considering the bridge was quite old. Overall the 30% change of the superstructure mass on average leads about 5% difference on the pier ductility.

5.2 Effects of the concrete compressive strength

The uncertainty of the material property is one of the major factors that need to be taken into account in the development of the fragility curves for bridges [1, 6, 20]. It is known that the concrete compressive strength (f_c') would affect the result of the moment-curvature or moment-rotation relationship of a given reinforced concrete section, which represents the hysteretic behaviour of the section under seismic loading. Given this, sensitivity analyses were conducted in this study for different f_c' for the column, such as 20 MPa, 25 MPa, 30 MPa, 35 MPa and 40 MPa. The responses corresponding to $f_c' = 30$ MPa were considered as a reference.

The results for the deck displacement and column curvature ductility are presented in Fig. 6a and b, respectively. The results in Fig. 6a show that smaller f_c' provides larger deck displacement and vice verse. However the variation in the responses is quite small, i.e. less than 5% with respect to the reference at the highest intensity level of 5.0Sa(0.99s). Such

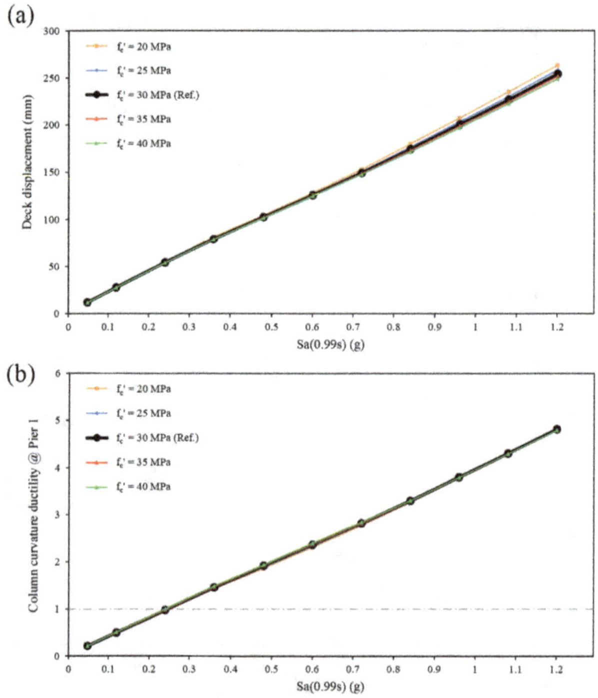

Figure 6: Responses due to change of the concrete compressive strength, (a) deck displacement; (b) column curvature ductility.

results were expected because the variation of the concrete strength mentioned above was only assigned to the column that affects the nonlinearity of the plastic hinge, and not to the deck and the girder. In addition, the superstructure should behave elastically according to the capacity design principle for bridges [13]. The small difference between the responses was due to the change of the stiffness of the substructure arising from different f_c'. The results in Fig. 6b indicate that f_c' does not have effect on the curvature ductility of the column at pier 1, where fixed bearings are used. This is because the bearing carries most of the seismic force given its relatively larger lateral stiffness.

5.3 Effects of the plastic hinge length

In order to evaluate the effect of plastic hinge length on the bridge response, the plastic hinge length (L_p) was varied by a factor of 1.1 to 1.5 times its normal value of 1 m determined according to CHBDC, and they are 1.0 m, 1.1 m, 1.2 m, 1.3 m, 1.4 m and 1.5 m. Figure 7a shows the results for the deck displacement, while Fig. 7b for the column curvature ductility. It can be seen clearly in Fig. 7a that the change of the plastic hinge length does not affect the deck displacement. Such observation is not surprising due to the mechanism of a bridge system, i.e. the displacement of the superstructure should be equal to the displacement of the substructure. Another observation of the results is that the increase of the maximum mean deck displacement is almost linearly proportional to the increase of the seismic excitation level.

As shown in Fig. 7b, the curvature ductility decreases with the increase of the plastic hinge length. For example, at the seismic excitation of 0.90 g, the column curvature ductility

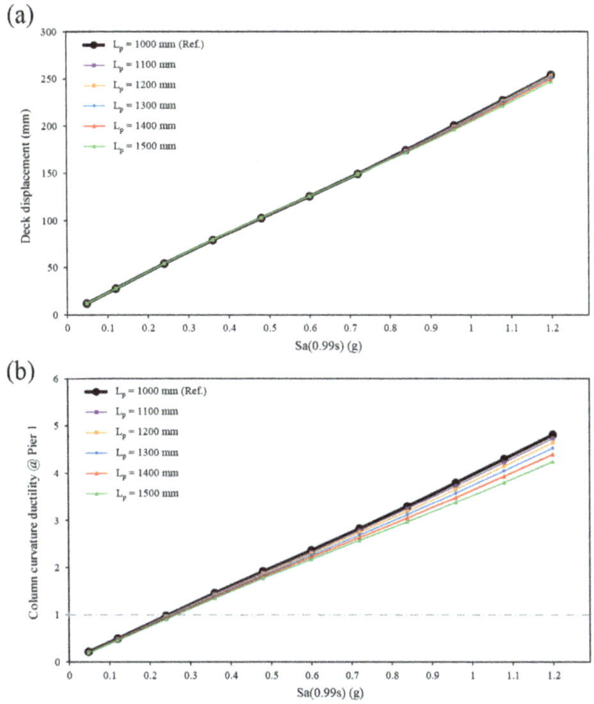

Figure 7: Responses due to change of the plastic hinge length, (a) deck displacement; (b) column curvature ductility.

corresponding to L_p = 1.0 m (i.e. the reference value) is about 2.8, while it is about 2.6 corresponding to L_p = 1.5 m, which gives a difference about 10%. It should be noted that similar results were observed from the results of pier 2 (under the expansion bearing) that are not presented in this paper. Furthermore, it was found in this study that the column curvature ductility of pier 1 (under the fixed bearing) was about 2.5 times larger than that of the pier 2. It indicates significant inelastic deformations occurred in pier 1 due to larger seismic force applied.

5.4 Effects of the damping

In the bridge modelling, Rayleigh damping was selected in order to take into account the bridge stiffness degradation during nonlinear response while a constant damping was assigned to all the 12 vibration modes. For the purpose of the investigation, the variation of damping ranged from 2% to 8%, in which the 5% damping was considered as a reference to compare the analysis results, as presented in Fig. 8.

As expected, smaller damping provides larger response, while larger damping creates smaller response. Detailed calculations showed that change of the damping by 1% would create variation on average about 3% in the deck displacement and column curvature ductility. In addition, the analysis results show that damping has significant effects on flexible columns at lower excitation levels. For example, the curvature ductility of the column at pier 2 resulting from 2% damping is about 17% larger than the reference value at the intensity level of 1.0Sa(0.99s). However, it is only 9% larger than the reference ductility at the excitation level of 5.0(0.99s).

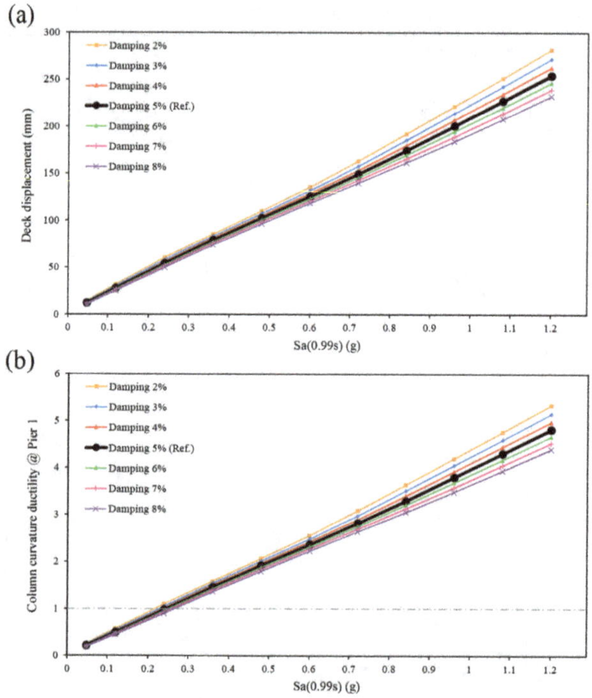

Figure 8: Responses due to change of the damping, (a) deck displacement; (b) column curvature ductility.

6 CONCLUSIONS

In this study, effects of the uncertainties of the superstructure mass, concrete compressive strength, plastic hinge length, and damping on the bridge seismic response were investigated. The sensitivity analysis on each modelling parameter was conducted on a 3-span reinforced concrete highway bridge located in Montreal, Canada. In total, 15 records obtained from the earthquakes around the world were used in the nonlinear time-history analysis. The deck displacement and the column curvature ductility were selected to represent the bridge response. The main conclusions from the study are summarized as follows:

- Changing the superstructure mass by 30% causes the deck displacement to change about 10% at the seismic excitation level five times the design earthquake while it does not have effect on the column curvature ductility.
- Changing the concrete compressive strength does not affect the deck displacement and column curvature ductility under the fixed bearing.
- Changing the plastic hinge length does not affect the deck displacement. However, it does affect the column curvature ductility, i.e. larger plastic hinge length provides smaller curvature ductility. Furthermore, increase in the plastic hinge length by 50% would reduce the ductility by about 10%.
- Damping has more effects on the response at the lower excitation levels especially for the column curvature ductility. The 5% damping is appropriate for use in the analysis.

REFERENCES

[1] Choi, E., DesRoches, R. & Nielson, B., Seismic fragility of typical bridge in moderate seismic zones. *Engineering Structure*, **26**(2), pp. 187–199, 2004. https://doi.org/10.1016/j.engstruct.2003.09.006

[2] Padgett, J.E. & DesRoches, R., Sensitivity of seismic response and fragility to parameter uncertainty. *Journal of Structural Engineering*, **133**(12), pp. 1710–1718, 2007. https://doi.org/10.1061/(ASCE)0733-9445(2007)133:12(1710)

[3] Dicleli, M. & Bruneau, M., Seismic performance of single-span simply supported and continuous slab-on-girder steel highway bridges. *Journal of Bridge Engineering*, **121**(10), pp. 1497–1506, 1995. https://doi.org/10.1061/(asce)0733-9445(1995)121:10(1497)

[4] Avşar, Ö., *Fragility based seismic vulnerability assessment of ordinary highway bridges in Turkey*, Ph.D. Thesis, Department of Natural and Applied Sciences, Middle East Technical University, Ankara, Turkey, 2009.

[5] Aviram, A., Mackie, K.R. & Stojadinovic, B., Effect of abutment modeling on the seismic response of bridge structures. *Journal of Earthquake Engineering and Engineering Vibration*, **7**(4), pp. 395–402, 2008. https://doi.org/10.1007/s11803-008-1008-3

[6] Tavares, D.H., *Seismic risk assessment of bridges in Quebec using fragility curves*, Ph.D. thesis, University of Sherbrooke, Sherbrooke, Quebec: Canada, 2011.

[7] Computers and Structures Inc., *SAP 2000 Integrated Software for Structural Analysis and Design*, V14, Berkeley, California, 2000.

[8] McKenna, F. & Feneves, G.L., *Open system for earthquake engineering simulation, 1.6.2 Edition*, Pacific Earthquake Engineering Research Center: Berkeley, California, 2005.

[9] ATC-32., *Improved seismic design criteria for California bridges: Provisional recommendations*. Applied Technology Council: Redwood City, California, 1996.

[10] Caltrans., Seismic design criteria, version 1.7. California Department of Transportation, Sacramento: California, U.S, 2013.

[11] Kelly, J.M., *Earthquake-Resistance Design with Rubber,* 2nd edn., Springer: London, 1997. https://doi.org/10.1007/978-1-4471-0971-6

[12] DesRoches, R., Leon, R.T. & Dyke, S., Response modification of bridges. MAE Center Project ST-12, *MAE report 30-08*. MAE Center: Illinois, U.S, 2003.

[13] Canadian Standards Association (CSA). *Canadian Highway Bridge Design Code (CHBDC), CAN/CSA S6-06*. Canadian Standards Association, Rexdale, Ontario: Canada, 2006.

[14] Mander, J.B., Priestley, M.J. & Park, R., Theoretical stress-strain model for confined concrete. *Journal of Structural Engineering*, **114**(8), pp. 1804–1826, 1988. https://doi.org/10.1061/(ASCE)0733-9445(1988)114:8(1804)

[15] Wilson, J.C. & Tan, B.S., Bridge abutments: formulation of simple model for earthquake response analysis. *Journal of Engineering Mechanics*, **116**(8), pp. 1828–1837, 1990. https://doi.org/10.1061/(ASCE)0733-9399(1990)116:8(1828)

[16] Kotsoglu, A. & Pantazopoulou, S., Modeling of embankment flexibility and soil-structure interaction in integral bridges. *Proceedings of the 1st European Conference on Earthquake Engineering and Seismology*, Geneva, Switzerland, 2006.

[17] Naumoski, N., Heidebrecht, A.C. & Rutenberg, A.V., Representative ensembles of strong motion earthquake records. *EERG Report, 93-1*, Department of Civil Engineering, McMaster University: Hamilton, Ontario, Canada, 1993.
[18] Naumoski, N., Heidebrecht, A.C. & Tso, W.K., Selection of representative strong motion earthquake records having different A/V ratios. *EERG report, 88-01*, Department of Civil Engineering, McMaster University: Hamilton, Ontario, Canada, 1988.
[19] Halchuk, S., Adams, J. & Anglin, F., Revised deaggregation of seismic hazard for selected Canadian cities. *Proceedings of the 9th Canadian Conference on Earthquake Engineering*, Ottawa, Ontario: Canada, pp. 420–432, 2007. https://doi.org/10.4095/223221
[20] Pan, J., Agrawal, A.K., Ghosn, M. & Alampalli, S., Seismic fragility of multispan simply supported steel highway bridges in New York State. I: Bridge modeling, parametric analysis, and retrofit design. *Journal of Bridge Engineering*, **15**(5), pp. 448–461, 2010. https://doi.org/10.1061/(ASCE)BE.1943-5592.0000085

SEISMIC RESPONSE ANALYSES OF RC PORTAL FRAMES WITH LARGE DEFORMABLE ELASTIC BRACES

KIICHIRO SAWADA
Shimane University, Japan.

ABSTRACT
Large deformable elastic braces (LDEBs) are devices which do never yield subject to large deformation under great earthquakes. In structures with LDEBs, elastic restoring force by LDEBs can improve seismic response under great earthquakes. In previous works, the effectiveness of LDEBs for the steel structures was confirmed by experimental tests and seismic response analyses. Here, the topology of LDEBs is determined by an optimization method, and seismic response analyses of RC portal frames with LDEBs are conducted. The effectiveness of LDEBs for the RC structures is discussed. RC portal frame are designed by Japanese seismic design code. LDEBs are equipped with the frame as knee braces. Seismic responses of not only the frame with LDEBs but also that without LDEBs are computed by dynamic nonlinear analysis software. In the analysis LDEBs are regarded as elastic bar elements. An input earthquake is JMA Kobe NS wave (1995 Kobe earthquakes). Hysteresis of story shear force and story drift, axial deformation of LDEBs, maximum and residual story drift are investigated. It is observed from computational results that LDEBs show remarkable improvements on maximum and residual story drifts of RC portal frame under a very large amplitude earthquake such as 1995 Kobe earthquake in Japan.
Keywords: an optimization method, large deformable elastic braces, RC frames, seismic response analyses.

1 INTRODUCTION

It is required for the seismic design in Japan that there are no damage under moderate earthquakes and no collapse under major earthquakes. Many buildings might experience a certain level of damage and deformation under major or great earthquakes. Recently various seismic control technologies [1–3] have been proposed. Especially, seismic upgradings by hysteresis dampers have been applied for many buildings in Japan. It is expected that elastic restoring force of a main frame and the hysteresis damping effect of the hysteresis damper reduce seismic response [4]. However, there are some posing problems for seismic upgrading design by hysteresis dampers. Firstly, increasing strength of the superstructure by seismic upgrading lead damage concentration of foundation structure under great earthquakes. Secondly, there is a possibility of plastic damage in main frame under great earthquakes. The plastic damage in main frame lead to large residual deformation because of lack of restoring force. Large deformable elastic braces (LDEBs) used in this study are devices which do never yield under great earthquakes [5]. It is expected that LDEBs realize reduction of seismic response with a slight increasing of strength of superstructure. This paper shows an example of morphogenesis of LDEBs by an optimization method. Moreover, seismic response analyses of portal RC frames with and without LDEBs are conducted, and the effectivenesses of LDEBs are discussed.

2 LARGE DEFORMABLE ELASTIC BRACES (LDEBS)

Figure 1 shows an example of large deformable elastic braces (LDEBs) [5]. This shape can realize large elastic deformation. It is confirmed from linear elastic FEM analysis subject to tensile load that yielding deformation and yielding force of the shape of Fig. 1 is 28.72 mm

DOI: 10.2495/CMEM-V6-N5-880-886

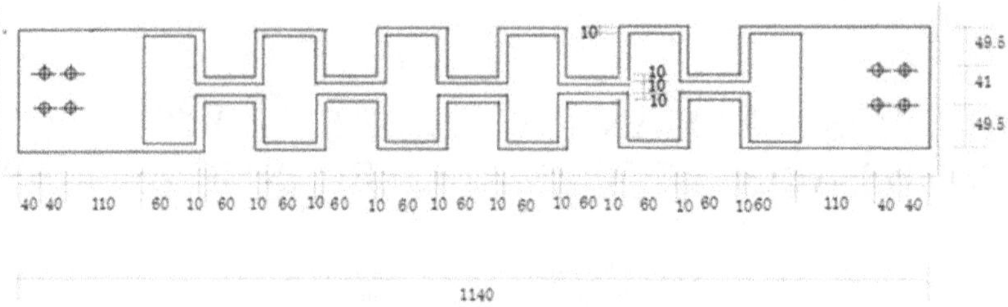

Figure 1: An example of LDEBs [5].

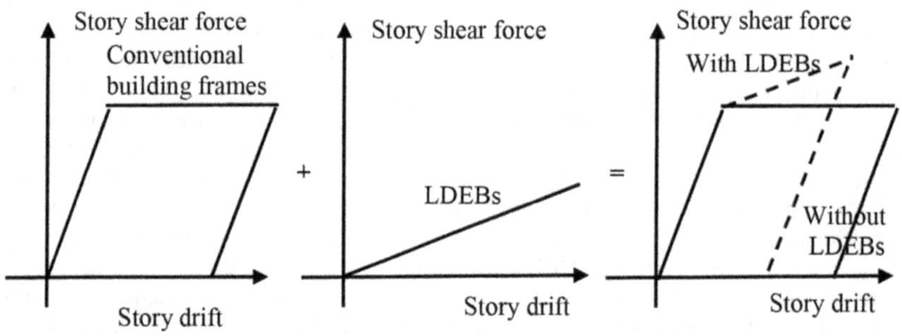

Figure 2: The relationship between story shear force and story drift of a conventional building frame and the frame with LDEBs.

and 15.20 kN, respectively. While seismic horizontal load–deformation relationship of conventional buildings are shown as Fig. 2, and that of buildings with LDEBs are shown as Fig. 3. It is expected that second stiffness by LDEBs can reduce maximum and residual story drift under major or great earthquakes. Here topology patterns of LDEBs with better performance than Fig. 1 are generated on a computer by an optimization method (genetic algorithm) as shown in the following section.

2.1 Topology optimization of LDEBs

A topology optimization problem for a plane stress two-dimensional plate is considered. An inverse Fourier transform function to represent the topological patterns of the plate is defined as follows.

$$F(x_k, y_k) = \sum_{u=-NFX}^{NFX} \sum_{v=-NFY}^{NFY} (a_{u,v} + b_{u,v} i) e^{i\pi(\gamma_1 u x_k / LX + \gamma_2 v y_k / LY)}, \tag{1}$$

where $u = -NFX, \ldots, -3, -2, -1, 1, 2, 3, \ldots, NFX$, $v = -NFY, \ldots, -3, -2, -1, 1, 2, 3, \ldots, NFY$, i represents an imaginary unit, e represents the Naperian base, and $a_{u,v}$ and $b_{u,v}$ are real-valued parameters to be computed in the optimization process, which are greater than –1 and less than 1. x_k and y_k represent x- and y-coordinates of the centre of the element k.

Since F should be a real function, the following complex conjugate conditions can be used.

$$a_{u,v} = a_{-u,-v},\ b_{u,v} = -b_{-u,-v}. \tag{2}$$

Each element is distinguished as either a solid or a void according to the value of F. If the value of F corresponding to the element k satisfies the following equation, the element k is a solid, otherwise it is a void.

$$F\min + MIN(\beta_1, \beta_2)\cdot(F\max - F\min) \leq F(x_k, y_k) \leq F\min + MAX(\beta_1, \beta_2)\cdot(F\max - F\min), \quad (3)$$

where F_{max} and F_{min} represent the maximum and minimum values of F, respectively, over all the elements. β_1 and β_2 ($0 \leqq \beta_1 \leqq 1, 0 \leqq \beta_2 \leqq 1$) represent parameters to be computed in the optimization process.

2.2 A formulation for topology optimization of LDEBs

A formulation for topology optimization of LDEBs can be formulated as follows.

Find $a_{u,v}$, $b_{u,v}$ and β_1 and β_2 which minimize

$$Z = 1/U+p[V/V0+MAX\{0,(Uy-U0)/U0\}+MAX\{P/Py,Py/P\}], \quad (4)$$

where U represents the displacement by given load P, V represents the plate volume for the current topology, V0 represents the plate volume where all the elements are solid, Uy represents yielding displacement at loading point, U0 represents the critical value of U and Py represents yielding load. Von Mises is used as yielding condition of plate elements.

2.3 Numerical results

The above mentioned optimization problem for a 5488 (196×28) elements meshed plate under tensile load is solved by real coded genetic algorithm with heuristic crossover [6]. Young's modulus is 205 GPa. The thickness, the width and the length of the plate is 9 mm, 210 mm and 1470 mm, respectively. Poisson's ratio is 0.3. Critical displacement U0 is set to 80 mm. 29000N and 58000N are given as load P. The Young's modulus of the void element is set to 10^{-4} times that of the solid element. The number of individuals is 50, mutation probability is 0.05, and crossover probability is 1.0. NFX and NFY in eqn (1) are both 2. γ_1 and γ_2 in eqn (1) are 36 and 1, respectively. Figure 3 represents the optimum topology of LDEBs until 100 generations. It is shown that this optimization method can realize very large elastic deformation, 2% or 2.5% of the plate length.

Figure 3: (a) Optimum topology of LDEBs (P = 29000N, Py = 28600N, Uy = 41.2 mm).

Figure 3: (b) Optimum topology of LDEBs (P = 58000N, Py = 56900N, Uy = 32.1 mm).

3 SEISMIC RESPONSE ANALYSES

Figure 4 shows a detail of an RC portal frame using LDEBs as knee braces for seismic response analyses. Each LDEB is sandwiched by two channels using non tension bolt connection. 12 re-bars (22 mm) are allocated in the column sections whose height and width are both 500 mm, and 8 re-bars are allocated in the girder section whose depth and width are 550 mm and 400 mm, respectively. Yield strength, Young's modulus and strain hardening ratio of the re-bars are 400 N/mm^2, 200 GPa, 0.005, respectively, and bilinear hysteresis rule is applied. Compressive strength of concrete is 30 MPa, and Mander *et al.* nonlinear concrete model [7] is used. Here, time domain analyses of not only a computational model of an RC portal frame without LDEBs shown in Fig. 5a, but also that with LDEBs (Elastic bar elements, axial stiffness 3.76 kN/mm), shown in Fig. 5b are conducted. An input earthquake is JMA Kobe NS wave (1995 Kobe earthquake). Peak ground acceleration (PGA) and duration time of the wave for computation is 818(gal) and 29(sec.), respectively. Seismostruct [8] is used as time domain analysis software. In the software, the sectional stress-strain state of beam-column elements is obtained through the integration of the nonlinear uniaxial material response of the individual fibres [8]. Figures 6 and 7 show the time history of story drift and

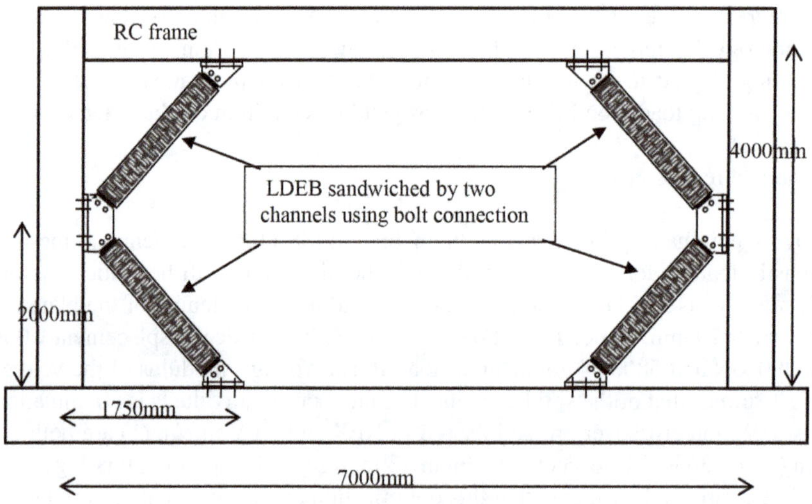

Figure 4: A detail of an RC portal frame with LDEBs.

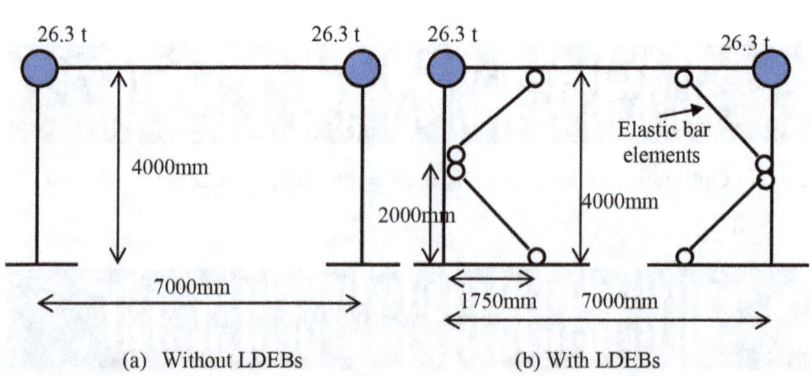

Figure 5: Computational models of an RC portal frame without LDEBs and that with LDEBs.

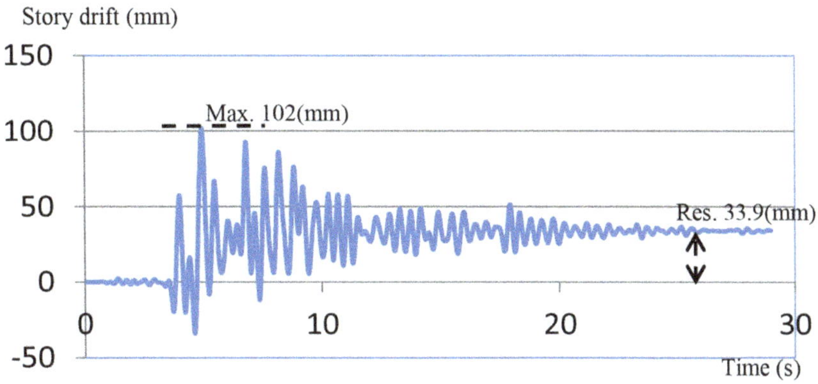

Figure 6: (a) Time history of story drift of the frame without LDEBs.

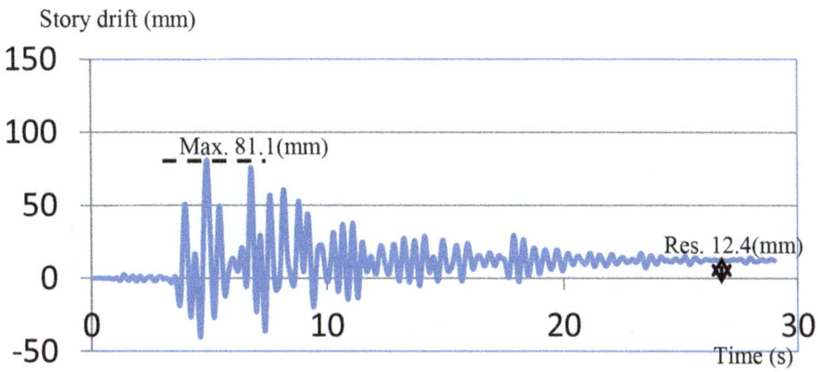

Figure 6: (b) Time history of story drift of the frame with LDEBs.

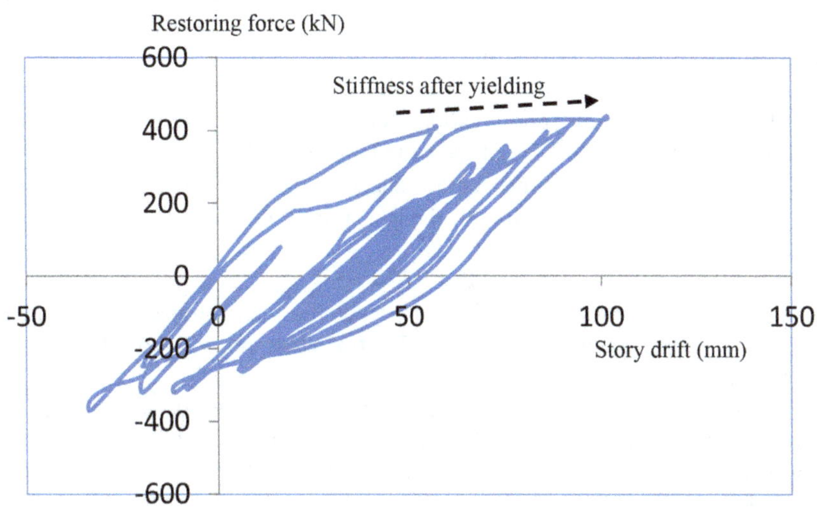

Figure 7: (a) The relationship between story restoring force and story drift of the frame without LDEBs.

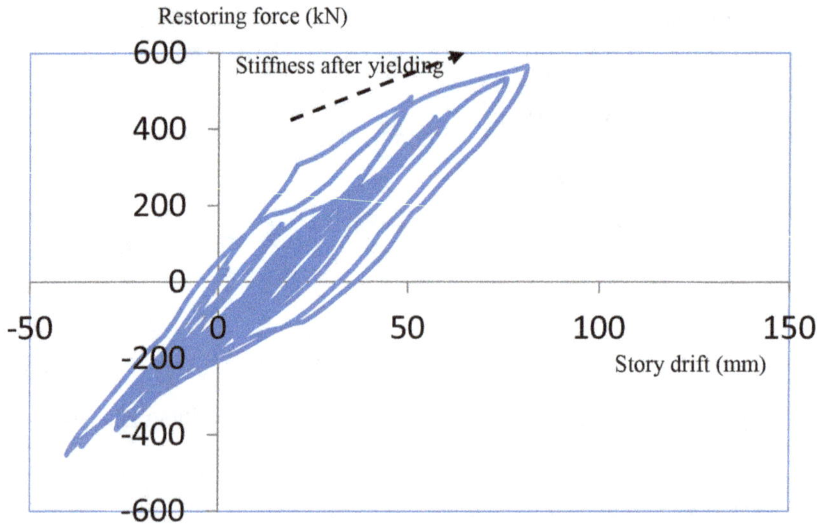

Figure 7: (b) The relationship between story restoring force and story drift of the frame with LDEBs.

the relationship between story restoring force and story drift obtained by time domain analysis. Broken arrows in these figures show that the RC frames with LDEBs have positive story stiffness after yielding to some extent, while the RC frame without LDEBs have just slight story stiffness after yielding. The maximum axial deformation of elastic bar elements was less than around 30 mm. It was also observed from these figures that LDEBs give improvements on maximum and residual story drifts.

4 CONCLUSIONS

This paper has shown an example of morphogenesis of Large Deformable Elastic Braces (LDEBs) by the optimization method. Moreover, seismic response analyses of portal RC frames with and without LDEBs have been conducted, and the effective-nesses of LDEBs have been discussed. Concluding remarks are as follows.

1. It has been shown that this optimization method presented here can realize very large elastic deformation, 2% or 2.5% of the plate length.
2. It has been observed from time domain analyses that LDEBs give improvements on maximum and residual story drifts for the RC portal frame.

REFERENCES

[1] Dargush, G.F. & Soong, T.T., Behavior of metallic plate dampers in seismic passive energy dissipation systems. *Earthquake Spectra*, **11**(4), pp. 545–568, 1995. https://doi.org/10.1193/1.1585827
[2] Takewaki, I., Murakami, S., Yoshitomi, S. & Tsuji, M., Fundamental mechanism of earthquake response reduction in building structures with inertial dampers. *Structural Control and Health Monitoring*, **19**(6), pp. 590–608, 2012. https://doi.org/10.1002/stc.457
[3] Takeya, N. & Shioya, S., Discussion for depressing a residual deformation of RC beam flexural yielded after Earthquake (Part.1). *Proceeding AIJ Annual Meeting*, 2011.

[4] Harada, Y. & Akiyama, H., Seismic design of flexible-stiff mixed frame with energy concentration. *Engineering Structures*, **20**(12), pp. 1039–1044, 1998. https://doi.org/10.1016/S0141-0296(97)00201-0

[5] Sawada, K., Nishida, G., Nakamura, K., Eto, H. & Tsuruda, M., Computational study on seismic steel structures with large deformable elastic braces. *Eighth International Conference on Steel and Aluminum Structures*, Hong Kong, China, December 2016 (USB).

[6] Sawada, K., Matsuo, A. & Shimizu, H., Randomized line search techniques in combined GA for discrete sizing optimization of truss structures. *Structural and Multidisciplinary Optimization*, **44**(3), pp. 337–350, 2011. https://doi.org/10.1007/s00158-011-0629-8

[7] Mander, J.B., Priestley, M.J.N. & Park, R., Theoretical stress-strain model for confined concrete. *Journal of Structural Engineering*, **114**(8), pp. 1804–1826, 1988. https://doi.org/10.1061/(ASCE)0733-9445(1988)114:8(1804)

[8] www.seismosoft.com/seismostruct

ADVANCES IN SEISMIC VULNERABILITY ASSESSMENT OF REINFORCED CONCRETE BUILDINGS APPLIED TO THE EXPERIENCE OF LORCA (SPAIN) 2011 EARTHQUAKE

J. L. RÓDENAS, A. TOMÁS & S. GARCÍA-AYLLÓN
Department of Civil Engineering, Universidad Politécnica de Cartagena (UPCT), Spain.

ABSTRACT
Despite the technical advances in seismic structural design, many regions still present a high level of seismic risk, principally due to the high vulnerability of their buildings. A modification of the empirical method for assessing the seismic vulnerability of reinforced concrete buildings in urban areas is proposed in this contribution. In the RISK-UE LM1 framework, the values of certain behaviour modifiers related to the typological, structural and urban parameters of the buildings have been modified according to a review and analysis of the currently available models and an evaluation of the actual seismic performance of buildings. This provides continuity to the progress of the previous works published to date. The proposal has been applied to the city of Lorca, Spain, for which ample knowledge of the damage occurred in the earthquake of May 11, 2011 is available. Less dispersion between actual observed and estimated damage in buildings is presented in comparison with the previous studies, with a statistical significance of 5%, thus achieving a more accurate evaluation of seismic risk. The new model also provides valuable information to be used in the planning and management of post-earthquake emergency situations when combining with GIS techniques, thus allowing for a better definition of several damage scenarios to enhance the development and urban preparedness in case of further seismic events.
Keywords: emergency management, empirical vulnerability assessment, observed damage, seismic vulnerability.

1 INTRODUCTION

Some of the most recent seismic disasters (L'Aquila 2009, Lorca 2011, Emilia Romagna 2012, Nepal 2015, Ecuador 2016, Amatrice 2016), have once again revealed the inadequate performance of many existing buildings against seismic action. This performance is observed not only in structures designed without seismic regulations – or with early versions of seismic codes –, but also in those designed with recent seismic codes.

Therefore, the analysis of the seismic behaviour of buildings continues to be one of the principal aims of Seismic Engineering [1, 2], not only from the perspective of the loss of human life and the physical damage originated in buildings, but also because of the damages caused by losses and inefficiencies in communication and services networks.

The seismic vulnerability of a structure can be defined as its intrinsic predisposition to damage at the occurrence of an earthquake of a certain severity [3]. This property, which is directly related to the design and construction characteristics of the building (structural typology, geometry, quality of the materials), therefore constitutes an internal factor, independent of the seismic action. Accordingly, seismic risk (the potential damage to a building, loss of human life and economic losses that may occur in a specified period of exposure) can be expressed as the convolution of vulnerability and seismic hazard (the probability of occurrence of a seismic event of certain severity in a specific site and during a determined period of exposure).

The availability of well-documented material on the damages caused by earthquakes (such as the Lorca earthquake in 2011), from on-site observations of the seismic effects as well as

DOI: 10.2495/CMEM-V6-N5-887-898

from comparison with previously estimated damages, provides an opportunity to improve the empirical methodologies, which can help evaluate the seismic response of buildings in urban areas and decrease the level of uncertainty associated with seismic risks.

2 RISK-UE METHODOLOGIES FOR SEISMIC VULNERABILITY ASSESSMENT OF BUILDINGS

RISK-UE Project [4] stated two methodologies for the evaluation of the levels of seismic risk in urban areas. The first one, named as methodology LM1 or the Level I method and based on empirical research, characterizes the seismic hazard of the urban area in terms of macro-seismic intensity EMS-98, and the vulnerability of the buildings through a vulnerability index obtained according to the Vulnerability Index Method (VIM). The second one, known as methodology LM2 or the Level II method, considers the seismic action in terms of the corresponding spectrum of demand and the seismic vulnerability according to the capacity spectrum of the analysed structure [5].

With respect to empirical methods, the VIM constitutes a procedure that combines the vulnerability classes of the EMS-98 scale and the Italian method [6] for the characterization of the seismic vulnerability of buildings. This methodology is based on basic vulnerability indexes I_{v_t} that characterize the seismic behaviour of the most usual structural typologies in urban areas. These indexes are included in the BTM (Building Typology Matrix) [7] and are shown in Table 1 for several building typologies defined in the BTM and have been modified subsequently by Lagomarsino and Giovinazzi [5].

In addition, since the seismic behaviour of a certain building also depends on factors such as its grade of conservation or the level of seismic design code, as well as on other structural and urban planning parameters such as the number of storeys above ground, the geometric and stiffness regularity or the relative position of the building in the block, the LM1 methodology modifies this basic index of vulnerability $I_{v\text{-}t}$ according to a series of values of penalty or improvement ΣM_C.

In this way, since the first version [7], a series of reviews have been developed in the definition and quantification of these behaviour modifiers [8–11], calibrating the influence of several dispositions (i.e. irregularity in plant or elevation, difference of height between nearby buildings, etc.), in the seismic performance of buildings.

Thus, the global vulnerability index of each building $I_{v\text{-}b}$ can be evaluated through the following expression [9]:

$$I_{\text{v-b}} = I_{\text{v-t}} + \sum\nolimits_{j=1}^{n} M_{\text{Cj}} + \Delta M_R. \tag{1}$$

where $I_{v\text{-}t}$ is the basic vulnerability index of the structural typology to which the building belongs (Table 1); ΣM_C are the different modifiers of behaviour that consider the structural and urban planning characteristics specifically for each building analysed; and ΔM_R is a

Table 1: Vulnerability indexes for RC building typologies defined in BTM [5, 7].

Typologies	Building type	I_v^{min}	I_v^-	I_v^*	I_v^+	I_v^{max}
RC1	Concrete moment frame	0.140	0.330	0.484	0.640	0.860
RC2	Concrete shear walls	0.140	0.210	0.384	0.510	0.700
RC3	Dual system	0.060	0.127	0.522	0.880	1.020

RC = Reinforced Concrete

regional modifier assigned depending on the characteristics and constructive dispositions specific to the zone of study, the seismic design codes, the date of construction, or the opinion of experts from actual observed damage.

Finally, to evaluate the level of seismic damage of every building, the LM1 method employed a semiempirical vulnerability function dependent on two parameters: the vulnerability index $I_{v\text{-}b}$, and the macroseismic intensity I in the EMS-98 scale [5]:

$$\mu_D = 2.5\left[1+\tanh\left(\frac{I+6.25\,I_{v\_b}-13.1}{Q}\right)\right]. \tag{2}$$

where Q is a ductility coefficient of value 2.3 for most of the building typologies of BTM [5], and μ_D is the mean damage grade defined in the EMS-98 scale for reinforced concrete (RC) buildings [12].

3 ADVANCES FOR EVALUATING THE SEISMIC VULNERABILITY OF REINFORCED CONCRETE BUILDINGS IN URBAN AREAS

The reviews proposed for the behaviour modifiers of the LM1 method [8–11], have focused principally on the analysis of those parameters related to the influence of diverse urban planning dispositions in the seismic response of the buildings, such as short columns and soft first storeys. Nevertheless, other parameters related to the level of the seismic design code that has been applied, or the influence of the natural period in the seismic response of the building need to be adjusted to the actual performance of RC buildings, in order to differentiate it from the behaviour of unreinforced masonry structures.

Thus, a proposal for the seismic vulnerability assessment of RC buildings according to the RISK-UE LM1 method is presented in this contribution, based on a thorough review of the current methods and on an analysis of the observed seismic performance of the buildings in recent earthquakes. This proposal provides continuity to the progress of previous works published to date. The modifiers proposed to obtain the vulnerability index $I_{v\text{-}b}$ are shown in Table 2.

Firstly, the modifiers related to the level of seismic design code and the construction characteristics have been revised. The modifiers defined in [8] have been adjusted according to the seismic response of buildings, since a worse performance is usually observed in the structures designed with older, and less accurate, seismic codes [13–15]. Thus, the modifier can adopt the values 0.16, 0.08 or 0.00:

a) 0.16: Case of pre-code buildings (designed without seismic regulations). In this type of buildings, the value of the modifier is based on the simultaneous met of the following considerations: (i) use of simplified analysis methods for the structure; (ii) the lack of seismic construction requirements and details for the structure; (iii) the lack of seismic construction requirements and details for the foundation; and (iv) the likely deficient condition and/or quality of materials.
b) 0.08: Case of low code buildings (designed with early versions of seismic codes). The incorporation of several seismic recommendations for the structure and foundation in the codes reduces this modifier from 0.16 to 0.08, since only the simplified analysis methods and the likely deficient condition and/or quality of materials are considered in this case.
c) 0.00: Case of medium code or high code buildings (designed with latter seismic codes). A value of 0.00 is assigned to this modifier because more accurate analytic methods have been employed and better condition and quality of materials have been used.

Table 2: Proposal for new values of modifiers for the RISK-UE LM1 method depending on the level of the seismic design code. A detailed explanation of each item is available in [9].

Behaviour modifiers		Seismic design code			
		Pre-code	Low code	Medium code	High code
Level of seismic and structural design code		0.16	0.08	0	0
N° storeys above ground	0 to 3	0.04	0.04	0.04	0.04
	4 to 7	0	0	0	0
	≥ 8	−0.04	−0.04	−0.04	−0.04
Plant irregularity		0.04 (RC < 0.5) 0.02 (0.5 ≥ RC < 0.7)			
Irregularity in elevation		0.04 ($d > 3$) 0.02 ($1 < d \leq 3$)			
Short column		0.08	0.08	0.08	0.08
Insufficient seismic structural joint		0.04	0.04	0.04	0
Slope of the ground		0.04	0.04	0.04	0.04
Relative position in the block of buildings	Intermediate	−0.04	−0.04	−0.04	−0.04
	Corner	0.04	0.04	0.04	0.04
	Full corner/Header	0.06	0.06	0.06	0.06
Relation with adjoining buildings		[−0.04, 0.04] *f* (relative height between buildings)			
Soft storey		0.2	0.2	0.2	0.2

In accordance with the considerations mentioned above, a categorization of the national seismic codes is required prior to the application of the method to any urban area. This modifier can therefore be obtained depending on the construction period of the building and the seismic code valid in that period, which can be known from the cadastral database.

Damage in low-rise buildings is generally identified with cracks in masonry infill walls at the ground floor and structural failures at the end of columns due to the higher stiffness of this type of structures [16, 17]. Therefore, the factor relating to the number of storeys above ground is assigned depending on the natural vibration frequency of the structure, which is represented in the response spectrums defined in the seismic codes. Thus, for low-rise buildings, whose fundamental natural vibration period is related to the top of response spectrums with constant acceleration, the vulnerability is penalized by increasing it by a value of 0.04. Conversely, for high-rise buildings, whose fundamental natural vibration period is related to the descent branch of acceleration in response spectrums, the vulnerability is diminished by a value of −0.04.

In practice, most of the buildings in urban areas, especially those placed on low-to-moderate seismicity regions, either have no seismic structural joint between buildings, or such a joint is not wide enough or has not been constructed properly. In this sense, cracks in masonry infill walls and even in structural elements can usually be observed in the joint between adjoining buildings [13, 16, 17]. Therefore, a penalty value of 0.04 has been conservatively considered, except for those buildings designed with a high-level code.

Thus, the proposed values enable to implement a more adjusted approach using cadastral survey information and making an on-site inspection of the evaluated zone, depending on the typological, structural and urban parameters of the buildings, similar to the methodology that has been performed in previous reviews of the RISK-UE LM1 method.

The two modifiers related to the typology of foundations and the conservation state of the building defined in the LM1 method are difficult to quantify in most cases. These parameters have been removed because they have already been considered in the modifier relating to the level of the seismic design code. Similarly, considering the definition of the modifier ΔM_R presented in eqn (1), it has also been removed from the evaluation of the vulnerability index $I_{v\text{-}b}$ because the aspects involved in the assessment have already been included in the different behaviour modifiers *Mc*. Thus, part of the uncertainty associated to the vulnerability assessment has been reduced and, moreover, the generalization of the proposal has been simplified by removing one parameter that requires calibration depending on the area analysed.

4 APPLICATION USING DAMAGE DATA FROM THE LORCA 2011 EARTHQUAKE

Although the earthquake on May 11, 2011 that struck Lorca had a moderate magnitude of 5.1 Mw, it finally caused several casualties (9 deaths and more than 300 injured) and important structural damage that meant that more than 10,000 people were unable to return to their houses [13, 14]. The earthquake was assigned with an EMS-98 intensity of VII in the main districts of the city [13, 14].

A field study to document the seismic behaviour in RC buildings damaged in the Lorca earthquake, 2011, was carried out by the Department of Civil Engineering of Universidad Politécnica de Cartagena. After a comparison and validation process from an initial sample of 1,050 buildings, 406 homogeneous cases were selected, after removing unusual data or those registers providing insufficient information.

The seismic damage and several urban characteristics that may influence the seismic structural response were registered. The collected data were divided into three blocks containing: (i) general data, such as the building typology or the address; (ii) the constructive features and urban configuration of the building, such as the presence of short columns or soft storeys; and (iii) the assessment of EMS-98 damage grade μ_D. In the second stage of the work, these data have been completed and compared with the cadastral data needed to obtain the aforementioned behaviour modifiers in the RISK-UE LM1 method, such as year of construction, number of storeys above ground, or aggregate building position.

Based on this input data, the observed damage in each building has been compared with the corresponding value from the LM1 method, according to the different reviews proposed in the literature. The construction period distribution of the buildings analysed, according to the Spanish seismic code, is shown in Table 3. The building typology RC1 (based on moment resistant RC frames of ordinary design with unreinforced masonry infill [7]) has been selected to obtain the vulnerability index. A similar categorization for the case of Spanish seismic codes has been used in [9, 11].

The EMS-98 damage grade μ_D observed in the buildings depending on the period of construction is shown in Fig. 1. Within the group of structures constructed prior to 1963, 40% presented high damage levels ($\mu_D = 3$), whereas this percentage decreased to 6.6% in the buildings constructed after 1996. On the contrary, only 6.7% of the buildings prior to 1963 presented a slight grade of damage ($\mu_D = 1$), whereas this percentage reached 39.6% in the more recently constructed buildings.

Table 3: Construction period distribution of the buildings analysed in Lorca according to the Spanish seismic code level.

Period of construction	Spanish seismic code	Code level	Number of buildings	Buildings %	BTM typology
Before 1963	-	Pre-code	45	11.1	RC1
1963–1970	Recommendation MV-101 (1962)	Pre-code	52	12.8	RC1
1971–1995	PGS-1 (1968) and PDS-1 (1974)	Low code	203	50.0	RC1
1996 until now	NCSE-1994 and NCSE-2002	Medium code	106	26.1	RC1

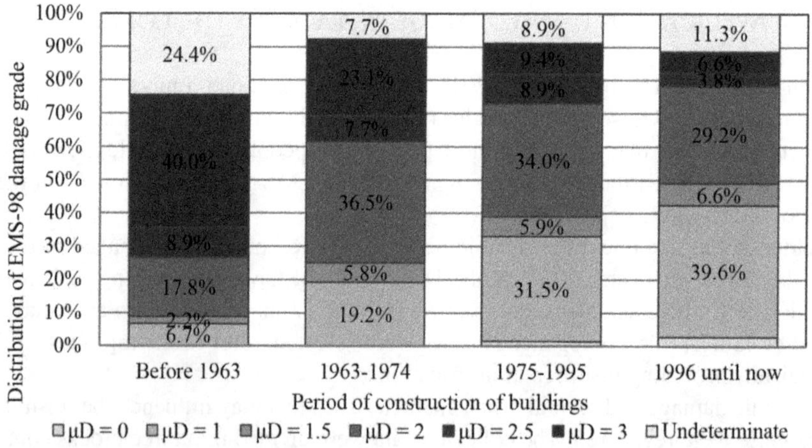

Figure 1: Distribution of EMS-98 mean damage grade observed on RC buildings after the Lorca earthquake, 2011, according to the period of construction.

The distribution of damage observed in different areas of Lorca after the 2011 earthquake is compared in Figs 2–6 with the calculated mean damage grade resulting from the RISK-UE LM1 method of [7] and the subsequent LM1 reviews of [8–11], respectively. For each building, the observed EMS-98 damage μ_D grade is plotted (as a cross mark) with its corresponding $I_{v\text{-}b}$ obtained from the respective LM1 version. Likewise, the simulated mean damage grades μ_D^* calculated from eqn (2) for an earthquake of EMS-98 intensity of VII and the same $I_{v\text{-}b}$ values are shown, along with the possible range $[\mu_D^-, \mu_D^+]$ associated with the interval $[I_v^-, I_v^+]$ defined in BTM (Table 1).

The high variability observed in the distribution of the cross marks is due to the behaviour modifiers considered by each author, which provide different $I_{v\text{-}b}$ values from the same building according to eqn (1). A wider range of $I_{v\text{-}b}$ values implies a better accuracy of the model, since a larger number of cases are considered. Anyway, the damage simulated by the LM1 method for RC buildings, and the subsequent revisions, is lower than the levels of the actual observed damage for an earthquake of EMS-98 intensity of VII. Therefore, these models would not totally represent the seismic behaviour of this type of buildings.

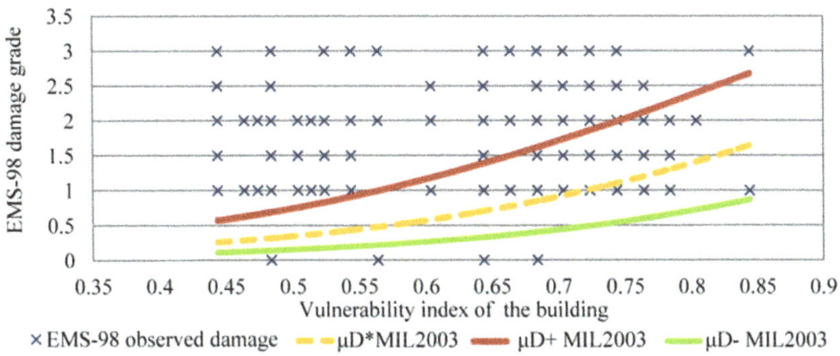

Figure 2: Comparison between EMS-98 observed damage on RC buildings from the Lorca earthquake, 2011, and the mean damage grade obtained from the RISK-UE LM1 method ([7], defined as MIL03), for an $I_{EMS\text{-}98}$ = VII earthquake, where μ_D* is the representative mean damage value, and $[\mu_D^-,\mu_D^+]$ the possible range of the estimated mean damage grade.

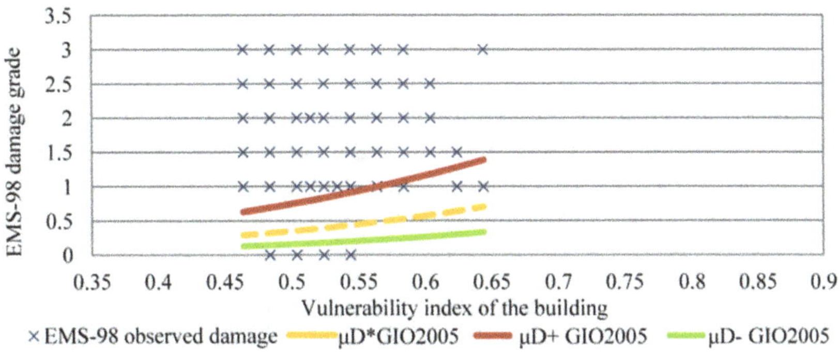

Figure 3: Comparison between EMS-98 observed damage on RC buildings from the Lorca earthquake, 2011, and the mean damage grade obtained from the revised RISK-UE LM1 method ([8], defined as GIO05), for an $I_{EMS\text{-}98}$ = VII earthquake, where μ_D* is the representative mean damage value, and $[\mu_D^-,\mu_D^+]$ the possible range of the estimated mean damage grade.

Finally, the comparison between the mean damage grade obtained according to the improved method and its correspondence with the observed damage is shown in Fig. 7. A better correlation between observed and simulated damage is appreciated, with a greater number of observations included within the range of probable values of the method.

In order to analyse the significance level of the results and determine the adjustment quality of the calculated damage, a new variable $Z_i = OD_i - \mu D_i$ has been defined, where OD_i is the EMS-98 observed damage for every building *i*, and μD_i is the corresponding mean damage grade obtained according to the different reviews of the LM1 method. In addition, the difference $Z_{ij} - Z_{i\_PRO}$ has been evaluated, where Z_{ij} is the value of the previous difference for the building *i* obtained from a certain review *j* of the LM1 method and Z_{i_PRO} is the value for the building *i* obtained from the new proposal.

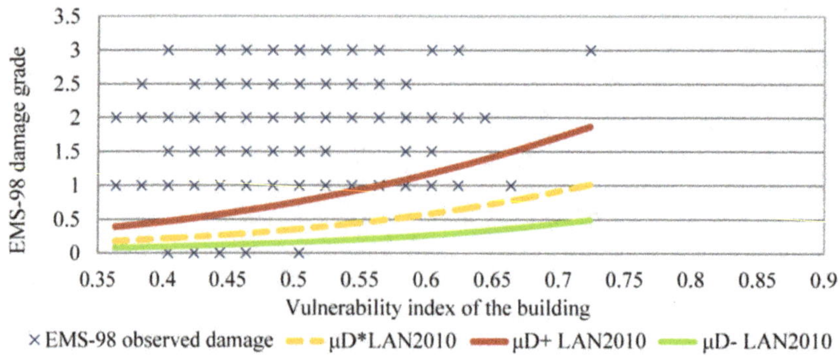

Figure 4: Comparison between EMS-98 observed damage on RC buildings from the Lorca earthquake, 2011, and the mean damage grade obtained from the revised RISK-UE LM1 method ([9], defined as LAN10), for an $I_{EMS\text{-}98}$ = VII earthquake, where $\mu_D{}^*$ is the representative mean damage value, and $[\mu_D{}^-,\mu_D{}^+]$ the possible range of the estimated mean damage grade.

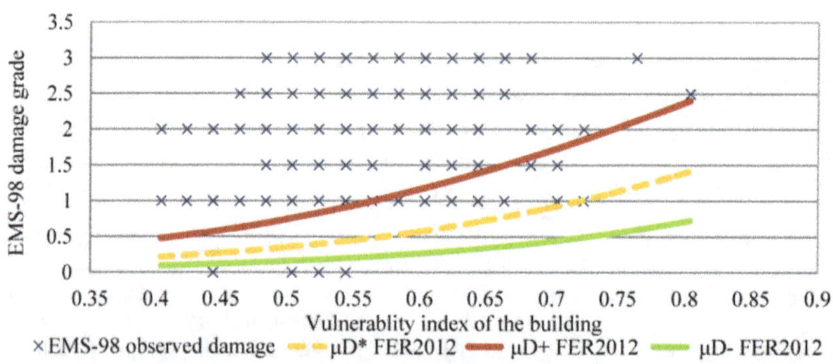

Figure 5: Comparison between EMS-98 observed damage on RC buildings from the Lorca earthquake, 2011, the mean damage grade obtained from the revised RISK-UE LM1 method ([12], defined as FER12), for an $I_{EMS\text{-}98}$ = VII earthquake, where $\mu_D{}^*$ is the representative mean damage value, and $[\mu_D{}^-,\mu_D{}^+]$ the possible range of the estimated mean damage grade.

When comparing the measures of central tendency and dispersion of the difference Z_{ij} - Z_{i_PRO}, some asymmetry arises in the distribution of the obtained data (Table 4). Consequently, a non-parametric Wilcoxon Signed-Rank Test for Paired Samples has been performed. For each of the different revisions of the LM1 method, a null hypothesis of equality of medians in the Z_{ij} - Z_{i_PRO} difference has been formulated. It implies that a significant difference would not be identified between the dispersion among observed and estimated damage obtained from the *j* review of the LM1 method and the dispersion obtained from the proposed model, for the case of the Lorca earthquake.

Thus, the median of the difference between observed and estimated damage obtained according to the proposed model would be less than the median obtained according to the currently available models of the LM1 method. By means of the value of the statistical parameter Z_{cal} – which follows a normal distribution according to the test methodology – the

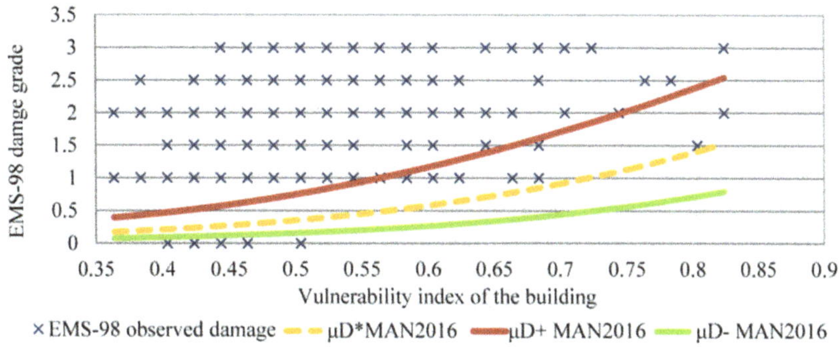

Figure 6: Comparison between EMS-98 observed damage on RC buildings from the Lorca earthquake, 2011, and the mean damage grade obtained from the revised RISK-UE LM1 method ([11], defined as MAN16), for an $I_{EMS\text{-}98}$ = VII earthquake, where μ_D^* is the representative mean damage value, and $[\mu_D^-, \mu_D^+]$ the possible range of the estimated mean damage grade.

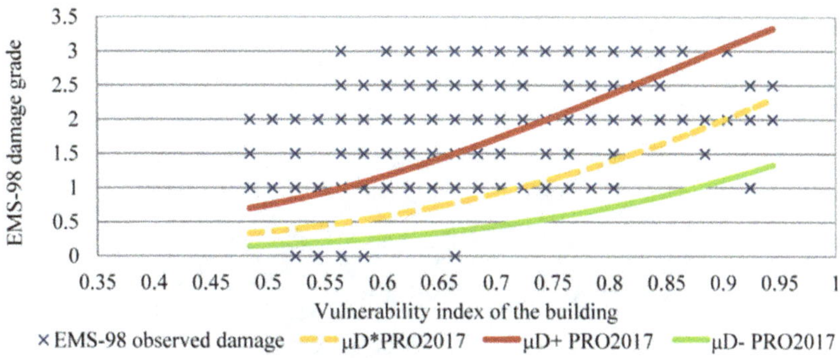

Figure 7: Comparison between EMS-98 observed damage on RC buildings from the Lorca earthquake, 2011, and the mean damage grade obtained from the proposed RISK-UE LM1 model (defined as PRO17), for an $I_{EMS\text{-}98}$ = VII earthquake, where μ_D^* is the representative mean damage value, and $[\mu_D^-, \mu_D^+]$ the possible range of the estimated mean damage grade.

Table 4: Central tendency measures of the difference Z_{ij} - Z_{i_PRO} comparing the proposed model with the current available review methods.

	MIL-03-PRO-17	GIO-05-PRO-17	LAN-10-PRO-17	FER-12-PRO-17	MAN-16-PRO-17
Sample mean	0.247	0.478	0.426	0.200	0.288
Sample standard dev.	0.367	0.378	0.382	0.285	0.249
Sample median	0.165	0.384	0.324	0.159	0.238

p-value can be obtained. With a p-value < 0.05, the null hypothesis of equality of medians is refuted for a statistical significance level of $\alpha = 0.05$ for each model (Table 5). Therefore, less dispersion between observed and estimated damage would be observed in comparison with the difference obtained according to the current reviews of the method, for the case of RC buildings analysed in Lorca.

The visualization of vulnerability indexes and expected damage by using GIS techniques from a geo-referencing process enables to identify the most vulnerable areas of the city and the planning and management of the different post-earthquake emergency situations. The geospatial distribution of the vulnerability indexes of RC buildings analysed in Lorca, obtained from the model of LM1 methodology proposed in this contribution, is shown in

Table 5: Development of Wilcoxon Signed-Rank Test for Paired Samples. *T* and *Zcal* correspond to statistical variables defined in the test methodology.

	MIL-03-PRO-17	GIO-05-PRO-17	LAN-10-PRO-1	FER-12-PRO-1	MAN-16-PRO-1
T=min(*T*+,*T*-)	11,071	159	2,070	7,501	1,550
Zcal	−10.89	−16.29	−15.42	−11.55	−14,99
Zα (One tail)	−1.64	−1.64	−1.64	−1.64	−1.64
p-value	0.000	0.000	0.000	0.000	0.000

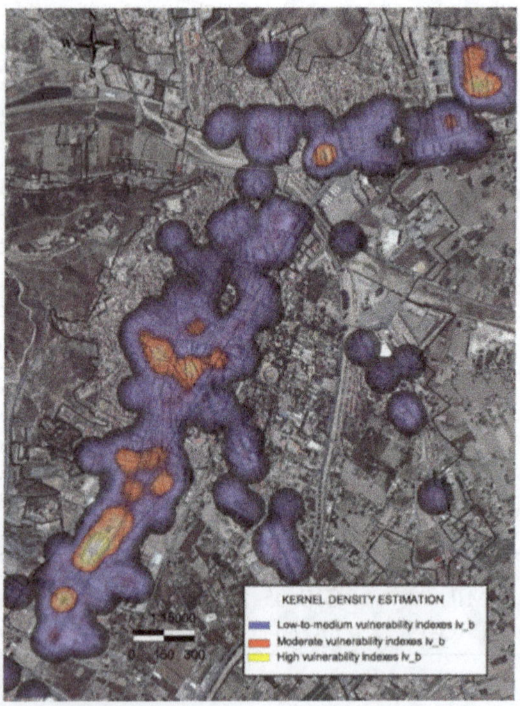

Figure 8: Kernel density estimation plot of the vulnerability indexes I_{v_b} of RC buildings analysed in the city of Lorca, obtained according to the proposed LM1 model.

Fig. 8, where a Kernel density estimation has been used. In the case of Lorca, the modelled damage reflects that the areas most vulnerable are mainly concentrated in the northeast, the central and the southwestern neighbourhoods of the city.

5 CONCLUSIONS

This article proposes to modify the empirical method of seismic vulnerability assessment in urban areas. In the RISK-UE LM1 framework, the behaviour modifiers related to the level of seismic code design, the number of storeys above ground, and the structural joint between adjoining buildings have been modified according to a review of the currently available models of LM1 methodology and an evaluation of the actual seismic performance of buildings.

This proposal provides continuity to the progress of the previous works published to date, simplifying the application of the methodology to any urban area. It can be implemented from cadastral information and an on-site inspection of the evaluated zone. The generalization of the proposal has been simplified by removing one parameter – the regional modifier – that requires calibration depending on the area analysed.

A better correspondence between mean damage grades estimated in RC buildings and the corresponding actual observed data reported from the Lorca (Spain) earthquake, 2011 is identified, with a significance level of 5%, thereby reducing the level of uncertainties associated with this type of methodology and providing a more accurate evaluation of seismic risk.

Despite the fact that such comparisons can hardly provide the basis for validation of an earthquake loss model, the observed tendency shows that the damage simulated by different methods remains below the actual observed damage in most cases, so the definition of the expected losses scenarios is not totally adequate. The existence of well-documented material on the damages caused by earthquakes (such as the Lorca earthquake in 2011), which does not happen too often, provides an opportunity to improve the methodologies of vulnerability assessment and help evaluate the seismic risk of buildings in urban areas.

By using GIS techniques, the proposal also enables for better planning and management of post-earthquake situations, characterizing the different damage scenarios more precisely.

REFERENCES

[1] Applied Technology Council, *FEMA P-58, Seismic Performance Assessment of Buildings*, FEMA: California, 2012.

[2] García-Ayllón, S. & Tomás, A., Actions, diagnosis and territorial management of the disaster in the city of Lorca after earthquake of May 11, 2011. In Brebbia, C. (ed.) *Risk Analysis IX. WIT Transactions on Information and Communication Technologies*, WIT Press: Southampton, pp. 381–392, 2013.

[3] Barbat, A., Carreño, M., Pujades, L., Lantada, N., Cardona, O. & Marulanda, M., Seismic vulnerability and risk evaluation methods for urban areas: a review with application to a pilot area. *Structure & Infrastructure Engineering*, **6**(1), pp. 17–38, 2010. https://doi.org/10.1080/15732470802663763

[4] Mouroux, P. & Le Brun, B., Presentation of RISK-UE Project. *Bulletin of Earthquake Engineering*, **4**(4), pp. 323–329, 2006. https://doi.org/10.1007/s10518-006-9020-3

[5] Lagomarsino, S. & Giovinazzi, S., Macroseismic and mechanical models for the vulnerability and damage assessment of current buildings. *Bulletin of Earthquake Engineering*, **4**(4), pp. 415–443, 2006. https://doi.org/10.1007/s10518-006-9024-z

[6] Benedetti, D., Benzoni, G. & Parisi, M., Seismic vulnerability and risk evaluation for old urban nuclei. *Earthquake Engineering & Structural Dynamics*, **16**(2), pp. 183–201, 1988.
https://doi.org/10.1002/eqe.4290160203
[7] Milutinovic, Z. & Trendafiloski, G., WP04: Vulnerability of current buildings. *RISK-UE project: an Advanced Approach to Earthquake Risk Scenarios with Applications to Different European Towns: Skopje*, 2003.
[8] Giovinazzi, S., *The Vulnerability Assessment and the Damage Scenario in Seismic Risk Analysis*, PhD thesis, Florence, 2005.
[9] Lantada, N., Irizarry, J., Barbat, A., Goula, X., Roca, A., Susagna, T. & Pujades, L. Seismic hazard and risk scenarios for Barcelona, Spain, using the RISK-UE Vulnerability Index Method. *Bulletin of Earthquake Engineering*, **8**(2), pp. 201–229, 2010.
https://doi.org/10.1007/s10518-009-9148-z
[10] Feriche, M., *Elaboration of Seismic Damage Scenarios in the City of Granada*, PhD thesis, Granada, 2012. (in Spanish).
[11] Martínez-Cuevas, S. & Gaspar-Escribano, J., Reassessment of intensity estimates from vulnerability and damage distributions: the 2011 Lorca earthquake. *Bulletin of Earthquake Engineering*, **14**(10), pp. 2679–2703, 2016.
https://doi.org/10.1007/s10518-016-9913-8
[12] Grünthal, G., *European Macroseismic Scale EMS-98*, Centre Européen de Géodynamique et de Séismologie, Luxembourg, 1998.
[13] National Geographic Institute (IGN), *Lorca Earthquake on May 11, 2011*, *Technical Report*, Madrid, 2011. (in Spanish).
[14] De Luca, F., Verderame, G., Gómez-Martínez, F. & Pérez-García, A., The structural role played by masonry infills on RC building performances after the 2011 Lorca, Spain, earthquake. *Bulletin of Earthquake Engineering*, **12**(5), pp. 1999–2026, 2013.
https://doi.org/10.1007/s10518-013-9500-1
[15] Andreini, M., De Falco, A., Girenisi, L. & Sassu, M., Structural damage in the cities of Reggiolo and Carpi after the earthquake on May 2012 in Emilia Romagna. *Bulletin of Earthquake Engineering*, **12**(5), pp. 2445–2480, 2014.
https://doi.org/10.1007/s10518-014-9660-7
[16] Romao, X., Costa, A., Paupério, E., Rodrigues, H., Vicente, R. & Varum, H., Field observations and interpretation of the structural performance of constructions after 11 May 2011 Lorca earthquake. *Enginering Failure Analysis*, **34**(1), pp. 670–692, 2013.
https://doi.org/10.1016/j.engfailanal.2013.01.040
[17] Manfredi, G., Prota, A., Verderame, G., De Luca, F. & Ricci, P., 2012 Emilia earthquake, Italy: reinforced concrete buildings response. *Bulletin of Earthquake Engineering*, **12**(5), pp. 2275–229, 2014.
https://doi.org/10.1007/s10518-013-9512-x

SEISMIC ANALYSIS AND DESIGN ASSISTED BY NUMERICAL SIMULATION OF SLENDER STEEL PORTAL FRAME STRUCTURES

[1]ALIA O. M. AHMED & [2] NIGEL D. P. BARLTROP
[1] Sudan University of Science & Technology, Sudan.
[2]University of Strathclyde, UK.

ABSTRACT

Lightweight (thin-walled or cold-formed) steel portal frame structure could be a popular and effective alternative to the traditional hot rolled structure and, with care to avoid buckling, could be used in earthquake areas owing to its economy and ease of fabrication and transportation, but no recommendations for seismic design of these structures is provided in the design codes. Accordingly, there is need for a lightweight design that is suitable for earthquake areas, which could be transported using lighter vehicles and erected quickly using smaller plant than is required for conventional hot rolled sections following an earthquake. The present paper shows some stages in the development of an earthquake-resistant frame, designed for by combining numerical finite element investigations with analytical check calculations based on EN 1993-1-3 for cold-formed steel members and EN 1993-1-1 for design of steel structures to estimate the loads on the frame structure within the use of EN 1998-1seismic design requirements. Although the initial buckling modes have been avoided, the frame still needs further modification to improve its ductility. It is planned to use this work to assist with the development of performance-based design recommendations for future structures that cover both thin-walled steel and cold-formed steel portal frame structures.
Keywords: cold-formed steel, portal frame, seismic design, steel, thin-walled structures

1 INTRODUCTION

Portal frame structures are widely used and have become the most often used structures for single storey non-residential buildings. However, there is a need for a lightweight design that is suitable for potential earthquake areas that could be easily transported and erected quickly following an earthquake and used for shelter, schools, hospitals, refugee camps etc. Likewise, they ought to be capable of resisting large aftershocks and could become a permanent building but might be removed and reused. Very light weight thin gauge steel construction is light but is likely to buckle under large deflections. Such a characteristic is not desirable during earthquakes.

In addition to that despite the establishment and development of codes for seismic design, they still lack design guidance for some types of structures. Lightweight steel portal frame structure systems could be an effective alternative to heavy steel, but no recommendations for seismic design of these structures is provided in the design codes. The design rules in the European code for cold-formed steel design [1] do not contain specific recommendations for seismic design of these structures. EN 1998-1 [2] does not specifically mention the use of thin-walled steel sections for seismic-resistant structures. However, it classifies thin gauge structures as low dissipative structures (e.g. low ductility) with a behaviour factor q of values from 1.5 to 2.0. Assuming that such type of structures are made by 'elastic' sections (e.g. class 3 or class 4), a behaviour factor q greater than 1.0 can be justified by overstrength and structural redundancy.

The present paper attempts to assist with the lack of seismic design guides of these structures by providing some recommendations for design rules for such type of structures.

DOI: 10.2495/CMEM-V6-N5-899-909

2 REVIEW ON SEISMIC PERFORMANCE OF THIN-WALLED PORTAL

Previous researchers have stated that light gauge steel structures, made with class 3 or class 4 sections, fabricated by cold forming or thin plate welding can be effectively used in seismic-resistant structures mainly due to their reduced weight/strength ratios [3] and [4].

Traditional capacity design based on equivalent elastic static analysis with reduction factors q of values $1< q < 2$ can be used provided the overstrength of joints and structural redundancy are available [3].

Seismic response of light-gauge steel framing can be significantly improved if shear walls are used to resist horizontal forces [3]. Both experimental and numerical results by researchers in the literature sustain to classify light-gauge steel structures as low-dissipative (e.g. class L, q=1.5–2.0), as specified in EN 1998-1 [2]. Previous research showed the possibility of using light gauge steel as earthquake-resistant structure but with recommendation to use shear walls as a lateral resisting system due to the problems of stability and imperfection of light gauge steel.

3 NUMERICAL INVESTIGATION

This section presents computational simulation techniques using the commercially available finite element software ANSYS [5]. The main objective of this study is to investigate the response of a frame model to horizontal displacements, which will be used for the seismic design of the frame combined with code calculations check.

3.1 Proposed models for the study

The two-pitched roof single-storey frame building models considered in this study have the same overall structural configuration of nine identical, equally spaced, moment resistant frames. The building is 40 m long by 20 m wide for all the models. The steel frames are fixed at their columns bases and have a span of 20 m and are spaced at 5 m, which leads to 8 bays or 9 frames (Fig. 1). The purlins are spaced equally at about 1.5 m and span between the steel frames. The columns are 6 m high and the distance from ground level to the apex of the frame is approximately 10 m for the first model (Fig. 2), while it is 6.875 m for the last model (Fig. 4).

The key features of this structural system were the use of creative built up cold-formed steel sections made from largest back-to-back lipped channel that can currently be rolled for the column and rafter members.

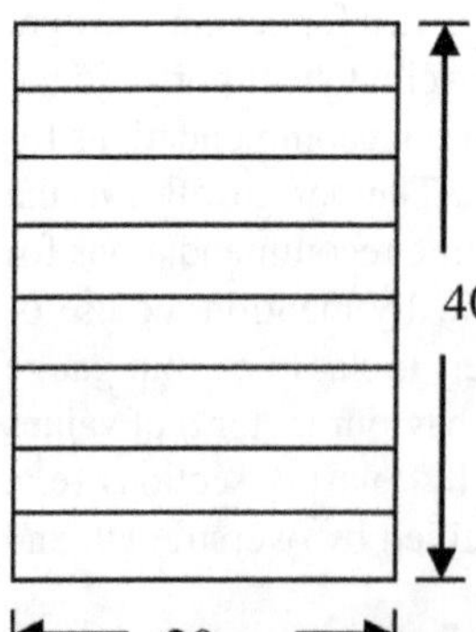

Figure 1: Plan of the building

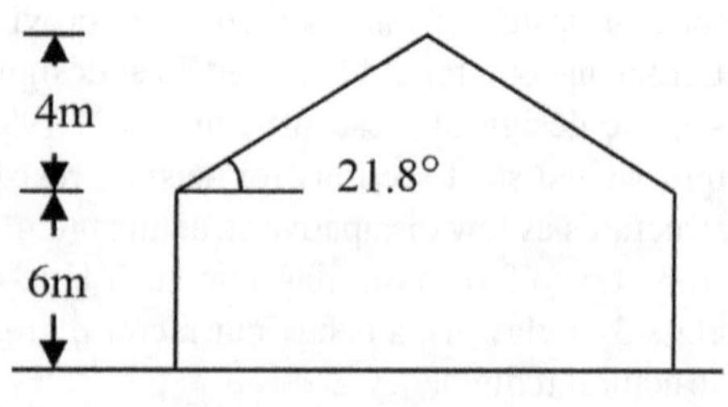

Figure 2: Portal frame for first model

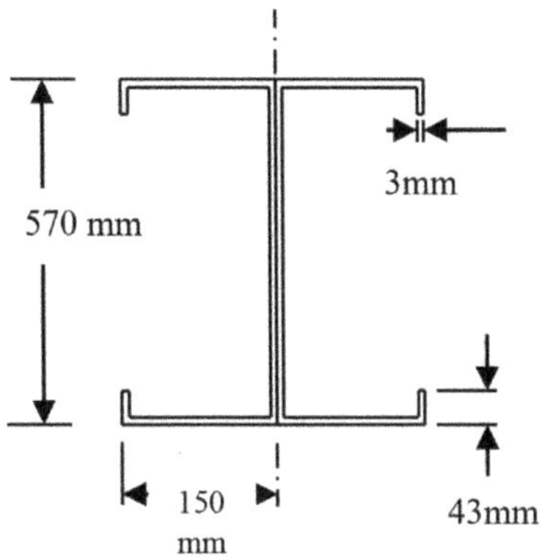

Figure 3: Dimensions of back- to back lip channel section used for column and rafter member for first model

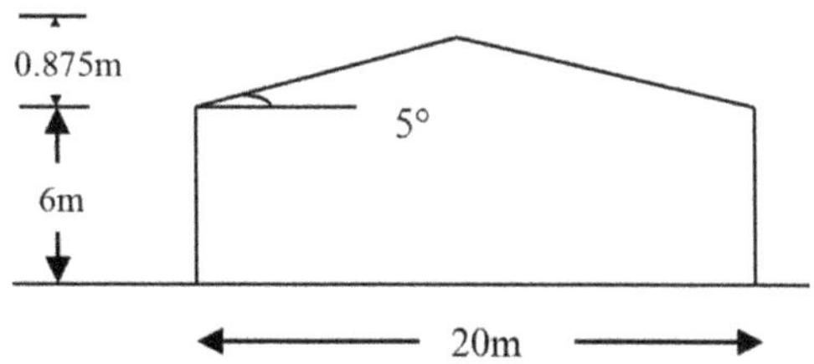

Figure 4: Portal frame for last model

The rafters in each frame for the first model have 570 mm total depth, the upper and lower flange width is 300mm, the thickness is 3mm and there is a lip of 43 mm. The columns of the frame have the same dimension of the rafters for the first model (see Fig. 3).

The rafters in each frame for the last modified model have 750 mm total depth, the upper and lower flange width is 500 mm with thickness 3 mm and lip of 53 mm (see Fig. 5).

The columns of the frames have 850 mm total depth and 6 mm web thickness, the upper and lower flange width is 650mm with 3mm thickness and 63 mm lip.

The first model used for the numerical investigation and used for the calibration of the seismic design is a frame of span 20 m with rigid joints and fixed at column bases, pitch 21.8° as was explained in previous section (see Fig. 2).

The brackets for the joints for ridge and knees are 5 mm thickness from cold-formed steel (see Figs 6 & 7). The proposed connections are bolted connections. The connected members are connected together through cold-formed steel plate using welded bracket elements

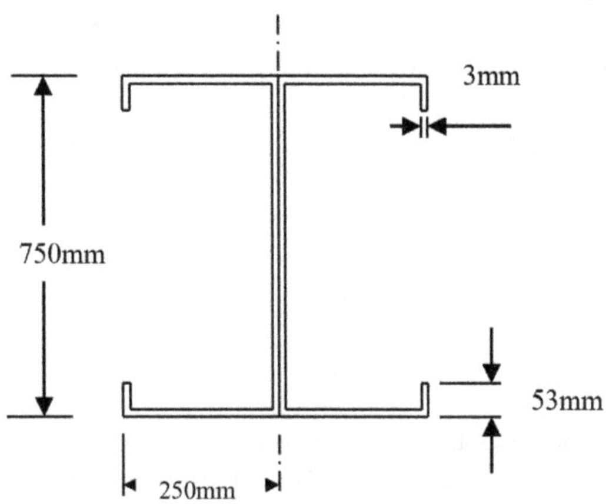

Figure 5: Dimensions of back to back lip channel section for rafter member for last model

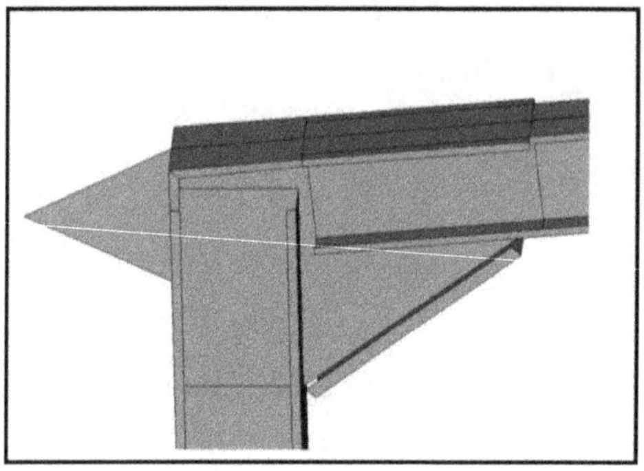

Figure 6: ANSYS model for knee connection

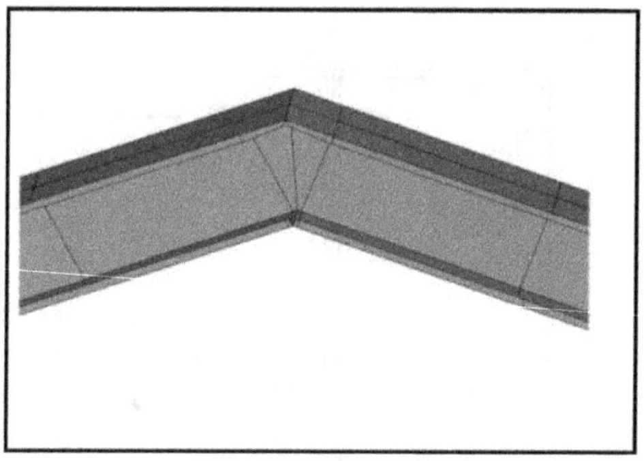

Figure 7: ANSYS model for ridge connection

(S355: Fy = 355 N/mm2) and M20 grade 8.8 bolts. To achieve a good performance of the connections, the plate was enhanced with folding stiffeners.

A gap in the lower part of the plate was applied to avoid the interaction between the folding stiffeners and the frame members for the finite element model. For simplicity, the connections for knees and ridge were modelled in the finite elements model as rigid.

Extra, artificial external triangular brackets were added at the eaves as shown in Fig. 8 to prevent concentration of the stresses when applying loads at the corner nodes.

3.2 Finite elements modelling of the frame

The commercially available finite element software ANSYS [5] was used for the analyses of the frame structure models in this research project. Rafters, columns and brackets are modelled using the four-noded thin shell element SHELL181. SHELL181 is well-suited for linear, large rotation, and/or large strain nonlinear applications. The finite element mesh used for the frame is shown in Fig. 8. The mesh size used in the finite element model for the shell elements was around 25x25mm.

The inelastic behaviour of the steel elements was considered using the von Mises yield criterion with isotropic yielding. Geometric nonlinearities: large displacement and P-δ effects were considered in the analyses. The stress-strain relationship is assumed to be elastic-perfectly plastic with a Young's modulus of 205E3 MPa up to the yield strength followed by strain hardening with value of 1% of the Young's modulus, i.e. the tangent modulus is 2.05E3 MPa and a yield strength of 400 MPa.

The analysis involves subjecting the frame to monotonically increasing static lateral displacements as it is an efficient tool to describe the behaviour of the frame beyond the plastic zone or within the strain hardening region. The lateral displacements were used first to test the response of the frame to lateral displacement until failure as the earthquake effect on the structure is assumed to be horizontal without considering the gravity load.

3.3 Some checking of the finite elements used

To validate the finite element models, the analysis was carried for a steel thin plate of 350 mm length, 350 mm wide, 6 mm thick with Young's modulus, E = 205 MPa and poison ratio 0.3. The plate was assumed to be made of cold-formed steel and modelled with type 181 shell elements.

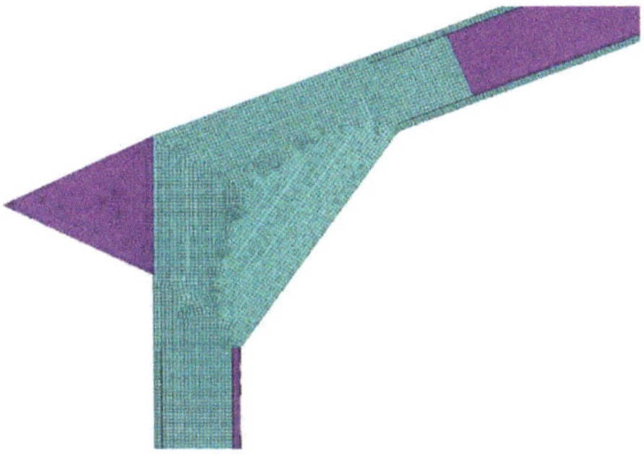

Figure 8: Finite element model for the frame

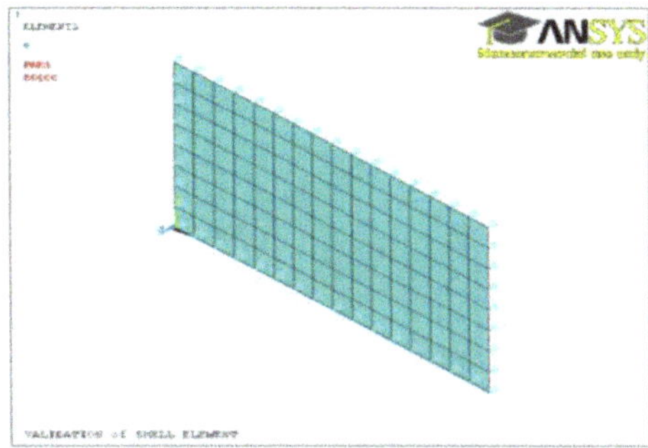

Figure 9. Plate model with shell elements (SHELL181).

The material was modelled as bilinear isotropic elastic-perfectly plastic. The inelastic behaviour of the steel elements was considered using the von Mises yield criterion as von Mises yield surface allows isotropic yielding. Geometric nonlinearities, large displacement and P-δ effects were considered in the analyses.

The plate was modelled with element mesh size about 12.5 mm for a uniform mesh of 128 SHELL181 shell elements, as shown in Fig. 9.

For verification of the model, a nonlinear buckling analysis was carried out to predict the critical load for buckling of a thin flat plate. Where the load is increased until the solution fails to converge, indicating that the structure cannot support the applied load (or that numerical difficulties prevent solution). In this case, the nonlinear geometry (nonlinear elastic) is considered without counting for nonlinear materials (nonlinear plastic).

This analysis resulted in a value of 71.0KN/m for the expected buckling load. This shows a good agreement with thin plate buckling theory that predicts it will buckle when the applied load is 75.9 KN/m.

The nonlinear method is the more accurate type of buckling analysis as it predicts the buckling load when the plate starts to buckle. This result in a smaller value for the expected buckling load obtained by nonlinear buckling analysis compared with the thin plate buckling theory.

3.4 Finite elements frame analysis results

The force displacement diagram is shown in Fig. 11, the structure had a considerable nonlinearity as the displacement after yielding value about 47 mm with yielding value about 33 mm. As shown in (Fig. 10) the structure appears to be buckling, in a lateral-torsional mode, at the right hand knee and failed due to the lateral-torsional buckling of the rafter and columns. The right side of the frame suffered worse torsional buckling than the left side because of the effect of the loading directions. The torsional buckling of structural members should be prevented in order to have a good seismic design. The improvement of the frame could be only in terms of frame members sections without considering the connections as the connections performed very well.

3.5 Concluding remarks

These results gave a general understanding for the behaviour of the frame structure under horizontal displacements, which should reflect the frame structure response and resistant to

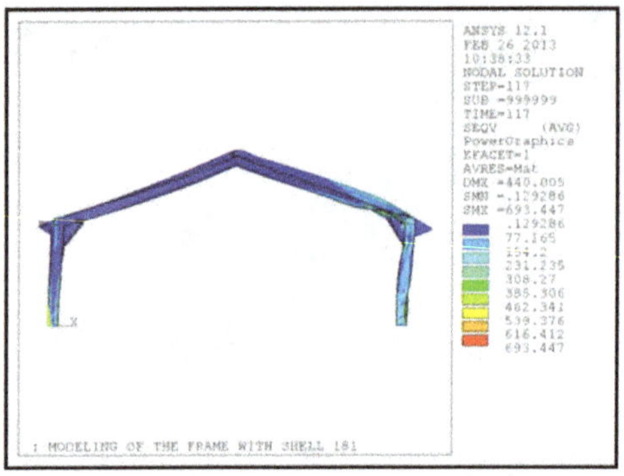

Figure 10. Von Mises stress (MPa)

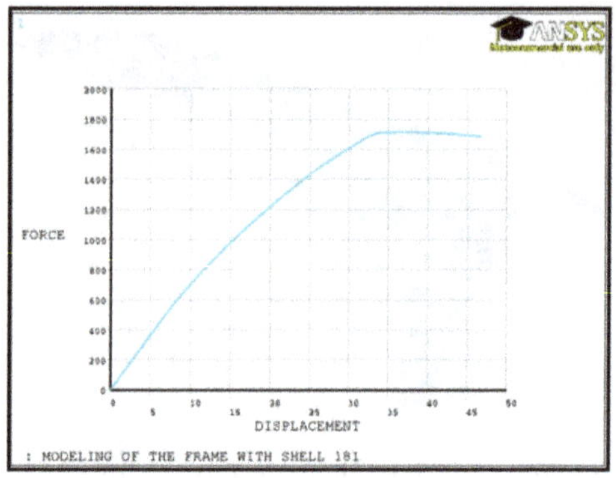

Figure 11. Force (N) vs. displacement (mm)

earthquake motion. The frame design in this study for the earthquake resistance will be the result of combining evaluation of structural performances by these numerical finite element investigations with check calculations based on EN 1993-1-3 [1] loading code.

The section used for the last model is built up cold-formed steel sections made from largest back-to-back lipped channel as detailed in present paper in section 3.1. The sections for column and rafter are chosen as result of the numerical finite element investigation, and then are checked by code calculation for different requirements to EN 1993-1-3 [1] and EN 1998-1 [2]. Use is made of EN 1993-1-3 [1] for cold-formed steel members and EN 1993-1-1 [6] for design of steel structures to estimate the loads on the frame structure within the use of EN 1998-1 [2] for the ductility class and behaviour factor for the proposed last model.

4 ANALYTICAL CALCULATIONS TO EUROCODE

4.1 Earthquake analysis of frame to eurocode

The elastic analysis of the buildings involved the analytical cold-formed light gauge steel model with respect to the steel strength and ductility class as it is steel grade S355 with ductility class low. The ductility class is used to estimate the behaviour factor, which is a reduction factor used for the estimation of the resistance to seismic forces.

The joints were assumed to be; fixed for column bases and rigid for beam to column. This model was analysed and the action effects for each individual member of the building were obtained for the two load combinations in eqns (1) and (2) according to Eurocode, using the commercially available structural analysis software (ANSYS).

$$Ed = (1.35G + 1.5Q) \tag{1}$$

$$Ed = \left(G + \sum \varphi_{2,i} + E_i\right) \tag{2}$$

The analysis for earthquake and the estimation for the forces for the initial design were carried out according to EC8 code. Both the equivalent lateral force procedure and the modal spectrum procedure were used for determination of base shear according to EN 1998-1 [2]. The shape of design spectrum obtained by EN 1998-1 [2] was as shown in Fig. 12. The value used was the value resulted from the static procedure after considering the torsional effects.

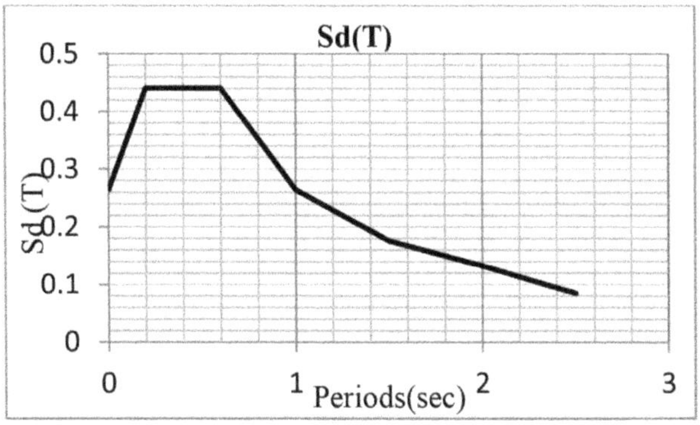

Figure 12. Design spectrum for elastic analysis to EN 1998-1 [3]

The imposed loads on the buildings were considered as a uniform load on the spans. Upon determination of the action effects per each element, the design of beam (rafter) or column section was carried out '"manually'.

The second order effects (P-δ effects) were ignored according to EN 1993-1-1 [6] as it has been found that the resulting value of α_{crest} is greater than 10, for that first order analysis used for the structure. Where α_{crest} is the factor by which the design loading would have to be increased to cause elastic instability in a global mode. Since the frame was sensitive to buckling in a sway mode, the global initial sway imperfection was allowed and according to EN 1993-1-1 [6].

4.2 Design of frame members to eurocode

In EN 1998-1 [2], light gauge steel or the thin-walled steel are classified as low dissipative structure, the ductility class, is recognised for low dissipative structural behaviour as low (DCL). According to EN (European Standard) EN 1998-1 [2], for low dissipative structures, the design forces may be calculated on the basis of an elastic global analysis without taking into account any significant non-linear material behaviour.

The frame structure, studied herein is a regular structural system which meets the criteria for regularity in plan and in elevation and satisfies the geometrical constraints of EN 1998-1 [2].

For dissipative zones in the case of columns, the normalised design axial force NEd for this frame structure conformed to $\frac{N_{Ed}}{N_{pl,Rd}} < 0.3$ which is well below the limiting values for the low DC building, as it is found to be 0.02 for the elastic design. Hand calculations for the design procedures of the frame structure was carried according to EN 1993-1-3 [1] and EN 1993-1-1 [6].

4.3 Column design and verification

The section 850x650x63x3 – back-to-back lip channel section – cold-formed steel grade S355 was selected for the column member and the verification of the member was carried out for the critical load combination. Since for design use of the critical load combination is made, the design actions for columns were those resulting from the seismic combination and, as pointed out earlier, the normalised design axial force N_{Ed} was found to be well below the limiting values specified in Eurocode 8 for the low ductility class. Many checks for the section of column were carried out to verify the ability of the column to carry the service load during the service life of the building such as resistance of the cross-section for shear, compression and bending moment in accordance with EN 1993-1-3 [1] for cold-formed steel and EN 1993-1-1 [6] for steel. Columns also have been verified as having sufficient resistance against lateral and lateral torsional buckling in accordance with EN 1993-1-3 [1] and EN 1993-1-1 [6].

4.4 Rafter Design & Verification

The section 750x500x53x3 – back–to-back lip channel section – cold formed steel grade S355 was selected for the rafter member and the verification of the member was carried out for the critical load combinations, as for the column member. The design of the rafters was performed for the gravity load combination Ed = (35G + 1.5Q), for the low ductility class examined, which yielded larger bending moments than the seismic load combination. Many checks for the section of rafter were carried out to verify the ability to carry the service load such as resistance of the cross section for shear, compression and bending moment in accordance with EN 1993-1-3 [1] for cold-formed steel and EN 1993-1-1 [6] for steel. Rafters also have been verified as having sufficient resistance against lateral and lateral torsional buckling in accordance with EN 1993-1-3 [1] and EN 1993-1-1 [6]. (These are of paramount importance for earthquake resistance).

5 THE PROPOSED DESIGNED MODEL OF PORTAL FRAME STRUCTURE

5.1 Model description

The model used herein is a result of calibration of the seismic design using numerical investigation and calculations to the design rules in the European codes specifications. The frame building model is frame of span 20 m with rigid joints and fixed at column bases, pitch 5°.The columns are 6 m high and the distance from ground level to the apex of the frame is approximately 6.875 m (see Fig. 4). The brackets for the joints for ridge and knees are 5 mm thickness from cold-formed steel (see Figs 13 and 14). The connection used here as same as

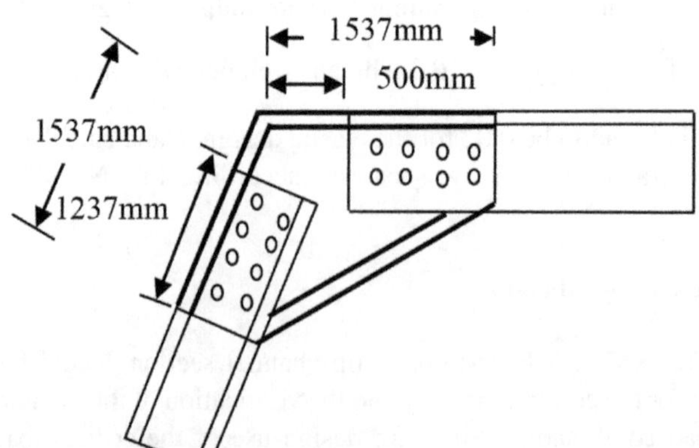

Figure 13. Knee arrangement for the last model

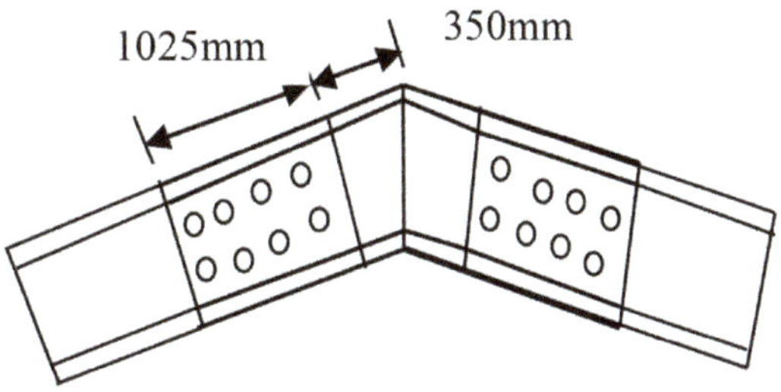

Figure 14. Ridge arrangement for the last model

the connections used in the first model in Figs 6 and 7 for both ridge and knees as this arrangement of connections showed good performance when the frame model was tested to lateral displacements loading. The model was defined with the same materials properties, type of element, material modelling and finite element idealisation like first model using ANSYS software; to check response of the new proposed designed modified model of the frame structure for the nonlinear static analysis under increasing lateral displacements until the failure of the frame structure.

5.2 Finite elements results of the proposed designed model

Since the frame members are sensitive to buckling, the frame needed to be restrained to prevent lateral and lateral torsional buckling by providing the frame members with restraint members in areas sensitive to buckling. These restraints include purlins as beam (rafter) restraints and rails as column restraints.

The restraints include lateral restraint, torsional restrains and intermediate restraints. These effects have been applied in the finite element model for the light gauge steel portal frame in ANSYS as spring element known in ANSYS as COMBIN40. This spring element has one degree of freedom at each node, either a nodal translation, rotation, pressure, or temperature.

The analysis resulted in the absence of the torsional buckling of the frame sections and the failure was seen to be by the local buckling of the flange upper part of the right column with yielding of the column base due to buckling of the flange and the web of the column as shown in Figs 15 and 16. This caused the failure of the frame structure. The force displacement

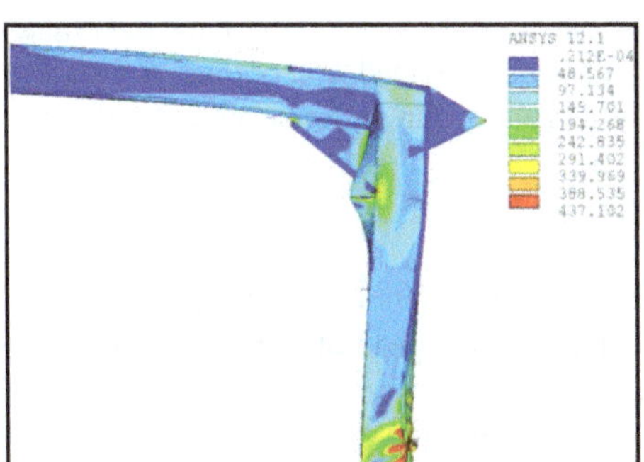

Figure 15: Von Mises stress (MPa) right.

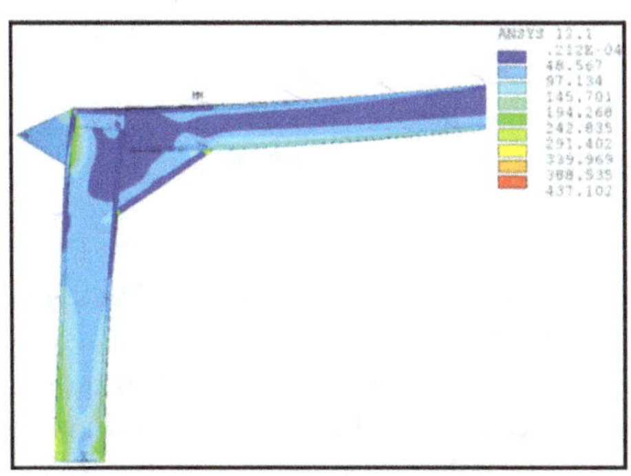

Figure 16: Von Mises stress (MPa) left.

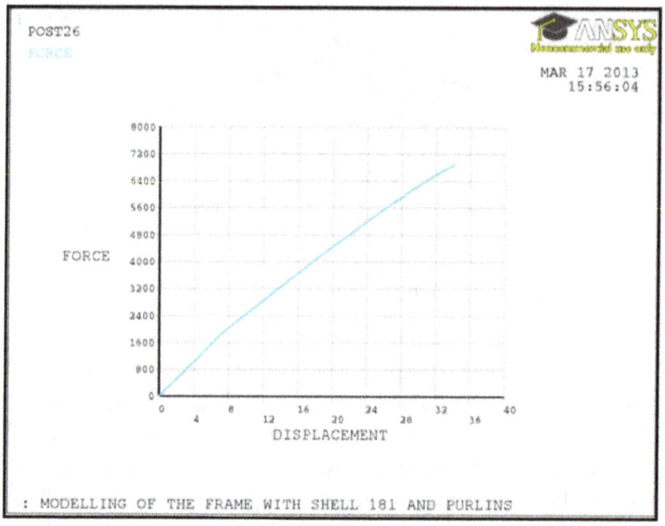

Figure 17: Force (N) vs. displacement (mm) diagram.

relationship showed ductility with value of 34 mm beyond the plastic zone and at the yielded point is 8 mm displacement as shown in Fig. 17. Due to the column buckling, a premature failure happened to the structure before yielding into plastic mechanism.

According to this result, the frame structure needs to be improved to be protected against this kind of premature failure which is not desirable in earthquake resistant structures.

6 ONCLUSIONS

From the first investigation, the importance of lateral restraints for this type of structure, when subject to horizontal forces or earthquake action, was noticeable: lateral torsional buckling is the main cause of failure for this type of structure, if it is not well restrained. Therefore, the design utilized lateral restraints directly or indirectly to the purlins.

The results of the numerical investigation combined with the code calculations checks resulted in an optimization of the model for the purpose of the study. The potential benefits of utilising light gauge steel or slender steel for the construction of portal frame buildings in earthquake prone areas were examined. The modes of failure of the designed frame structure were investigated with a three-dimensional detailed finite element model. This demonstrated that, as was expected, premature failure of the frame is due to buckling of the slender frame section profiles. Such a mode of failure is not desirable in an earthquake-resistant structure as the capacity drops very suddenly when the structure is overloaded. In contrast, a structure with more ductility has the capability to absorb energy and can better survive the shaking from the ground accelerations.

REFERENCES

[1] EN 1993-1-3. "Eurocode 3: Design of Steel Structures. Part 1-3: General Rules. Supplementary Rules for Cold-Formed Thin Gauge Members and Sheeting", CEN- European Committee for Standardization, 2006.

[2] EN 1998-1. "Eurocode 8: Design of Structures for Earthquake Resistance. Part 1: General rules, Seismic Actions and Rules for Buildings", CEN - European Committee for Standardization, 2004.

[3] Dubina, D., Ductility and seismic performance of thin walled cold-formed steel structures. *International Journal of Steel Structures*, **4**, pp. 209–222, 2004.

[4] Ono, T. & Suzuki, T., *Inelastic behaviour and earthquake-resistance design method for thin-walled metal structures*. Proceedings on IABSE Coll. on Thin-Walled Metal Structures in Building, Stockholm, pp. 115–122, 1986.

[5] ANSYS, Inc. *Structural analysis guide*, Release 12.1, 2009.

[6] EN 1993-1-1., *Eurocode 3: Design of Steel Structures. Part 1-1: General Rules and Rules for Buildings*. CEN- European Committee for Standardization, 2005.

IMPACT OF SPATIAL VARIABILITY OF EARTHQUAKE GROUND MOTION ON SEISMIC RESPONSE OF A RAILWAY BRIDGE

RACHID DERBAL[1,2], NASSIMA BENMANSOUR[2] & MUSTAPHA DJAFOUR[2]
[1]Department of Civil Engineering, University centre of Ain Temouchent, Algeria.
[2]Department of Civil Engineering, University of Tlemcen, Tlemcen, Algeria.

ABSTRACT
This paper studies the impact of spatially varying ground motions on the responses of a railway bridge. The evaluation of the seismic hazard for a given site is to estimate the seismic ground motion at the surface. This is the result of the combination of the action of the seismic source, which generates seismic waves, the propagation of these waves between the source and the site, and the local conditions of the site. Firstly, the seismic ground motions are modelled by assuming the base rock motions of the same intensity and modelling them with a filtered Tajimi-Kanai power spectral density function and a spatial ground motion coherency loss function. Then, the power spectral density function of ground motion on surface is derived by considering the site amplification effect based on the one-dimensional seismic wave propagation theory. A comparison between the bridge responses to uniform ground motion, to spatial ground motions with and without considering local site effects is established. Discussions on the seismic ground motion spatial variability and local site conditions effects on structural responses of railway bridge are made.
Keywords: bridge responses, coherency loss, site effects, spatially varying ground motions, Tajimi-Kanai power spectral

1 INTRODUCTION

The seismic analysis of the structures requires a good understanding of the seismic loading. For extended structures, such as bridges, there is the appearance of a phenomenon called spatial variability of ground motions. A good representation of the seismic input must take into account this phenomenon. The sources of this variation have been described [1, 2] as: wave passage effect, wave coherency loss, and site effects.

Many studies have investigated the spatial variability of ground motion and assume that the site is uniform and homogeneous. The only variations are loss of coherency and a phase delay. This hypothesis can not be retained because the soil is heterogeneous. Therefore, the seismic wave propagation is affected in terms of intensity and frequency content.

For this purpose, a good representation of the seismic loading must take into account all the factors of the spatial variability of the ground motion, namely the loss of coherency loss effect, the waves passage effect and, in particular, local site effect.

Some researchers have tried to model the effect of local site conditions on earthquake ground motion spatial variations.

Kaiming *et al.* [3] have proposed an approach where the spatial ground motions are modelled in two steps. Firstly, the base rock motions are assumed to have the same intensity and are modelled with a filtered Tajimi-Kanai power spectral density function and an empirical spatial ground motion coherency loss function. Then, power spectral density function of ground motion on surface of the canyon site is derived by considering the site amplification effect based on the one dimensional seismic wave propagation theory. A discussion on the ground motion spatial variation and local soil site amplification effects on structural responses

DOI: 10.2495/CMEM-V6-N5-910-920

are made. They conclude that the effects of neglecting the site amplifications in the analysis as adopted in most studies of spatial ground motion effect on structural responses are highlighted.

Konakli and Der Kiureghian [4, 5] presented a simulation of the seismic ground motions by two approaches: The conditional approach where the simulated motion is conditioned by an observed acceleration at a site and the unconditional approach where the ground motion is compatible with an estimated spectral density function

The work of Benmansour *et al.* [6–8] evaluated the method proposed by the RPOA (the Algerian bridge seismic regulation code) and compared it with more refined approaches and with the provisions of EC8. They have developed a method of generating of asynchronous seismic ground motion in the sense that the displacement signals can be obtained directly without going through the double integration. Based on the study of several bridges, the results obtained of this work show that the simplified RPOA method overestimates the seismic demand. They proposed to modify some provisions, and the new approach gives results which are in better

Kaiming and Hao [9] developed a method for generating asynchronous seismic ground motions that takes account of local site effects. This method is based on the theory of wave propagation presented by WOLF 1985. As a hypothesis, the motion at the rock base is composed of SH wave (off-plan) or the combination of P and SV waves with an incident angle given. The simulated movements are compatible with the spectral density function of the target response spectrum. They concluded that the proposed method leads to a more realistic modeling of the asynchronous seismic ground motions in sites with different characteristics compared to the hypothesis of identical intensity of the movements.

In this study, the generation model of seismic ground motions described by Benmansour *et al.* [8, 10–13] is developed to take account local site effect. Several recent studies have shown that the local conditions of the site should not be neglected when interpreting the spatial variability of the seismic ground motions. Consequently, the model proposed in this study takes into account all the effects of the spatial variability of the spatially varying ground motion, in particular the site effect. The aim of this work is to evaluate the influence of the site effect on the dynamic response of bridges. The generation of asynchronous seismic ground motion is made for all supports of bridge. Structural responses to uniform ground motion and to spatial ground motion with considering the site effect are calculated and compared. Discussions on the ground motion spatial variation and site effect in terms of the site properties on structural responses are made.

2 BRIDGE AND SPATIAL GROUND MOTION MODEL

2.1 Railway bridge model

Figure 1 illustrates the bridge model on the site, in which CA, P1, P2, P3 and CB are the supports on ground surface. The corresponding points at the base rock are CA', P1', P2', P3'and CB', respectively.

ρ_i, v_i, ξ_i and h_i are the density, shear wave velocity, damping ratio and depth of the soil under support, respectively, where i represents CA, P1, P2, P3, CB, respectively. The corresponding parameters on the base rock are ρ_R, v_R and ξ_R Table 1 gives the values of these parameters.

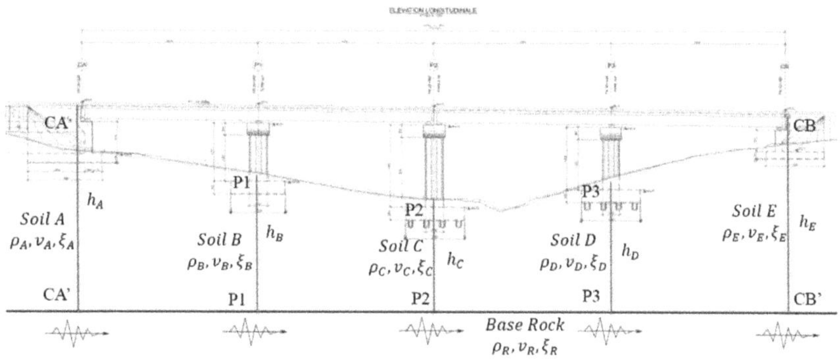

Figure 1: Longitudinal view of a bridge crossing a canyon site.

Table 1: Parameters of soil foundation.

Soil A		**Soil B**		**Soil C**		**Soil D**		**Soil E**	
ρ_A 2500	v_A 800	ρ_B 1500	v_B 300	ρ_C 1500	v_C 200	ρ_D 1800	v_D 400	ρ_E 2000	v_E 500
ξ_A 5%	h_A 59.6	ξ_B 5%	h_B 40.5	ξ_C 5%	h_C 45.6	ξ_D 5%	h_D 50.6	ξ_E 5%	h_E 57.9

2.2 Base rock model

Assume the amplitudes of the power spectral densities at different locations on the base rock are the same and in the form of the filtered Tajimi-Kanai power spectral density function [3, 9]:

$$S_g(\omega) = |H_P(\omega)|^2 S_0(\omega) = \frac{\omega^4}{\left(\omega_f^2 - \omega^2\right)^2 + \left(2\omega_f \omega \xi_f\right)^2} \cdot \frac{\omega_g^4 + 4\xi_g^2 \omega_g^2 \omega^2}{\left(\omega_g^2 - \omega^2\right)^2 + 4\xi_g^2 \omega_g^2 \omega^2} \tag{1}$$

in which $|H_P(\omega)|^2$ is a high pass filter [13, 15], ianai power spectral density function [8], ω_g ωg and ξ_g ξg are uency and damping ratio of the Tajimi-Kanai power spectral density function, Γ is a scaling factor depending on the ground motion intensity, and ω_f and ξ_f are the central frequency and damping ratio of the high pass filter.

I study, it is assumed that $\omega_f = 0.5\pi\ Hz$, $\xi_f = 0.6$, $\omega_g = 6\pi\ Hz$, $\xi_g = 0.6$and $= 0.00565\ \mathrm{m}^2 / s^3$ (see Fig. 1). These values correspond to a peak ground acceleration (PGA) 0.2 g with duration *T* = *20 s* (Der Kreiughian 1980) [3]. Figure 2 shows the power spectral density of the base rock motion.

2.3 Coherency model

The complex coherency function describes the correlation between the amplitudes and phase angles of two ground motion time histories in the frequency domain. This function is defined as (Der Kiureghian, 1996) [3, 14, 15]:

$$\gamma_{jk}(\omega) = \frac{S_{jk}(\omega)}{\sqrt{S_j\left(\omega S_k(\omega)\right)}} \tag{2}$$

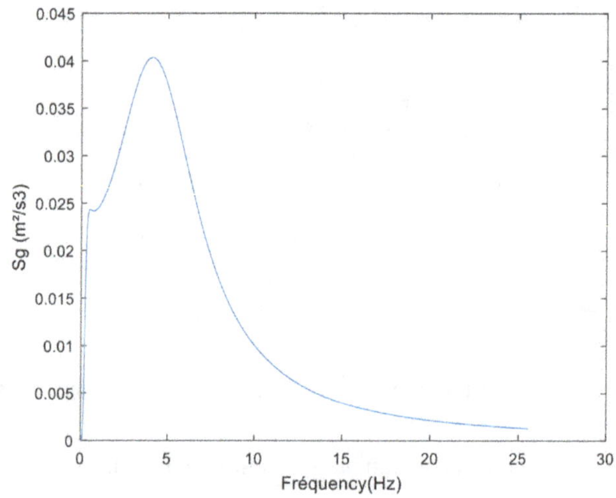

Figure 2: Filtered ground motion power spectral densfution on the base rock.

where ω is the circular frequency; $S_j(\omega)$, Sk(ω) $S_k(\omega)$ are the power spectral density functions of the time histories $g_j(t)$ and $g_k(t)$, respectively; and $S_{jk}(\omega)$ is the cross-power spectral density of the considered time histories.

The coherency function can be written as (Der Kiureghian, 1996) [7]:

$$\gamma_{jk}(\omega) = |\gamma_{jk}(\omega)| exp\left(i - \frac{\omega d_{jk}}{v}\right) \tag{3}$$

where d_{jk} is the projected horizontal distance along the direction of propagation of the waves, which is from station j to station k, and v ishe surface apparent velocity of waves, considered as constant over the frequency range of the wave.

The Hindy and Novak coherency loss model is used to model the seismic ground motion at the base rock:

$$|\gamma(f,\lambda)| = exp(-\alpha(2\pi f \lambda)^{\beta}) \tag{4}$$

The evolution of the loss coherency as a function of the frequency is described in Fig. 3. The parameters of this model are: α =3,007 10^{-4} and β =0.9 [16].

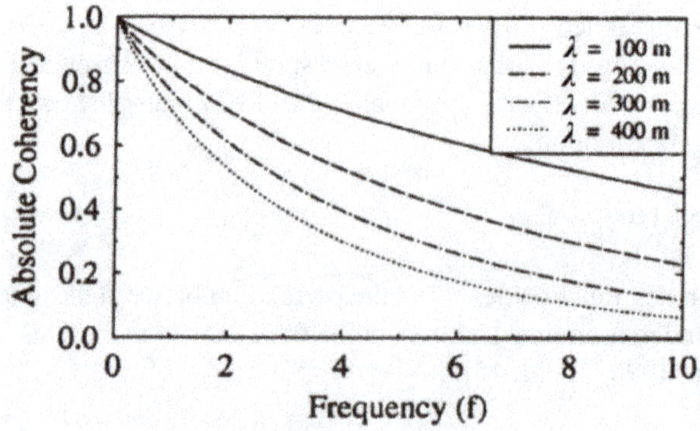

Figure 3: Hindy and Novak coherency loss model [16].

3 SPATIAL GROUND MOTION MODEL

The stationary time series are simulated using the method described by Deotatis (1997) [14], which is as follows.

From eqn (2), the cross-spectral density matrix $S_0(\omega)$ for the stationary process $g_j(t)$; j = 1,2,...,n. is given by [3, 9, 16]:

$$S_{jk}(\omega) = \left[S_j(\omega) S_k(\omega) \gamma_{jk}(\omega)\right] ; j,k = 1,2,\ldots,n. \tag{5}$$

In order to simulate samples of the n-variant stationary stochastic process $g_j(t)$; $j,k = 1,2,\ldots,n$. Its cross-spectral density matrix $S_0(\omega)$ given in eqn (5) is factorized into the following product using Cholesky's decomposition method:

$$S_0(\omega) = H_0(\omega) H^{T^*}(\omega) \tag{6}$$

The elements of $H_0(\omega)$ can be written in polar form as:

$$H_0(\omega) = \left|H_{jk}(\omega)\right| exp\left(i\theta_{jk}(\omega)\right) ; j > k. \tag{7}$$

where:

$$\theta_{jk}(\omega) = tan^{-1}\left(\frac{Im\left[H_{jk}(\omega)\right]}{Re\left[H_{jk}(\omega)\right]}\right) \tag{8}$$

Using eqns (6) and (7) the stationary stochastic vector process $g_j(t)$; $j,k = 1,2,\ldots,n$, can be simulated by the following series as $N \to \infty$ [6, 8]

$$g_j(t) = 2\sum_{m=1}^{n}\sum_{l=1}^{N}\left|H_{jm}(\omega)\right|\sqrt{\Delta\omega} cos\left(\omega_l - \theta_{jm}(\omega_l) + \phi_{ml}\right) \tag{9}$$

where:

$$\omega_l = l\Delta\omega ; l = 1,2,\ldots,N \tag{10}$$

$$\Delta\omega = \frac{\omega_u}{N} \tag{11}$$

N represents the number of the frequency step $\Delta\omega$ needed to reach the upper cut-off frequency ω_u.

The $\{\phi_{ml}\}$; $m = 1,2,\ldots,n$; $l = 1,2,\ldots,N$ appearing in eqn (9) are n sequences of independent random phase angles distributed uniformly over the interval $[0,2\pi]$.

4 NUMERICAL RESULTS AND DISCUSSION

4.1 Finite element model of railway bridge

In this paper, we considered a linear dynamic analysis. A finite element model in 3D was created using a finite element code (see Fig. 4). The bridge, composed of four equal spans, has a total length of 160m. The pile heights are 11.9 m, 17.4 m and 12.9 m respectively. The superstructure of the bridge is connected to the piers by rigid elements (rigid link elements). The deck is supported by four beams of metal frame PRS2300. The deck and the piers are

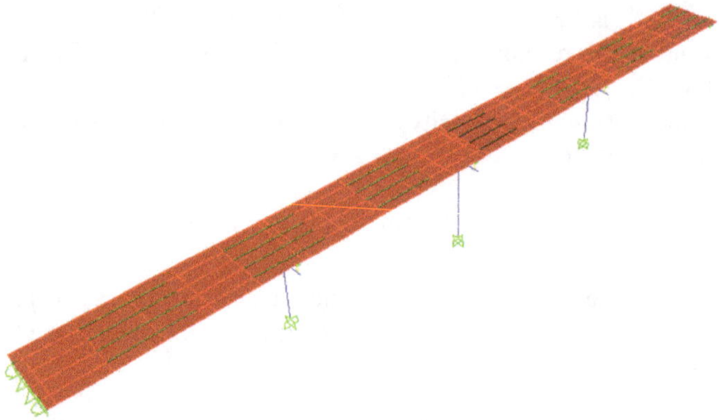

Figure 4: Finite element model of bridge.

assumed to have elastic behavior. The bridge piers are modelled by beam elements at two nodes (Beam element). The bearing devices are modelled in link elements. The conditions of support of the two ends of the bridge are modelled according to the details of fixing the beams provided by the execution plans. The damping ratio of the structure is assumed to be 5%.

4.2 Modal analysis

The first three periods and frequency of the railway bridge are given in Table 2. Figures 5–7 show the shape of the Eigen modes of vibration.

Table 2: Parameters of the three first Eigen modes of vibration.

	Mode 1	Mode 2	Mode 3
Period (s)	0.90561	0.89320	0.73683
Frequency (Hz)	1.10423	1.11957	1.35716

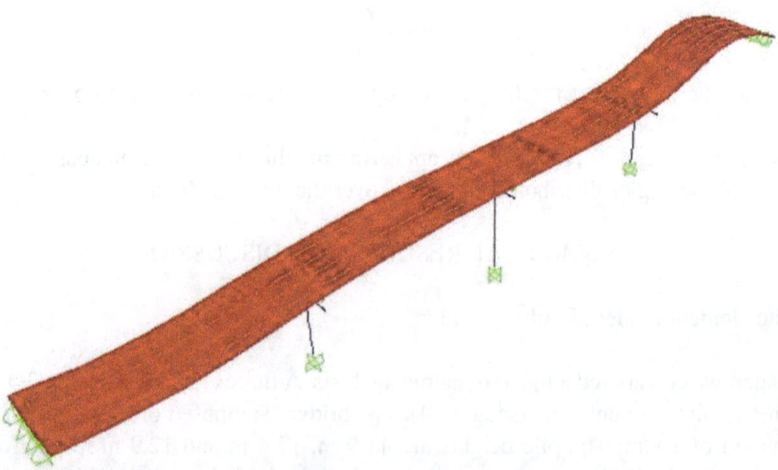

Figure 5: Mode 1: T = 0.90561 s.

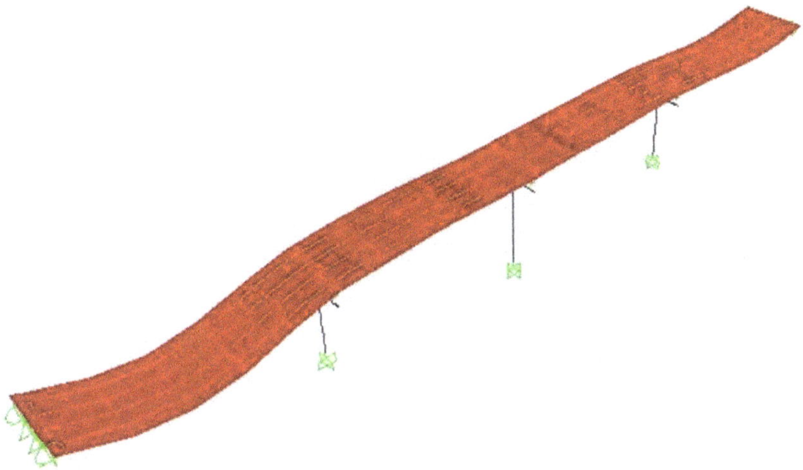

Figure 6: Mode 2: T = 0.89320 s.

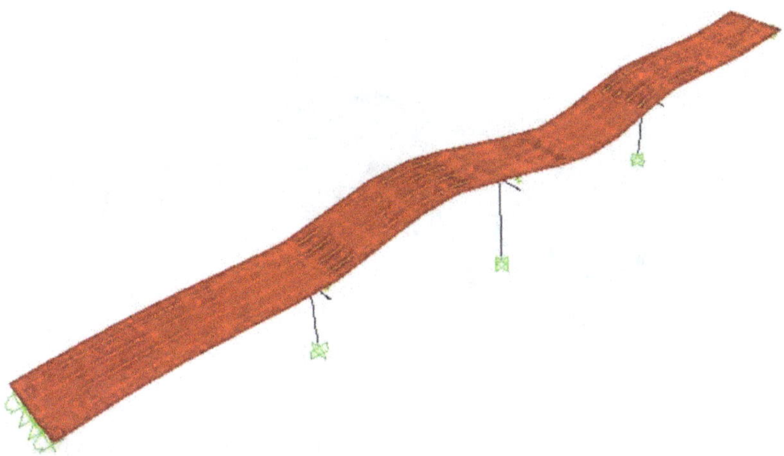

Figure 7: Mode 3: T = 0.73683 s.

4.3 Seismic loads generated

Note that the seismic loads are applied as time history of displacements at the bottom of piers. These displacements were generated using the simulation model described above (see Figs 8 and 9).

4.4 Dynamic analysis

To show the impact of the spatial variability of ground motions, taking into account local site conditions, four analyses were performed. The first analysis considers that load is uniform, i.e. that the seismic loading generated for the first bridge support (using the simulation approach described above) is applied to all of the bridge supports by taking into account local

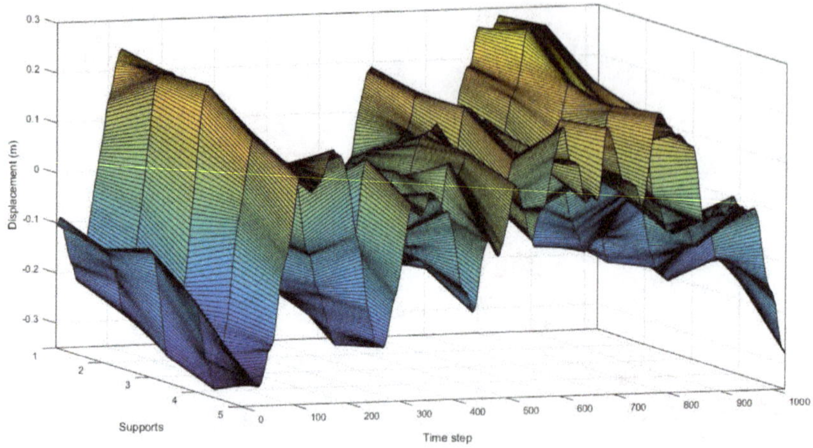

Figure 8: Displacements generated by considering local site effects.

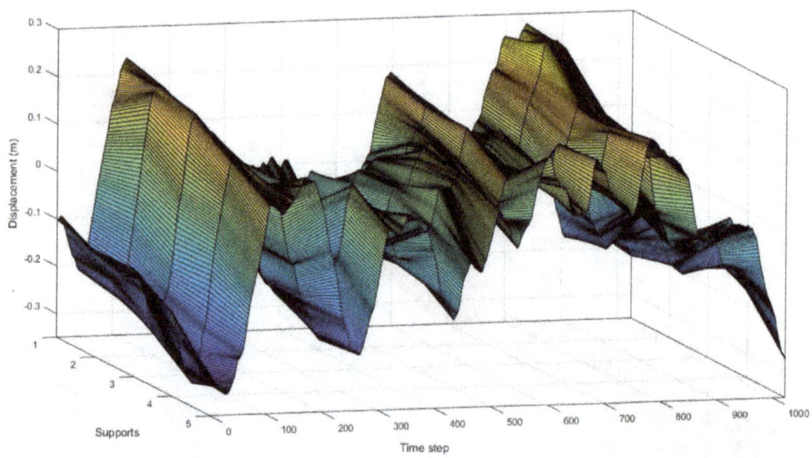

Figure 9: Displacements generated without considering local site effects.

site effects (denoted UNIF-WSE). The second analysis is performed in the same way of the first but neglecting local site effects (denoted UNIF-WOSE).

In the third analysis, each support has its own seismic loading (using the simulation approach) by considering local site effects (denoted SVGM-WSE). The last analysis is carried out in the same way as the third but neglecting local site effects (denoted SVGM-WOSE). The generated seismic loads are defined as time history displacements applied at the piers bottoms.

The results of the structural dynamic analysis, subjected to the four cases of excitations, are compared in terms of bending moment and shear force in piers. The maximum values of the bending moments and shear force obtained at each pier are illustrated in Figs 8 and 9.

We observe that the SVGM-WSE analysis provides higher internal forces than those produced by UNIF-WSE analysis and this is for the entire piers (see Fig. 10).

For these two analyses which take into account site effect, the decrease of bending moments vary between 4% for the pier P1 and 22% for the pier P3.

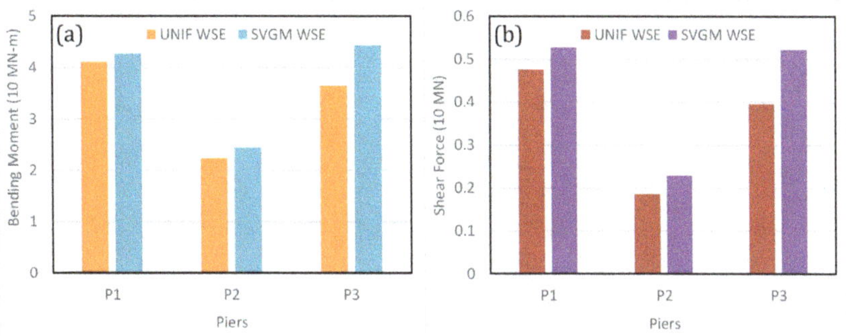

Figure 10: Dynamic analysis considering local site effects (a) Maximum bending moment. (b) Maximum shear force.

Figure 11 shows that bending moments outcome from SVGM-WOSE analyses are higher than those of UNIF-WOSE analysis with 3% for pier 1 and 17% for pier 3. Whereas for pier P2, there is a decrease of 6% in the bending moment. Noting that these analyses neglect the site effect. It is important to note that SVGM-WSE and SVGM-WOSE analyses caused seismic demand increase. These observations are consistent with previous studies results (Monti *et al.*, 1996) [8].

Results of seismic analysis of a bridge under variable seismic load, which take into account the site effect (SVGM-WSE), are higher than those produced from the analysis neglecting the site effect (SVGM-WOSE). Bending moments increase with 4% for pier P1 and 20% for pier P2. Shear forces values show an increase which vary between 4% for pier P1 and 18% for the pier P2 (see Fig. 12).

Note that site local condition variation leads to an increase in seismic loading especially for the piers, whereas the coherence model gives rise to pseudo-static displacements.

A seismic loading realistic representation must take into account these two parameters; coherency loss and site effect. Neglecting the site effect, case of several studies, can leads to an under estimation of internal forces.

The results presented in the case of variable seismic loading taking into account site effect correspond to foundation soil parameters (fixed above). In the case where soil parameters are divergent between bridge supports, an increase in the amplification factor is possible. Hence, the local site effect can lead to an important seismic demand.

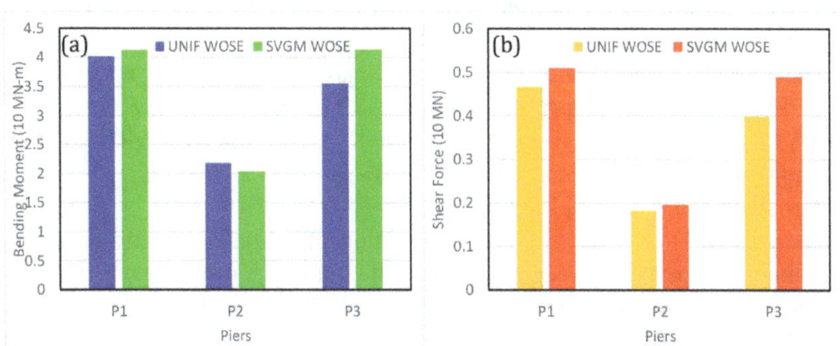

Figure 11: Dynamic analysis neglecting local site effects. (a) Maximum bending moment. (b) Maximum shear force.

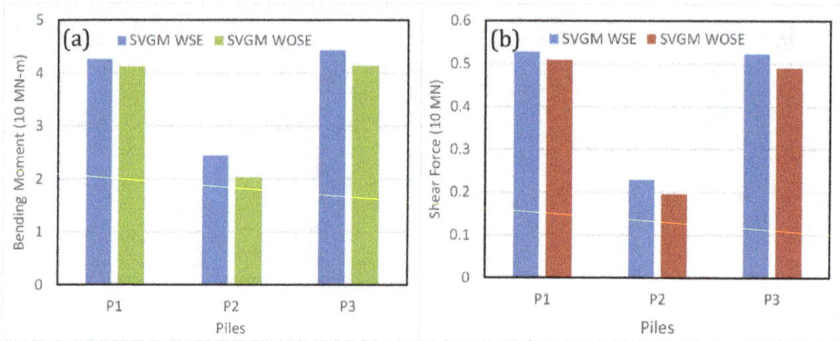

Figure 12: Comparison between the internal forces resulting from SVGM-WSE and SVGM-WOSE. (a) Maximum bending moment. (b) Maximum shear force.

5 CONCLUSION

A study was carried out to evaluate the impact of a spatial variability of earthquake ground motions, taking into account the local conditions of the soil foundation on the dynamic behavior of a railway bridge. A seismic simulation model has been developed to take account of local site effect.

Four loading cases are considered in this study. Two cases of spatial variability of ground motions, the first considers a coherency loss model the site effect. The second assumes that the spatial variability ground motion is defined only by a coherency loss function. Two other analyses under uniform seismic loading are carried out. The seismic loading was applied in the form of time history of displacements imposed on the base of each pier.

Results of this work show that the dynamic analysis under spatial variable seismic loading taking into account the site effect leads to an increase of the seismic demand. A dynamic analysis under spatial variability ground motions neglecting the site effect may lead to an underestimation of the seismic demand of the bridge.

The soil parameters of the site can generate an amplification of the seismic excitation while the coherency loss model leads to pseudo-static motions. Thus, a realistic representation of seismic loading must take into account these two effects.

REFERENCES

[1] Deodatis, G., Non-stationary stochastic vector processes: seismic ground motion applications. *Probabilistic Engineering Mechanics*, **11**, pp. 149–168, 1996. https://doi.org/10.1016/0266-8920(96)00007-0

[2] Der Kiureghian A., A coherency model for spatially varying ground motions. *Earthquake Engineering and Structural Dynamics*, **25**(1), pp. 99–111, 1996. https://doi.org/10.1002/(sici)1096-9845(199601)25:1<99::aid-eqe540>3.3.co;2-3

[3] Kaiming, B., *Effects of ground motion spatial variations (PHD thesis)*. University of western Australia, Crawley, 2011.

[4] Konakli, K. & Der Kiureghian, A., Simulation of spatially varying ground motions including incoherence, wave, passage and differential site response effects. *Earthquake Engineering & Structural Dynamics*, **41**(3), pp. 495–513, 2012. https://doi.org/10.1002/eqe.1141

[5] Zerva, A., *Spatial Variation of Seismic Ground Motions: Modeling and Engineering Applications*, CRC Press, Group, Taylor & Francis, 2009.

[6] Benmansour, N., Djafour, M., Bekkouche, A., Zendagui, D. & Benyacoub, A., Seismic response evaluation of bridges under differential ground motion: a comparison with the new Algerian provisions. *European Journal of Environmental and Civil Engineering*, **16**(7), pp. 863–881, 2012.
https://doi.org/10.1080/19648189.2012.681951

[7] Benmansour, N., *Étude du comportement dynamique des barrages voûtes face au mouvement sismique différentiel* (Master Thesis). University of Tlemcen, Algeria, 2004.

[8] Benmansour, N., *Effet de la variabilite spatiale du mouvement sismique sur le comportement dynamique des ponts* (Doctoral Thesis). University of Tlemcen, Algeria, 2013.

[9] Kaiming, B. & Hong, H., Modelling and simulation of spatially varying earthquake ground motions at sites with varying conditions. *Probabilistic Engineering Mechanics*, **29**, pp. 92–104, 2012.
https://doi.org/10.1016/j.probengmech.2011.09.002

[10] Djafour, M., Meddane, N., Derbal, R., Megnounif, A., Zendagui, D. & Bekkouche, A., *Response of a gravity arch dam to spatially varying earthquake ground motion.* 8th National Conference on Earthquake Engineering, California, 2006.

[11] Derbal, R., *Analyse dynamique de barrages poids - voûtes sous chargement sismique différentiel* (Master thesis). University Abou Bakr Belkaid, Tlemcen, Algeria, 2005.

[12] Djafour, M., Meddane, N., Derbal, R., Zendagui, D. & Bekkouche, A., Étude du comportement dynamique d'un barrage poids-voûte face au mouvement sismique différentiel. 18ème Congrès Français de Mécanique, 27–31 août 2007.

[13] Zendagui, D., Berrah, M.K. & Kasusel, E., Stochastic deamplification of spatially varying seismic. *Soil Dynamics and Earthquake Engineering*, **1**, pp. 409–421, 1999.
https://doi.org/10.1016/s0267-7261(99)00015-9

[14] Dumanoglu, A.A. & Soyluk, K.A., Stochastic analysis of long span structures subjected to spatially varying ground motions including the site-response effect. *Engineering Structures*, **25**(10), pp. 1301–1310, 2003.
https://doi.org/10.1016/s0141-0296(03)00080-4

[15] Harichandran, R. & Vanmarcke, E., Stochastic variation of earthquake ground motion in space and time. *Journal of Engineering Mechanics, ASCE*, **112**, pp. 154–174, 1986.
https://doi.org/10.1061/(asce)0733-9399(1986)112:2(154)

[16] Chen, M.T. & Harichandran, R.S., Response of an earth dam to spatially varying earthquake ground motion. *Journal of Engineering Mechanics ASCE*, **127**, pp. 932–939, 2001
https://doi.org/10.1061/(asce)0733-9399(2001)127:9(932)

SENSITIVITY OF STRUCTURAL DAMAGE TO EARTHQUAKE GROUND MOTION SCENARIOS. THE TORREVIEJA EARTHQUAKE CASE STUDY

N. AGEA-MEDINA[1], S. MOLINA-PALACIOS[1], D. H. LANG[2], I. FERREIRO-PRIETO[3], J. A. HUESCA[4], J. J. GALIANA-MERINO[5] & J. L. SOLER-LLORENS[1]
[1] Dpto. Ciencias de la Tierra y del Medio Ambiente, Universidad de Alicante, Spain.
[2] NORSAR, Department of Earthquake Hazard and Risk, Kjeller, Norway.
[3] Dpto. Expresion Grafica y Cartografia, Universidad de Alicante, Spain.
[4] Dpto. Construcciones Arquitectónicas, Universidad de Alicante, Spain.
[5] Dpto. Fisica, Ingenieria de Sistemas y Teoria de la Señal, Universidad de Alicante, Spain.

ABSTRACT

Structural damage computation using analytical methods requires the knowledge of the ground motion distribution in the urban area caused by a given earthquake. In this manuscript, the ground motion estimates (i.e. PGA and spectral acceleration values) are obtained through simulation of the 1829 Torrevieja earthquake using the NGA ground motion prediction equations (GMPE). The building stock under consideration has been classified according to the methodology presented in RISK-UE. The computations have been done using the last version of the software SELENA. The epistemic uncertainties of the analysis are accounted for by means of a logic tree computation scheme. The logic tree has two branches for the uncertainty in the earthquake scenario, two branches for the GMPEs and three branches to consider the uncertainties in average shear wave velocity Vs30 (soil conditions). Results indicate large differences derived for the different earthquake loss scenarios (ELE) obtained following each branch of the logic tree.

The greatest structural damages and losses are obtained when the earthquake is located in the Bajo Segura fault zone, using Campbell and Bozorgnia GMPE and for soft soil conditions. This article has allowed us to see how the different possible input parameters for ELE should be carefully analyzed for each case study and the importance of providing ELE results in terms of mean values with corresponding uncertainty ranges.

Keywords: analytical method, earthquake damage, epistemic uncertainty, seismic risk, seismic vulnerability

1 INTRODUCTION

The south of Alicante province (Spain) is characterized by a relatively high seismic activity. In the past, several earthquakes have seriously affected its municipalities. The most important earthquake was the Torrevieja earthquake on March 21, 1829. This event reached a macroseismic intensity X following the EMS-98 intensity scale [1] as reported by Martinez Solares and Mezcua [2]. According to Albini and Rodriguez de la Torre [3], it can be stated that this event represented the mainshock of a seismic sequence of at least 42 events that occurred between 1828 and 1830, with EMS-98 intensities reaching IX–X. It is expected that regional earthquake ground motions were highly amplified by the soft soils covering the areas closest to the Segura River, thereby causing significant structural damage in a wide area.

The city of Guardamar del Segura was also seriously affected by this earthquake. According to Larramendi [4], 397 buildings suffered extensive and complete damage, while 132 buildings experienced moderate damage. With respect to social losses, 8 persons died and 14 suffered injuries of moderate severity. The seismic activity of the Bajo Segura fault zone was studied by Alfaro *et al.* [5] and it was predicted that the maximum moment magnitudes range between 6.6 and 6.8 for individual fault segments and between 6.9 and 7.1 for a complete rupture of the Bajo Segura fault. These values are also consistent with the estimated magnitude of the

DOI: 10.2495/CMEM-V6-N5-921-932

1829 Torrevieja earthquake, which was considered to be in the range between Ms6.3 [6] and Ms6.9 [7]. Additionally, the focal depth was supposed to be shallow (approximately 5 km).

The United Nations Disaster Relief Organization defines the risk connected to a certain hazard as the expected physical damage and connected losses that are computed from the convolution of probability of occurrence of hazardous events and the vulnerability of the exposed elements to this hazard.

For a deterministic analysis, seismic hazard – the first component – refers to the shaking effects at a certain site caused by a scenario earthquake. While the term exposure represents the availability and inventory of buildings, infrastructure facilities and people in the respective study area subjected to a certain seismic event, the structural (i.e. physical) vulnerability stands for the susceptibility of each individual element (building, infrastructure, etc.) to suffer damage given the level of earthquake shaking. This results in structural (and non-structural) damages, which directly implicate economic losses as well as casualties (social losses).

Mena [8] carried out the first global study of vulnerability in various urban areas of Spain by using also Geographic Information System (GIS), using the Italian method [9] and the vulnerability-damage functions proposed by Barbat *et al.* [10] and Yepez [11]. The study was done for the two predominant model building types in the city of Barcelona, i.e. unreinforced masonry and reinforced concrete frame buildings.

The European project RISK-UE represented a step forward regarding vulnerability assessment and seismic risk computation in Europe. Under this framework, various damage functions were developed: vulnerability curves for the macroseismic method and capacity and fragility curves for the analytical/mechanical method that could be applied to the typical building types in Europe [12]. Later, two methods were proposed by RISK-UE: the vulnerability index method (VIM) and the Capacity Spectrum Method (CSM) were applied to residential buildings in Barcelona (Spain) in order to evaluate the seismic risk [13–16].

In 2009, Molina *et al.* [17] applied the capacity spectrum method, as implemented in SELENA, in order to compute earthquake loss scenarios for the urban area of Almoradi.

Recently, Serrano-Lanzarote and Temes-Córdovez [18] carried out a seismic vulnerability study of the residential buildings in the Valencia autonomous region using the vulnerability index method. However, the damage and losses were computed assuming a hypothetical earthquake and intensity estimates, so a detailed study should be done to improve their results.

Therefore, the goal of the present study is to conduct detailed earthquake loss estimation scenarios for the city of Guardamar del Segura, which consider not only physical ground motion parameters (i.e PGA and spectral accelerations by simulation a repetition of the 1829 Torrevieja earthquake) but also a detailed vulnerability model (using analytical capacity curves and fragility functions). The computation will be done using the last version of the software SELENA [19]. The epistemic uncertainties of the analysis will be considered by means of a logic tree. We will include two branches of the logic tree for the earthquake source parameter, two branches for the soil conditions represented by Vs_{30} values, and two branches corresponding to the ground motion prediction equations (GMPE) used to simulate the PGA in the city. The most recent Next Generation Attenuation (NGA) relationships ASK14 [20] and CB14 [21] will be used. Therefore, this paper will not only provide a mean ELE and corresponding uncertainties, but also a sensitivity study of the importance of each branch in the seismic risk results.

2 METHODOLOGY

The damage probability for each model building type will be obtained using the analytical methodology implemented in SELENA. The performance of each building typology is represented by its performance point, i.e. the spectral displacement that is experienced by the building typology under the respective seismic demand. This spectral displacement is further connected to a lognormal distribution of damage that the respective building typology will experience, which is, considering the building stock inventory model, computed into absolute values of damage and loss.

The method used to estimate the performance point has been recommended by Eurocode 8 and generally known as N2 method [22].

The process consists of obtaining the appropriate inelastic response, i.e. performance point (defined as point of intersection of idealized SDoF system based capacity curve with the inelastic demand spectrum) for a given earthquake record. Once the performance point has been obtained, the conditional probability of being in, or exceeding a damage state *ds* given by the spectral displacement S_d (or other seismic demand parameter) is defined by the following equation:

$$P\left[ds|S_d\right] = \Phi\left[\frac{1}{\beta_{ds}} ln\left(\frac{S_d}{\bar{S}_{d,ds}}\right)\right] \tag{1}$$

where:

$\bar{S}_{d,ds}$: median value of spectral displacement at which the building reaches the threshold of damage state *ds*; β_{ds}: standard deviation of the natural logarithm of spectral displacement for damage state *ds* and Φ: standard normal cumulative distribution function.

The structural damage is classified into four levels, i.e. slight, moderate, extensive and complete damage states. This damage classification concept was initially provided in the report 'Expected Seismic Performance of Buildings' [23] and later adopted for the HAZUS methodology [24].

In order to apply this methodology, first, the deterministic earthquake is defined. Secondly, the vulnerability model (damage functions in terms of capacity and fragility curves) is proposed, and finally the population distribution and the cost of repair and reconstruction to compute human and economic losses results are presented.

In order to better represent the ground motion in the study area, and to simplify the computation, we have divided the city into ten geographical units (Geounits; Fig. 1), that will be used as a reference unit in terms of computation and representation of damage and loss results.

2.1 Deterministic earthquake scenario parameters

As stated before, our study will simulate the repetition of the 1829 Torrevieja earthquake. We have assumed the fault parameters of the Bajo Segura fault zone and the Torrevieja fault zone given by the active fault catalogue in the Iberian Peninsula QAFI [25] (Table 1). As it can be seen in Fig. 1, two possibilities for the repetition of the earthquake are assumed, each one linked to one of the main faults.

2.2 Soil Conditions and Ground Motion Prediction Equations.

The regional geology of the study area is composed by sedimentary materials, mainly soft soils, which are likely to amplify the seismic ground motion. We have identified two main soil

Figure 1: Distribution of geounits and fault traces: the Bajo Segura fault (E1) and the Torrevieja fault (E2).

Table 1: Main source parameters of the earthquake scenarios used as input for SELENA.

Earthquake Scenario:	E1	E2
ID QAFI	ES620	ES616
Fault Name	Bajo Segura (3/3)	Torrevieja
Latitude	38.08	38.03
Longitude	-0.68	-0.7
Focal Depth (km)	5	5
Mw	6.6	6.6
Fault orientation (degrees from North)	77	116
Dip (degrees)	60	90
Fault Mechanism	Reverse	Strike-Slip

types: soil type B (deposits of very dense sands, gravels, or very stiff clays characterized by a gradual increase of mechanical properties with depth and an average $V_{s,30}$ ranging from 360 to 800 m/s) and soil type C (deep deposits of dense or medium-dense sand, gravel or stiff clay with thickness from several tens to many hundreds of meters and an average $V_{s,30}$ ranging from 180 to 360 m/s), following the soil classification scheme of EN 1998-1 [26]. It was decided to use soil type B (360–800 m/s), instead of a subdivision as B1 (500–800 m/s) and B2 (360–500 m/s), in order to point out the current uncertainty of the $V_{s,30}$ distribution of the city.

Regarding the selection of GMPEs, we have followed the recommendations of the RISMUR II Project working group that has carried out a revision of the seismic risk in the Murcia Region. They proposed the use of the following GMPEs: ASK [20] and CB14 [21]. Both GMPEs include the soil amplification in the attenuation relationship by means of some terms, which are dependent on the $V_{s,30}$ value. Therefore, all geounits will be assigned three different $V_{s,30}$ values: the lower bound of the Eurocode 8 range, i.e. 360 m/s and 180 m/s, the mean value, i.e. 580 m/s and 270 m/s, and the upper bound, i.e. 800 m/s and 360 m/s for soil types B and C, respectively.

2.3 Vulnerability and Exposure Model

The building stock in Guardamar comprises 4,078 buildings and a residential population of 15,589. This data has been obtained through a thorough analysis of the Spanish Cadastral database (SEC), the Spanish Institute of Statistics (INE), the local council of Guardamar del Segura, the ARGOS website, and several field trips conducted in 2015.

Following previous studies, [17] and [27], the building stock has been classified into model building types taking into consideration the main structural system, primary work materials, year of construction (i.e. the seismic design level), and the building's height range. Each model building type has been assigned a set of capacity and fragility curves [12].

Table 2 shows the model building types observed in Guardamar del Segura. The table contains the ID given in this study, the corresponding label [12], a brief description of the structural system and primary work materials, its height (represented by number of stories), the period of construction used as an indicator of the seismic design level, and the total number of buildings that were assigned to each class. As it can be seen, the city is highly vulnerable because approximately 80% of the building stock was built previously following the provisions of the NCSE-94 [28]. Only 9.5% of the buildings were designed according to NCSE-94 and 10.5 % according to NCSE-02 [29].

Table 2: Classification of the building stock in Guardamar del Segura into model building types (MBT).

L-G Label	Id	FRAME/ HORIZONTAL STRUCTURE			Period of construction	Nº of stories	Nº of buildings
M3.w_L	UM.w_L	Unreinforced Masonry	Wooden slabs		< 1950	1 – 2	16
M.3v_L	UM.v_L		Masonry vaults		< 1950	1 – 2	490
M.3v_M	UM.v_M				3 – 5	1	
M.3sm_L	UM.sm_L		Composite steel and masonry slabs		1950 – 1970	1 – 2	1191
M.3sm_M	UM.sm_M				3 – 5	99	
RC1_L	RC1.A_L	Reinforced Concrete	Concrete moment frame	PGS – 68	1969 – 1975	1 – 3	92
RC1_M	RC1.A_M					4 – 7	66
RC1_H	RC1.A_H					8+	5
RC1_II_L	RC1.B_L			PDS – 74	1976 – 1995	1 – 3	1097
RC1_II_M	RC1.B_M					4 – 7	305
RC1_I_L	RC1.C_L			NCSE – 94	1996 – 2003	1 – 3	150
RC1_I_M	RC1.C_M					4 – 7	176
RC1_I_L	RC1.D_L			NCSE – 02	2004 – Act.	1 – 3	167
RC1_I_M	RC1.D_M					4 – 7	181
RC1_I_H	RC1.D_H					8+	42

The predominant typology is RC1.B_L, corresponding to low rise (1 to 3 stories) reinforced concrete frame structures built between 1976 and 1994, using the regulations given PGS-68 [30] and PDS-74 [31].

The economic loss model depends on reconstruction values (complete damage) which follow the construction costs for new buildings given by the Spanish cadastral database in Euro/m^2. These construction values depend on the model building type, the quality of the building, the year of construction and the building's use. The cost of repair for extensive, moderate and slight damage will be assumed as 50%, 10% and 2% of the reconstruction cost, respectively [32].

2.4 Accounting for epistemic uncertainties: The logic tree

Figure 2 shows the designed logic tree computation scheme for our study. For example, Branch 1 corresponds to the ELE obtained by using an earthquake located in the Bajo Segura fault zone, using the ASK14 GMPE and assuming that soil types B and C will have the lower bound $V_{s,30}$. At the end this branch will be labelled as E1 – SMIN – ASK. In total, the logic tree has allowed us to obtain 12 ELE scenarios for Guardamar and, after weighting these branches, SELENA will provide a mean value and corresponding standard deviation range.

The weights assigned to each branch of the logic tree are the following: 0.8 to E1 and 0.2 to E2; 0.33 to SMIN, 0.34 to SMED and 0.33 to SMAX; 0.7 to ASK14 and 0.3 to CB14. As we can see, the E1 scenario has been given a higher probability because many authors [5] related it to the 1829 Torrevieja earthquake. Regarding the GMPE, the RISMUR II team [33] carried out a comparison of various GMPE models and they concluded that ASK14 should have a higher weight than CB. The herein assigned weights to the GMPEs follow the recommendations of the RISMUR II team.

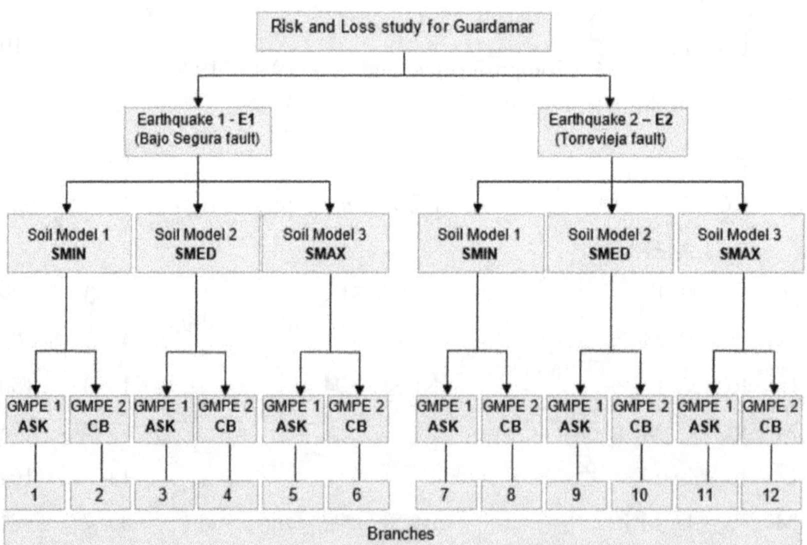

Figure 2: Branches of the logic tree of the ELE for Guardamar.

3 RESULTS AND DISCUSSION

3.1 Uncertainty of the structural damage scenarios

Figure 3 represents the structural damage distribution in terms of percentage of the total building stock for the entire city for both scenarios, while concentrating on typology RC1.B_L. As it can be seen, the structural damage is higher in the case of scenario E1 (Bajo Segura fault zone) than in the case of scenario E2 (Torrevieja fault). This is obviously due to the geometry of the fault which is crossing the city and causing a smaller rupture distance.

The higher probability of complete damage is observed for E1 in combination with CB14 GMPE. It ranges from 7% when using upper bound Vs_{30} values and the ASK14 attenuation relationship to 77% when using lower bound $V_{s,30}$ values and the CB14 attenuation relationship. On the other hand, for scenario E2, the probability of complete damage is lower and it ranges from 0% to 17%.

Therefore, again we can observe how the difference in $V_{s,30}$ can lead to important changes in the structural damage, so a detailed microzonation is always recommended when ELE has to be obtained for emergency and management planning.

3.2 Uncertainty in economic losses

As indicated before, the differences in the performance points may be the reason for the differences in structural damage. Therefore, once structural damage is converted into economic losses (Fig. 4), a similar behaviour is expected. Again, the economic losses are higher when scenario E1 and CB14 attenuation relationships are used and when the soil conditions are represented by the lower bound $V_{s,30}$ values. In this case, the total losses are 1,074 million Euro.

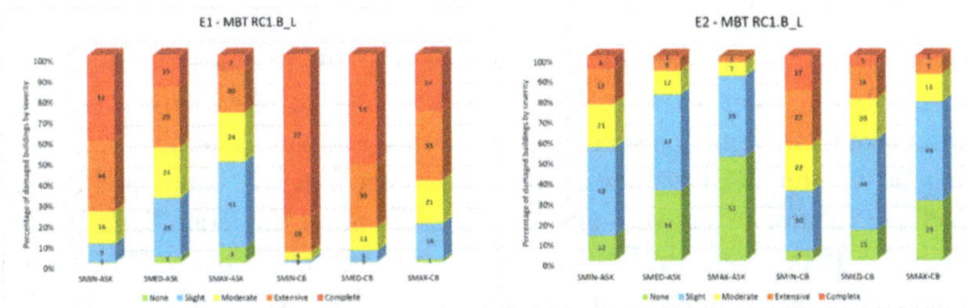

Figure 3: Summary of the damage distribution for the entire city: E1 (left); E2 (right). Note that only model building type RC1.B_L is represented.

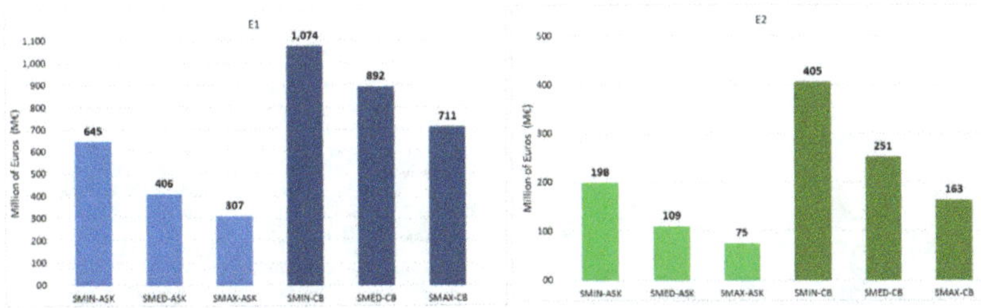

Figure 4: Total economic loss distribution for both scenarios.

However, when using the scenario E2, the upper bound $V_{s,30}$ values and the ASK14 attenuation relationship the economic losses are reduced to 93% approximately. As explained before the economic model used to quantify the repair/reconstruction costs is not very accurate because we have only used general values; therefore, this is only an approximation of the economic losses but it also helps to see how different the losses can be depending on the chosen input parameter.

3.3 Uncertainty in human losses

Figure 5 shows the human losses (injured and deaths) in the city for the two earhquake scenarios and the rest of the branches of the logic tree. The results have been computed assuming a nighttime scenario in which most of the population are inside the residential buildings, thereby serving some sort of a worst-case scenario. The behaviour of the human loss distribution is the same as previously observed with structural damage.

The maximum number of people seriously affected (i.e. heavy injuries and deaths) are 375 for the scenario E1 and 128 for the scenario E2 while the minimum is 63 for the scenario E1 and 9 for the scenario E2. If we add all the people injured, we obtain a maximum of 3165 for the scenario E1 and 1193 for the scenario E2.

3.4 Mean ELE for the city of Guardamar del Segura.

Finally, as explained before small variations in the input parameters can result into big differences amongst the obtained damage and loss results are depending on small variations in the input parameters. In our case, and after reviewing the literature, we found that both scenarios E1 and E2 could be possible for a repetition of the 1829 Torrevieja earthquake. Additionally, if there is not any detailed information about the $V_{s,30}$ distribution in the city, a first approach using the $V_{s,30}$ ranges provided in Eurocode 8 could be a way of representing the different possibilities. Since no specific attenuation relationships are available for the respective region, two of the NGA have been selected that were initially proposed for the update of the Spanish seismic hazard map. Therefore, we always should provide the local authorities with a mean earthquake loss estimation scenario with corresponding uncertainties.

Regarding the typology RC1.B_L, we have obtained that 276 ± 78 buildings suffer complete damage, 268 ± 35 suffer extensive damage, 193 ± 26 suffer moderate damage and finally 275 ± 58 and 83 ± 45 buildings suffer slight or none damage, respectively. On the other hand, if we add all the different model building types, the damaged number of buildings will increase until 1646 ± 253 building that suffer complete damage, 847 ± 113 suffer extensive

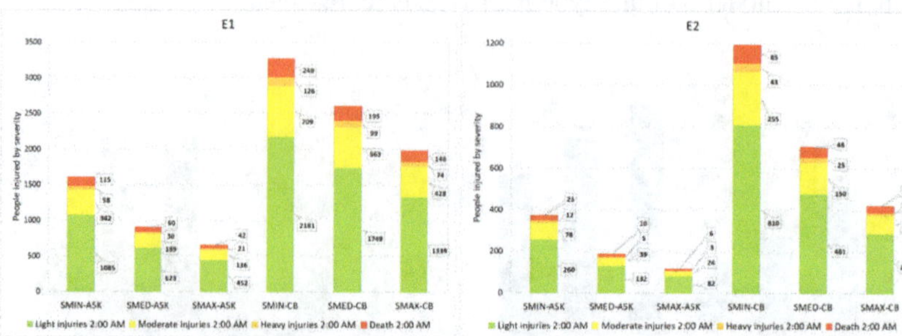

Figure 5: Total human losses distribution for both scenarios.

damage, 648 ± 108 moderated damage and 656 ± 129 and 263 ± 110 buildings suffer slight or none damage, respectively. The mean economic losses due to this structural damage will be 515 ± 79 million Euros and the mean number of people with heavy injuries will be 47 ± 11 and death will be 93 ± 22.

4 CONCLUSIONS

This paper has allowed us to see how the different but possible input parameters that are required to compute an ELE for the city of Guardamar should be carefully analyzed for each case study. In our case, we have developed a logic tree with two different earthquake scenarios, E1 and E2, two different ground motion prediction equations and three different soil conditions. All this branches were assumed to be valid if an ELE is going to be computed for the city, therefore all are realistic.

The obtained results allow us to state the following conclusions:

a) The earthquake scenario source parameters are very important even in a situation like the proposed in this paper where with a similar magnitude but associated a different fault types and mechanism, we obtain structural damage differences up to a 60%.
b) The selection of a correct GMPE is always a difficult task. In our case the CB14 GMPE provides the higher damage that can be up to a 36% when compared with ASK14.
c) Finally, the $V_{s,30}$ conditions in the city also have an important influence on the computed damage. The softer the soil, the higher the damage, usually. In our case, it can be observed differences up to 50% between the damage computed with the minimum and the maximum $V_{s,30}$ values.

Therefore, the steps that we will follow for the next detailed ELE to be computed in the city will be the following: a) First, as the $V_{s,30}$ values provide higher differences, we will carry out a detailed microzonation in order to get a better characterization of the soil behaviour; b) secondly, we will try to analyze if the earthquake scenario can be better defined or complemented with other possibilities; c) thirdly, we will better analyze the behaviour of the attenuation relationships in order to choose those which can simulate our ground motion scenarios in a better way, and d) finally, the vulnerability of Guardamar del Segura's building stock of will be represented through a more detailed vulnerability model [34].

Anyway, the use of the logic tree given in SELENA has been useful in order to obtain a mean ELE scenario, which we consider reliable because all the input parameters are applicable to the case study and with corresponding uncertainties, so stakeholders and local governments can use it in order to establish the first draft of a management and emergency plan for the city.

ACKNOWLEDGMENTS

The present research has benefited from funding of the Ministerio de Economía, Industria y Competitividad through research project CGL2016-77688-R and the Generalitat Valenciana through the research project AICO/2016/098.

REFERENCES

[1] Grünthal, G. (Ed.), European macroseismic scale 1998. *Cahiers du Centre Européen de Géodynamique et de Séismologie*, **7**, pp. 1–99, 1998.
[2] Martinez Solares, J.M. & Mezcua, J., *Catálogo sísmico de la Península Ibérica (880 a.C.–1900)*, Monograph nº 18. Instituto Geográfico Nacional, Madrid, publication, p. 303, 2002.

[3] Albini, P. & Rodriguez de la Torre F., The 1828–1829 earthquake sequence in the provinces of Murcia and Alicante (Southeastern Spain): a reappraisal of the historical sources, *The use of historical data in natural hazards assessments*, eds. Glade T, Albini P, Frances F, Kluwer: Dordrecht, pp. 35–54, 2001.

[4] Larramendi, J.A., *Memoria y relación circunstanciada de los estragos que la terrible catástrofe de los terremotos de 2,1 de marzo y siguientes, principalmente el del sábado Santo l 8 de abril hasta el presente día, han causado en Torrevieja y demás pueblos de la gobernación de Orihuela y sus inmediaciones, en la ciudad de Murcia y algunos pueblos de la provincial.* Biblioteca Nacional de España. Imprenta Real, Madrid, 1829.

[5] Alfaro, P., Bartolomé, R., Borque, M.J., Estevez, A., García-Mayordomo, J., García-Tortosa, F.J., Gil, A.J., Gràcia, E., Lo Iacono, C. & Perea, H., The Bajo Segura fault zone: active blind thrusting in the eastern betic cordillera (SE Spain). *Journal of Iberian Geology*, **38**(1), pp. 271–284, 2012.
https://doi.org/10.5209/rev_jige.2012.v38.n1.39217

[6] Delgado, J. & López Casado, C., El terremoto de 1829. *La catástrofe sísmica de 1829 y sus repercusiones,* ed. G. Canales, Excma. Dip. Provincial de Alicante, Excmo. Ayto. Almoradí and Universidad de Alicante, 1999.

[7] Muñoz, D. & Udías, A., Three large historical earthquakes in Southern Spain. *Seismicity, Seismotectonics and Seismic Risk of the Ibero-Maghrebian Region*, ed. J. Mézcua and A. Udías, Publ. Inst. Geogr. Nac, 8, pp. 175–182. Madrid, 1991.

[8] Mena, U., *Evaluación del Riesgo Sísmico en Zonas Urbanas*, PhD. Thesis, Barcelona. Dpto. Ingeniería del Terreno, Cartográfica y Geofísica, Universidad Politécnica de Cataluña, 2002.

[9] Benedetti, D. & Petrinni, V., Sulla vulnerabilitá sísmica di edifici in muratura. Proposte di un método di valutaziones. *L'industria delle Construccioni*, **149**, pp. 66–78, 1984.

[10] Barbat, A.H., Pujades, L.G. & Lantada, N., Performance of buildings under earthquakes in Barcelona, Spain. *Computer-Aided Civil and Infrastructure Engineering,* **21**(8), pp. 573–593, 2006.
https://doi.org/10.1111/j.1467-8667.2006.00450.x

[11] Yépez, F., *Metodología para la evaluación de la de la vulnerabilidad y riesgo sísmico de estructuras aplicando técnicas de simulación,* PhD. Thesis, Barcelona, Dpto. Ingeniería del Terreno, Cartográfica y Geofísica, Universidad Politécnica de Cataluña, 1996.

[12] Lagomarsino, S. & Giovinazzi, S., Macroseismic and mechanical models for the vulnerability and damage assessment of current buildings. *Bulletin of Earthquake Engineering,* **4**, pp. 415–443, 2006.
https://doi.org/10.1007/s10518-006-9024-z

[13] Lantada, N., *Evaluación del riesgo sísmico mediante métodos avanzados y técnicas GIS. Aplicación a la ciudad de Barcelona*, PhD. Thesis, Dpto. Ingeniería del Terreno, Cartográfica y Geofísica. Universidad Politécnica de Cataluña, Barcelona, 2007.

[14] Pujades, L.G., Barbat, A.H. & Lantada, N., Evaluación del riesgo sísmico en zonas urbanas: desarrollo de escenarios. *Revista internacional de ingeniería de estructuras,* **12**(1), pp. 1–28, 2007.

[15] Barbat, A.H., Pujades, L.G. & Lantada, N., Seismic damage evaluation in urban areas using the capacity spectrum method: application to Barcelona. *Soil Dynamics and Earthquake Engineering,* **28**, pp. 851–865, 2008.
https://doi.org/10.1016/j.soildyn.2007.10.006

[16] Lantada, N., Pujades, L.G. & Barbat, A.H., Vulnerability index and capacity spectrum based methods for urban seismic risk evaluation. A comparison. *Natural Hazards,* **51**(3), pp. 501–524, 2009.
https://doi.org/10.1007/s11069-007-9212-4

[17] Molina, S., Lang, D.H., Galiana-Merino, J.J., Jiménez-Delgado, A., Zaragoza-Martínez, A., *et al., Seismic risk scenarios for the urban area of Almoradi (Southeast Spain).* Presented at 8th International Workshop on Seismic Microzoning and Risk Reduction, Almería, Spain, 2009.

[18] Serrano-Lanzarote, B. & Temes-Córdovez, R., Vulnerabilidad y riesgo sísmico de los edificios residenciales estudiados dentro del Plan Especial de evaluación del riesgo sísmico en la Comunidad Valenciana. *Informes de la construcción,* **67**(539), e104, pp. 1–11, 2015.
https://doi.org/10.3989/ic.13.182

[19] Molina, S., Lang, D.H., & Lindholm, C.D., SELENA – An open-source tool for seismic risk and loss assessment using a logic tree computation procedure. *Computers and Geosciences,* **36**(3), pp. 257–269, 2010.
https://doi.org/10.1016/j.cageo.2009.07.006

[20] Abrahamson, N.A., Silva, W.J. & Kamai, R., Summary of the ASK14 ground motion relation for active crustal regions. *Earthquake Spectra,* **30**(3), pp. 1025–1055, 2014.
https://doi.org/10.1193/070913eqs198m

[21] Campbell, K.W. & Bozorgnia, Y., NGA-West 2 ground motion model for the average horizontal components of PGA, PGV, and 5% damped linear acceleration response spectra. *Earthquake Spectra,* **30**(3), pp. 1087–1115, 2014.
https://doi.org/10.1193/062913eqs175m

[22] Fajfar, P., Capacity spectrum method based on inelastic demand spectra. *Earthquake Engineering & Structural Dynamics,* **28**(9), pp. 979–993, 1999.
https://doi.org/10.1002/(sici)1096-9845(199909)28:9<979::aid-eqe850>3.0.co;2-1

[23] EERI Ad Hoc Committee on Seismic Performance, *Expected seismic performance of buildings*, Oakland, CA, 1994.

[24] Federal Emergency Management Agency FEMA, *HAZUS99 Earthquake Loss Estimation Methodology, User Manual,* Federal Emergency Management Agency, Washington, DC, United States, 1999

[25] García-Mayordomo, J., Insua-Arévalo, J.M., Martínez-Díaz, J.J., Jiménez-Díaz, A., Martín-Banda, R., Martín-Alfageme, S., Álvarez-Gómez, J.A., Rodríguez-Peces, M., Pérez-López, R., Rodríguez-Pascua, M.A., Masana, E., Perea, H., Martín-González, F., Giner-Robles, J., Nemser, E.S., Cabral, J. & QAFI compilers. The Quaternary Active Faults Database of Iberia (QAFI v.2.0). *Journal of Iberian Geology,* **38**(1), pp. 285–302, 2012.
https://doi.org/10.5209/rev_jige.2012.v38.n1.39219

[26] CEN, *EN 1998-1, Eurocode 8—design of structures for earthquake resistance, part 1: general rules, seismic actions and rules for buildings*. European Committee for Standardization, Brussels, 2004.

[27] Mouroux, P. & Le Brun, B., Risk-UE Project: An Advanced Approach to Earthquake Risk Scenarios with Application to Different European Towns. *Assessing and Managing Earthquake Risk,* ed. Oliveira, C.S., Roca, A. and Goula, X., Springer Netherlands, pp. 1–14, 2006.

[28] Ministerio de Obras Públicas, Transporte y Medio Ambiente, *Real Decreto 2543/1994, de 29 de diciembre de 1994, por el que se aprueba la NCSE-94 Norma de Construcción Sismorresistente: parte General y de Edificación,* Boletín Oficial del Estado, nº 33, pp. 3935–3980, España, 1995.

[29] Ministerio de Fomento, *Real Decreto 997/2002, de 27 de septiembre de 2002, por el que se aprueba la NCSE-02 Norma de Construcción Sismorresistente: parte General y de Edificación,* Boletín Oficial del Estado, nº 244, pp. 35898–35967, España, 2002.

[30] Presidencia del Gobierno, *Decreto 2987/1968, de 20 de septiembre de 1968, por el que se aprueba la EH-68. Instrucción para el proyecto y la ejecución de obras de hormigón en masa o armado*, Boletín Oficial del Estado, nº 290, pp. 17257–17291, España, 1968.

[31] Ministerio de Planificación del Desarrollo. (1974). Decreto 3209/1974, de 30 de agosto de 1974, por el que se aprueba la *PDS-1-1974 Norma Sismorresistente: parte A*. Boletín Oficial del Estado, nº 279, pp. 23585–23601, España, 1968.

[32] Federal Emergency Management Agency FEMA, *HAZUS-MH MR4 Technical Manual.* Washington, DC, 2003.

[33] Benito, B., Rivas, A., Pérez, M. Quirós, L.E., Ruiz, S., *et al.*, *Servicio de actualización del análisis de riesgo sísmico (RISMUR) en la región de Murcia (RISMUR II)*, Memoria Técnica, 2015.

[34] Pitilakis, K., Crowley, H. & Kaynia, A.M. (Eds), *SYNER-G: Typology definition and fragility functions for physical elements at Seismic Risk: Buildings, Lifelines, Transportation Networks and Critical Facilities,* Springer, 1st edition, 2014.

TOWARDS DAMAGE-CONSISTENT PERFORMANCE-BASED DESIGN OF CRITICAL INFRASTRUCTURES

JENS-UWE KLÜGEL & SUNAY STÄUBLE-AKCAY
NPP Goesgen-Daeniken, Switzerland.

ABSTRACT

The objective of earthquake-resistant design of critical infrastructures like nuclear power plants or lifelines is to ensure the prevention of catastrophic disasters. Experience from recent past like the earthquake of Amatrice (2016) or the Napa earthquake of 2014 have shown that traditional code requirements based on probabilistic seismic hazard maps are not able to prevent disasters. The purpose of probabilistic hazard assessment is to support risk analysis. The latter is used to separate tolerated residual risks from non-tolerable, more frequent risks. Therefore, these methods do not intend to provide protection against extreme events. Additionally, it is proven that the traditional hazard parameter used in probabilistic seismic hazard maps, peak ground acceleration (PGA), is not very suitable for the description of the physical impact of earthquakes on structures, systems and components. The only hazard parameter describing physical effects of earthquakes at least on macroseismic scale is intensity or in engineering units, intensity factors. The actual EMS-98 scale correlates reasonably well with the damage of structures classified into vulnerability classes.

The availability of large databases of registered earthquake time-histories covering a wide range of site intensity values allows to model earthquake impact directly using dynamic time-history analysis methods. On this basis a methodology was developed that allows to design critical infrastructures for certain levels of seismic intensity directly.

The methodology and some applications are presented.

Keywords: disaster prevention, earthquake engineering, performance-based design, seismic hazard analysis

1 INTRODUCTION

The big Japan earthquake of March 11, 2011, the most costly natural hazard ever recorded, but also the earthquakes of Amatrice (2016) and the Californian Napa earthquake (2014) demonstrated that modern seismic hazard analysis and earthquake engineering methods as applied today have significant limitations in preventing disasters. Civil code regulations in most countries of the world are based on probabilistic seismic hazard (PSHA) maps with only few exceptions. Performance-based design methods as developed by NEHRP in the U.S. [1,2] are based on these maps. These methods define a performance level for structures and systems for different seismic use groups ensuring a low probability of failure for a given hazard level. The latter is defined in terms of the probability of exceedance for a certain period of time, e.g. 10% in 50 years for ordinary buildings. Typically three service levels are defined [3]:

- Serviceability.
- Damage control.
- Collapse prevention.

With different corresponding performance criteria:

- Near-elastic response.
- Limited inelastic response.
- Large inelastic response.

DOI: 10.2495/CMEM-V6-N5-933-943

Different design procedures were developed to ensure the intended performance level as:

1. Linear static.
2. Linear dynamic.
3. Nonlinear static.
4. Nonlinear dynamic.

Similarly, for critical infrastructures, like nuclear power plants, an analogous approach was developed distinguishing different performance categories in dependence on the technological function of the structure [4], system or component (SSC) under consideration. This approach defines five different seismic design categories [5] and provides detailed design procedures for the top three categories with the highest functional requirements.

Nevertheless experience from past disasters questions whether this approach is able to guarantee the intended level of safety to the public. Although there are many debates related to seismic design procedures and structural dynamics the root cause of the problem has to be seen in the missing link between seismic hazard analyses and engineering design. The key issues known are:

1. The purpose of probabilistic hazard assessment is to support risk analysis. The latter is used to separate tolerated residual risks from non-tolerable, more frequent risks. Therefore, these methods do not intend to provide protection against extreme events and thus cannot prevent disasters.
2. The limit states of different SSCs have to be defined in terms of different engineering parameters. Both probabilistic and traditional deterministic seismic hazard analysis do not provide the full level of information needed by engineers as long as they are focussing on simple ground motion parameters [6].

This paper presents an approach and the associated methodology that allows to move towards a truly damage-consistent performance-based design with some applications for critical infrastructures. It can easily be expanded to other structures and it makes use of the recent developments in structural dynamics and computer sciences allowing for the application of advanced simulation technics.

2 DESCRIPTION OF METHOD

2.1 Main objectives and general requirements to performance-based methods

The main and primary objective of any seismic design method with respect to disaster prevention is to ensure a very low probability of critical functional failure of the object to be designed even in case of very extreme rare earthquakes.

Additionally, it is reasonable to formulate a secondary objective. The methods and techniques applied shall allow to optimize the economic resources assigned to pursue the primary objective. The objective of optimization of economic resources puts a very strict requirement to the methods applied for the seismic design. They have to be as much as possible realistic and as such, in compliance with empirical information from earthquake recordings. The concept of performance-based design allows allocating resources in dependence on the functional importance of the design object and therefore, in principal has the capability to support an optimization of resources. Unfortunately the actual performance-based methods significantly deviate from the requirement of realism. They suffer from several deficiencies, which may even lead to a violation of the primary objective, to ensure a robust design.

The most relevant deficiencies are:

1. Performance-based criteria are expressed in terms of absolute values for the probability of exceedance [5]. This expresses the belief that the results of a PSHA reflect the recurrence of earthquakes in a realistic way. This belief is not justified. In a short cut this is not correct for the following main reasons:
    a. The stochastic process of occurrence of earthquakes is not ergodic and it is certainly not a homogeneous poisson process [7, 8]
    b. PSHA is based on uniform hazard spectra (UHS). The latter do not reflect the true ground motion response of single earthquakes. An UHS represents the weighted combination of many earthquakes with different damaging characteristics. As the result an UHS cannot be directly related to the physical effects of earthquakes.
    c. Modern PSHAs and their hazard curves consider two main sources of uncertainty: aleatory variability (reproducible in an experiment) and epistemic uncertainty (knowledge-based uncertainty), the latter being subjective. Thus results of PSHA by definition deviate from empirical observations due to the subjectivity associated with the treatment of uncertainty [9].
2. The main focus on acceleration spectra instead of (as for example) tripartite spectra in PSHA limits the applicability of hazard assessment results. The governing seismic failure modes for many SSCs are not controlled by peak values of accelerations.

As a consequence, the performance-based methods in use today are not related to the observed physical effects of earthquakes. Therefore, the quality of a seismic design based on these methods cannot be judged at all. The current methods cannot be recommended for the development of a graded approach to the seismic design of SSCs.

As general requirements for the development of an alternative approach to performance-based design one can define:

- Compliance with empirical information on the physical effects of earthquakes
- Flexibility to allow for the use of different engineering parameters suitable for the design of different structures.

The performance-based method presented below is meeting these requirements.

2.2 Selection of seismic hazard parameter

A realistic performance-based design is only possible if the seismic hazard assessment is performed in parameters that are closely linked to the physical damage in earthquakes. The only hazard parameter that is directly connected to the physical effects of earthquakes and that is used in seismic hazard assessment is intensity [6]. The intensity scale EMS-98 [10] is directly linked to the physical effects of earthquakes. It is very well calibrated against observed damage for buildings of different vulnerability classes. Therefore, intensity in EMS-98 provides the perfect basis for the development of a performance-based design.

Certainly, engineers pose the challenging question that they have not learned to perform calculations in intensities or in intensity factors (if integer site intensities are converted into intensity factors). Therefore, intensities have to be converted into engineering parameters.

The progress in the registration of time–histories from earthquakes as well as the large progress in computational simulations allows to solve this issue that for a long time has prevented the use of intensities as the basis of seismic design of SSC. Big databases similar to Ref. [11] that were prepared as part of the PEGASOS refinement Project (PRP) [12] contain a large number of time-histories that can be classified in terms of site intensity and site conditions. They can be used directly for structural analyses and the design of SSCs. For performance-based design it is mandatory that a set of time-histories is used to cope with the uncertainty associated with time-histories leading to the same site intensity and therefore to the same structural damage. Figure 1 illustrates the range of uncertainty of time-history recordings for intensity VIII (EMS-98) on the example of records from 11 different earthquakes.

In terms of PGA we observe a spread of data from 0.06 g (earthquake #8, Kerman, 22.02.2005, Derwood station (Iran)) to 0.37g (earthquake#1, Friuli, 6.05.1976, station Tolmezzo Centrale – Diega Ambesta 1). It is worth to note that this spread cannot be explained by site conditions lonely.

The database [11] contains time-histories that very well cover the range of intensities up to VIII, containing a few records of intensity IX. For critical infrastructures such as nuclear power plants or river dams the recommended range of seismic site conditions corresponds to this range of recordings. The direct use of time-histories provides the flexibility required for a performance-based design. If necessary they can be converted into response-spectra or tripartite spectra so that traditional response based engineering design methods can be applied. For critical infrastructures or for the design of aseismic foundations (base-isolated systems), the direct use of time histories is strongly recommended.

For high seismic areas there it may be necessary to cover higher intensity ranges the lack of recorded time-histories can be compensated by different types of waveform modelling, reaching from synthetic seismograms like in the neodeterministic method [13] to kinematic or dynamic modelling. A method successfully applied in high seismic regions is the 'Japanese recipe' used for the design of critical infrastructures [14] including nuclear

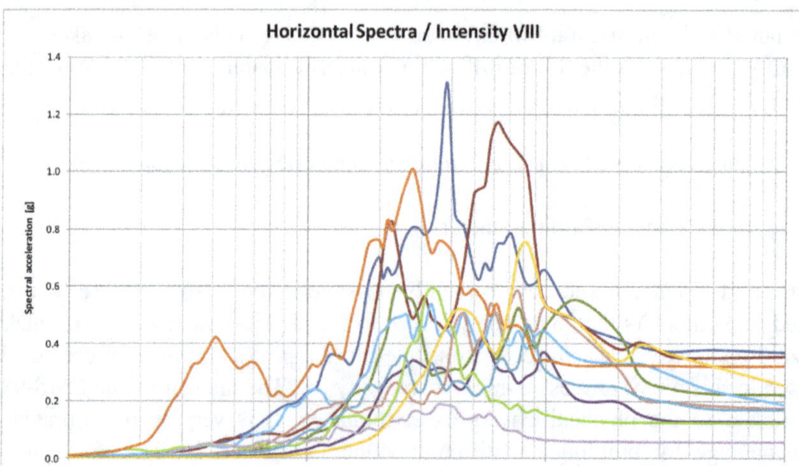

Figure 1: Uncertainty range of horizontal response spectra for intensity VIII.

power plants. Waveform modelling allows to develop all engineering parameters wanted by engineers including the development of tripartite spectra used in the more traditional design methods.

2.3 Adaptation of seismic hazard analysis results

National standards in many countries require the implementation of PSHA as the basis for the seismic design or the review of the design of structures, systems and components including critical infrastructures. Some countries require a comparison with deterministic hazard assessment results or even prefer a deterministic, physics-based approach. All types of hazard analysis can be performed directly in terms of intensity or can be converted into intensity scale. For example, in Switzerland the large scale PSHA-Study PEGASOS [15] was converted into site intensities (site intensity factors) using the results of hazard deaggregation [16]. Hazard deaggregation and conversion into intensity scale frequently allows to understand the true meaning of a study results in terms of the damaging effects expected. Because the uncertainty of hazard assessment results is frequently bounded within a single intensity unit the conversion into integer site intensity values is simple and the range of uncertainty can be maintained and propagated through the analysis.

2.4 Performance-based design goals for critical infrastructures

Due to the difficulties to calculate the frequency of earthquakes in a realistic way it is suggested to define performance criteria in terms of conditional probability of failure for a specified design earthquake. A review of the performance-based approach used in [5] shows that for many SSCs even for increased service levels (seismic design categories SDC 3 to SDC 5) the conditional probability of failure is rather high, reaching from 0.25 to 0.025. The requirements in [5] are significantly lower than in many other countries of the world. The approach suggested here is closer to the international practice. It is based on the traditional double earthquake concept distinguishing between an operability earthquake and a safety earthquake. The categorization of performance levels for SSCs follows the traditional approach in the industry by distinguishing between serviceability, damage control and collapse prevention. The seismic design categorization distinguishes between three categories corresponding to different levels of serviceability. The categories are:

- SDC 1 – damage control for the operational earthquake OE and collapse prevention for the safety earthquake SE are required
- SDC 2 – serviceability for the operational earthquake OE and damage control for the safety earthquake SE is required
- SDC3 – serviceability for the safety earthquake SE is required

The following table describes the performance goals suggested in terms of conditional probability of significant deviation from linear-elastic response.

For the calculation of the conditional probability of significant deviation from linear-elastic response it is possible to use the well-established methods for developing fragility functions in terms of damage indices as needed by engineers [3]. This allows for a structure or component specific approach and ensures a high level of flexibility with respect to the engineering parameters selected for the design process.

Table 1: Performance goals for seismic design categories.

Seismic design category	Categorization criteria	Operational earthquake	Safety earthquake
SDC 1	SSCs needed to prevent losses of lives without active function	0.05	0.1
SDC 2	SSCs that shall maintain operability of active function during the operational earthquake or passive SSCs that shall not fail during the safety earth-quake	0.01	0.05
SDC 3	SSCs that shall maintain operability of active functions after the safety earthquake	Close to zero (0.005)	0.01

2.5 Summary of the procedure

The procedure that was developed can be subdivided into the following steps:

1. Convert existing seismic hazard assessment results into intensity scale if they are not yet readily available in this format.
2. Define the design site intensity level for the operational earthquake (OE) and for the safety earthquake (SE).
3. Select time-histories matching the site intensity levels and the site conditions from registered earthquakes. If needed complete the available number of records by wave modelling approaches (synthetic seismograms, kinematic models etc.).
4. Design calculations and verification
    a. Classify structures, systems and components (SSCs) to be designed against earthquakes into seismic design categories according to Table 1.
    b. Develop the design requirements according to the performance-based criteria in table 1.
    c. Construction design process aimed at meeting the requirements of step 4b.
    d. Verification of the design.

3 PRACTICAL EXAMPLES

The application of the procedure is demonstrated on the practical example of the new build of a nuclear power plant. It is assumed that a second unit for the nuclear power plant Goesgen shall be constructed at the same site. As the existing plant the new unit shall be a pressurized water reactor (PWR) from the same vendor. For simplicity it is assumed that the reactor building design will be the same as for the existing unit. Based on the available seismic hazard assessment study PEGASOS [15] it is required to develop the seismic design requirements for structures, systems and components that are categorized into the three different service levels SDC 1 to SDC 3. For simplicity, it is assumed that the design requirements shall be developed in terms of design level PGA and floor response spectral acceleration at 10 Hz (a typical value for the first natural frequency of many pieces of nuclear equipment).

3.1 Seismic hazard analysis results

In Switzerland the safe shutdown earthquake (the safety earthquake according to the nomenclature of the procedure) is defined as an earthquake with a mean annual frequency of exceedance of 10^{-4}/a. The operational earthquake is defined as 50% of the safe shutdown earthquake. Additionally, the seismic design shall envelope the largest historical event. For Goesgen this is the historical earthquake of Basel (1356) that would lead to a site intensity at the Goesgen site of VII-VIII. In [17] the detailed process of hazard deaggregation and conversion to the intensity scale is described. Here the final results are reproduced.

3.2 Design site intensity levels

Based on the results in Table 2 the safe shutdown earthquake (safety earthquake) has to be assigned to a site intensity of VIII and correspondingly, the operational earthquake to site intensity VII.

3.3 Selection of time-histories and conversion into engineering parameters

The seismological design parameters have to be converted into PGA based on empirical observations. Selecting the 11 records presented in Fig. 1 for site intensity VIII we obtain a mean PGA of 0.21g and a standard deviation of 0.1g. From the database [11] we also restore the corresponding time-histories. It is possible to approximate the hazard by a lognormal distribution as it is common for performance-based approaches [3]. Similarly, we proceed for the operational earthquake with intensity VII. For this intensity level we selected 17 records from [11] with a mean PGA of 0.08g and a standard deviation of 0.047g. The hazard assessment results in engineering parameters are shown below:

For the conversion of ground motion parameters into spectral accelerations a dynamic structural analysis was performed. For simplicity, it is assumed that the in-structure floor response at elevation 18.00 m of the reactor building can be regarded as bounding for the design of systems and components of the new nuclear power plant. Therefore, we present here only the results for this building level.

The time-histories selected from the database [11] were used for a dynamic linear analysis of the reactor building using a three-dimensional model in SASSI 2010. Figure 2 shows the resulting linear-elastic floor response spectra and the quantiles of the associated discrete probability distributions for the reactor building level 18m for site intensity VIII.

3.4 Development of performance-based design requirements for SSCs

For the development of performance-based design requirements in the example we consider two typical buildings and three different groups of components. Regarding buildings, we consider the reactor building with the containment and the turbine building. Due to its high

Table 2: Conversion of the PEGASOS hazard to site intensity (EMS-98) – Goesgen site.

Frequency of exceedance, [1/a]	Mean site intensity factor (from PEGASOS hazard deaggregation) (Intensity (EMS-98))
10^{-3}	VI.6 (VII)
10^{-4}	VII.5 (VII–VIII)
10^{-5}	VIII.5 (VIII–IX)
10^{-6}	VIII.9 (IX)

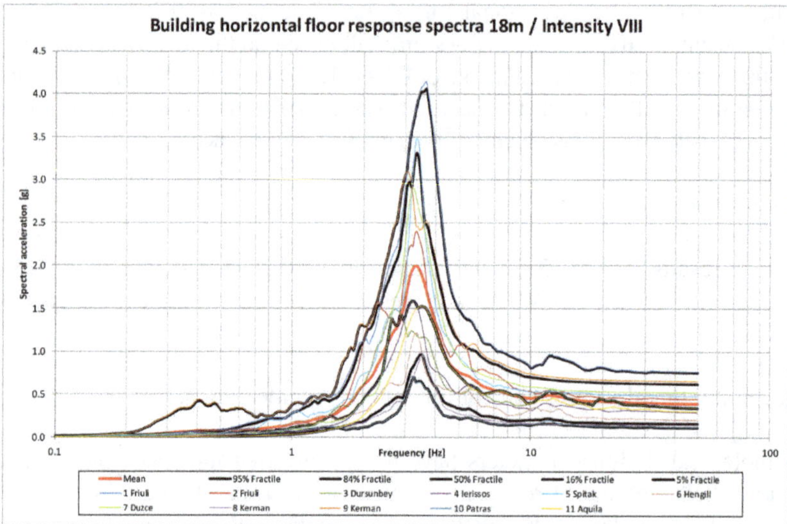

Figure 2: In-structure floor response spectra, intensity VIII, 18m, reactor building.

importance for the safety of the plant, especially with respect to the retainment of radioactivity after an accident, the reactor building is assigned to seismic design category SDC3. The turbine building of a PWR has significantly lower importance (for the vendor AREVA (KWU Germany)). Therefore, it is assigned to seismic design category 2. As additional components tanks, pumps and motor-operated valves (MOVs) are selected for this example. For these groups of components, the performance-based design requirements for all three seismic design categories are developed.

The capacity functions for SSCs are frequently described in the form of double lognormal distributions as [18]:

$$A = A_m \varepsilon_R \varepsilon_U. \tag{1}$$

Here ε_R and ε_U are log-normally distributed with unit medians and standard deviations β_R and β_U respectively. They represent the inherent randomness (aleatory variability) about the median and the epistemic uncertainty of the median value respectively. In some cases, the composite variability is used, defined by:

$$\beta_c = \sqrt{\beta_R^2 + \beta_U^2} \tag{2}$$

A is the seismic capacity and A_m is the unknown median capacity for the component. This parameter is to be defined based on the performance-based requirements in Table 1. The capacity of a component is frequently characterized by its HCLPF (high confidence of low probability of failure) value, which is defined as:

$$HCLPF = A_m e^{-1.645(\beta_R + \beta_U)}. \tag{3}$$

This value approximately corresponds to a probability of failure of 1%.

To calculate the performance objectives given as conditional probabilities of significant deviation the seismic hazard is represented in form of a lognormal distribution with mean and standard deviation as shown in Table 3 and Table 4.

Table 3: Seismic design basis, PGA.

Earthquake	Mean PGA, [g]	Standard deviation, [g]
Safety earthquake	0.21	0.10
Operational earthquake	0.08	0.047

Table 4: Seismic design basis for components, SA (10Hz), 18m.

Earthquake	Mean SA(10Hz), [g]	Standard deviation, [g]
Safety earthquake	0.450	0.216
Operational earthquake	0.146	0.089

$$H = H_m \gamma_R \gamma_U . \tag{4}$$

Here γ_R and γ_U are log-normally distributed with unit medians and standard deviations ϑ_R and ϑ_U similarly as defined for the capacity functions. The combined variability is denoted as ϑ_c.

With capacity and hazard expressed by lognormal probability distributions (using the combined variability as in eqn (2)) the probability of significant deviation from linear elastic behaviour can be calculated analytically from the standard normal distribution.

$$P = 1 - \int_{-z}^{\infty} \varphi(z) dz. \tag{5}$$

Here z is defined as

$$z = -\frac{\ln(A_m) - \ln(H_m)}{\sqrt{\beta_c^2 + \vartheta_c^2}}. \tag{6}$$

Using standard values for the component capacity variabilities and for the hazard parameters (converted to logarithmic space) we obtain the results shown in Table 5. The results are expressed in terms of the median design capacity and of the HCLPF value following the standardized approach to fragility analysis for nuclear power plants [18]. Where applicable, the results are presented in terms of PGA and of SA (10 Hz) at the location of the component. The design requirement expressed in spectral acceleration is of practical interest because it defines the boundary conditions for qualification tests. Such tests are required for safety classified components of nuclear power plants.

The analysis of the results shows that the safety earthquake controls the requirements for the design of the pant for all seismic design categories. Therefore, it is feasible to use a single earthquake design for nuclear power plants. Note, that for other industries the situation might be different. It is also worth to note that the application of performance-based methods for nuclear power plants as in use in the USA [4,5] today would lead to a significantly weaker seismic design due to the permissible higher probabilities of failure.

Table 5: Performance-based seismic design requirements for NPP SSCs.

SSC	Perfor-mance level	Capacity variability parameters		OE median PGA (SA), [g]	SE median PGA (SA), [g]	Final design, [g]	
		β_R	β_U			Median (A_m)	HCLPF
Reactor Building	SDC3	0.21	0.41	0.80 (-)	0.95 (-)	0.95 (-)	0.34 (-)
Turbine Building	SDC2	0.21	0.41	0.37 (-)	0.61(-)	0.61 (-)	0.22 (-)
Pump	SDC1	0.3	0.23	0.22 (0.41)	0.44 (0.94)	0.44 (0.94)	0.18 (0.37)
Pump	SDC2	0.3	0.23	0.35 (0.64)	0.54 (1.16)	0.54 (1.16)	0.23 (0.45)
Pump	SDC3	0.3	0.23	0.66 (1.21)	0.80 (1.71)	0.80 (1.71)	0.33 (0.67)
Tank	SDC1	0.23	0.35	0.23 (0.43)	0.45 (0.98)	0.45(0.98)	0.18 (0.34)
Tank	SDC2	0.23	0.35	0.38 (0.70)	0.57 (1.22)	0.57 (1.22)	0.22 (0.43)
Tank	SDC3	0.23	0.35	0.75 (1.37)	0.9 (1.93)	0.9 (1.93)	0.35 (0.68)
MOV	SDC1	0.3	0.23	0.22 (0.41)	0.44 (0.94)	0.44 (0.94)	0.18(0.37)
MOV	SDC2	0.3	0.23	0.35 (0.64)	0.54 (1.16)	0.54 (1.16)	0.23 (0.45)
MOV	SDC3	0.3	0.23	0.66 (1.21)	0.80 (1.71)	0.80 (1.71)	0.33 (0.67)

4 SUMMARY AND CONCLUSIONS

A procedure for damage-consistent performance-based seismic design of critical infrastructures was developed. The method allows to achieve the main objective of disaster prevention, a robust seismic design of critical infrastructures, in combination with an optimization of economic resources due to the definition of graded performance goals. The performance goals are defined in terms of conditional probabilities of significant deviation from linear-elastic response of SSCs. The implementation of the procedure is demonstrated for the postulated new build of a nuclear power plant unit. The procedure is applicable for all types of critical infrastructures. The categorization of SSCs into seismic design categories can be adjusted in dependence on the problem and the performance objectives to be addressed.

REFERENCES

[1] FEMA-273, *NEHRP guidelines for seismic rehabilitation of buildings. building seismic safety council*. FEMA, Washington, DC, 1997.

[2] FEMA-445, *Next-generation performance-based seismic design guidelines. program plan for new and existing buildings*. FEMA, Washington, DC, 2006.

[3] Elnashai, A.S. & Di Sarno, L., *Fundamentals of earthquake engineering. from source to fragility*. Second Edition., John Wiley & Sons, Ltd.: Chichester, West Sussex, United Kingdom, 2015.

[4] Budnitz, R.J., *Opportunities for advancing Technology-neutral and performance-based design methods for the seismic design and Regulation of Nuclear Power Plant Structures, Systems and Components*. Lawrence Berkeley National Laboratory, Berkeley, CA, 2010.

[5] American Society of Civil Engineers (ASCE), *Seismic design criteria for structures, systems, and components in nuclear facilities, ASCE/SEI 43-05*. ASCE, Reston, VA, USA, 2005.

[6] Klügel, J.-U. & Mualchin, L., Earthquake engineering needs and seismic hazard assessment, ERES 2013. *WIT Transactions on The Built Environment,* **132**, pp. 47–67, 2013.
[7] Klügel, J.-U., Uncertainty analysis and expert judgment in seismic hazard analysis. *Pure and Applied Geophysics,* **168**, pp. 27–53, 2011. https://doi.org/10.1007/s00024-010-0155-4
[8] Klügel, J.-U., Error inflation in probabilistic seismic hazard analysis. *Engineering Geology,* **90**, pp. 186–192, 2007. https://doi.org/10.1016/j.enggeo.2007.01.003
[9] Bommer, J.J. & Abrahamson, N.A., Why do modern probabilistic seismic-hazard analyses often lead to increased hazard estimates. *BSSA,* **96**(6), pp. 1967–1977, 2006. https://doi.org/10.1785/0120060043
[10] Conseil de L'Europe, G. Grünthal., (Editor)., *European Macroseismic Scale 1998.* Luxembourg, 1998.
[11] European-Mediterranean Seismological Center. *Resorce reference database for seismic ground-motion in Europe*. EMSC, France, 2013.
[12] Renault, P., Abrahamson, N.A., Coppersmith, K.J., Koller, M., Roth, P. & Hölker, A., *PEGASOS Refinment Project, Rev. 1.* swissnuclear, Olten, 2014.
[13] Hassan, H.M., Romanelli, F., Panza, G.F. & ElGabry, M.N., Update and sensitivity analysis of the neo-deterministic seismic hazard assessment for Egypt. *Engineering Geology,* **218**, pp. 77–89, 2017. https://doi.org/10.1016/j.enggeo.2017.01.006
[14] Irikura, K. & Miyake, H., Recipe for predicting strong ground motion from crustal earthquake scenarios. *PAGEOPH,* **168**, pp. 85–104, 2011. https://doi.org/10.1007/s00024-010-0150-9
[15] Abrahamson, N.A., Coppersmith, K. & Sprecher, C., *Probabilistic seismic hazard analysis for swiss nuclear power plant sites* (PEGASOS project), vol. **1–6**. Nagra, Wettingen, 2004.
[16] Klügel, J-U., Understanding site-specific PSHA results by hazard deaggregation into site intensities, Poster. *EGU General Assembly Meeting, Natural Hazards*, Vienna, 2016.
[17] Klügel, J-U., Attinger, R., Rao, S. & Vaidya, N., *Adjusting the fragility analysis method to the seismic hazard input, Part I. The intensity-based method, paper 1567.* 20th International Conference on Structural Mechanics in Reactor Technology (SmiRT 20), Espoo, Finland August 9–14, 2009, Espoo, 2009.
[18] EPRI, *Methodology for developing seismic fragilities, TR-103959.* EPRI, Palo Alto, 1994.

SEISMIC FRAGILITY CHARACTERISTICS OF STRUCTURAL POPULATIONS WITH IRREGULARITIES

JUNWON SEO
Department of Civil and Environmental Engineering, South Dakota State University, USA.

ABSTRACT

Earthquakes result in substantial structural damage of a number of structures across a region. Many studies have made an effort to examine regional seismic damage and vulnerability to different structure types, e.g. buildings and bridges, using varying computational methodologies. This paper focuses on the use of Response Surface Metamodels (RSMs) in conjunction with Monte Carlo Simulations (MCSs) to quantify probabilistic seismic performance for different classes of structural populations with irregularities, including irregular steel buildings and steel girder bridges. As part of the regional vulnerability study, each of the selected classes is constructed based upon the appropriate experimental design technique, i.e. the Central Composite Design (CCD), and the responses of each class subjected to multiple ground motions are captured during the nonlinear time history analyses of an individual computational model. Then, a RSM for each class is established by performing a least-square regression analysis within the considered CCD space. Seismic fragility curves are generated by means of the joint RSM-MCS enabling to treat uncertainties regarding overall configuration irregularities and additional structural parameters considered significant for each class. The influence of the irregularity parameters on seismic vulnerability for each class is investigated by comparison of the resulting fragilities. Results reveal that the RSM-MCS is able to efficiently assess seismic vulnerability of each class and directly examine the parameters' influence on corresponding behaviours.

Keywords: irregularities, metamodel, seismic response, structural populations, vulnerability.

1 INTRODUCTION

A number of structures across susceptible regions to earthquakes have been substantially damaged, resulting in economic and human losses. To make wise decisions for enhanced earthquake mitigations, it is required that seismic losses of structures with regional characteristics data (e.g. structural dimensions) be appropriately predicted. It is well known that the most efficient methodology to estimate seismic losses of structures accounting for uncertainty in seismic loadings and structural capacities is the seismic fragility analysis [1–4]. Numerous studies have been performed to investigate regional seismic damage and fragility to various structure types, e.g. buildings and bridges, via different computational or statistical modelling techniques [5–8]. Different metamodels created with seismic responses computed by performing nonlinear time history analyses of structural models have been utilized to estimate seismic fragilities of structures with an aid of reliability techniques [4, 9]. Recently, metamodels have been used to estimate seismic restoration costs of structural populations [10].

This paper is intended to evaluate the feasibility of Response Surface Metamodels (RSMs) classified as a metamodel type for probabilistic seismic performance quantification for structural populations with irregularities. RSMs coupled with Monte Carlo Simulations (MCSs) were applied to two different classes for irregular structures: 1) steel moment resisting frame buildings and curved steel girder bridges. As part of the study, sample combinations for each structure class were constructed based upon the Central Composite Design (CCD), and the responses of each combination subjected to ground motions were obtained during the nonlinear time history analyses. Then, a RSM for each class is established by performing a least-square regression analysis within the considered CCD combination space. Seismic fragility curves for each class were generated by means of the joint RSM-MCS enabling to treat uncertainties

DOI: 10.2495/CMEM-V6-N5-944-954

regarding structural irregularities and other parameters considered significant on seismic behaviours. The influence of the considered irregular parameters on seismic vulnerability per class was investigated by comparison of the resulting fragility curves.

2 APPROACH

A seismic RSM is typically defined as a polynomial regression function that can be created by fitting a dataset consisting of structural characteristics and seismic responses. The RSMs can be used not only to forecast seismic responses of structure(s), but also to generate their fragility curves at low-computational costs in conjunction with reliability techniques. Details on the RSM-based seismic response prediction and fragility curve generation are provided in the following subsections.

2.1 Seismic response prediction

The use of a second-order polynomial RSM function is a traditional approach to predict seismic response quantities. The RSM can be expressed as indicated below [11].

$$y = \beta_0 + \sum_{i=1}^{k} \beta_i x_i + \sum_{i=1}^{k} \beta_{ii} x_i^2 + \sum_{i=1}^{k-1} \sum_{j>i}^{k} \beta_{ij} x_i x_j + \varepsilon \tag{1}$$

where y is the estimated seismic response; x_i, x_j are input variables, such as irregularity parameters; β_0, β_i, β_{ij}, β_{ii} are coefficients to be determined from experimental sets of seismic response quantities and input variables; and k is the number of input variables. It should be noted that the minimum number of experimental sets required for determining RSM coefficients is normally equivalent to the number of the input variables [11].

To predict seismic response quantities of structures, RSMs consisting of input and output variables should be created at first. The input variables are composed of structural, material, and geometrical parameters necessary for the construction of analytical models associated with their variation in structures located across a region. A range of values for each input variable can be determined by examining design plans for structures or reviewing findings from the past studies. Experimental designs, including CCD, can be employed to establish possible combinations, which are structural models.

Based on each input combination with the corresponding values, a structural model can be developed with multiple frame elements using Finite Element Modelling (FEM) software. To account for uncertainty in potential earthquake scenarios associated with seismic resistance which varies depending on structural characteristics, multiple ground motions need to be considered. One ground motion at a time can be applied to each of the generated FEM model, and then the corresponding seismic response quantities can be determined by performing its seismic analysis. Note that the quantities can be categorized as output variables. The outputs can be incorporated into the inputs to complete a matrix. The matrix is used as the basis to be run by a least square regression analysis for the development of RSMs enabling the seismic response prediction.

2.2 Seismic fragility curve generation

Prior to providing a procedure for the generation of seismic fragility curves through RSMs, it is worth noting its physical definition. The seismic fragility curve for structures is a conditional probability, providing the probability of failure or exceedance that a structure will exceed a particular state of damage given ground motion intensity levels [1–4]. This fragility curve is mathematically expressed in Equation (2):

$$P_f = \Phi\left(\frac{\ln\left(\frac{S_D}{S_C}\right)}{\sqrt{\beta_d^2 + \beta_c^2}}\right) \tag{2}$$

where P_f is the exceedance probability, S_c is the median value of the seismic resistance at a given damage state, βc is the dispersion or lognormal standard deviation of the seismic resistance, S_d is the computed seismic demand at a chosen ground motion intensity parameter, β_d is the logarithmic standard deviation for the demand and Φ is the standard normal distribution function [5].

Based on the notion for the generation of the standard fragility curves, seismic fragility curves can be created using RSMs in conjunction with MCSs. MCSs with randomness within a probabilistic characteristic for each input variable can be conducted to compute all possible seismic responses using RSMs. The probabilistic characteristics can be quantified in terms of probability density functions. The likelihood of the response exceeding a particular damage state is determined from numerous seismic response outputs obtained from the MCS with RSMs. This procedure provides a single probability value which is part of a fragility curve at a given ground motion intensity level; thus, this can be iterated for the other earthquake intensity levels to generate a complete fragility curve. More information on the RSM-based fragility curve generation can be discovered in the past studies [1, 4].

3 APPLICATION TO IRREGULAR STRUCTURE POPULATION

The RSM-MCS approach was applied to irregular structure populations: 1) steel moment-resisting building structures and 2) curved steel girder bridge structures. To assess the seismic vulnerability of each of the structure populations, fragility curves considering geometrical and structural characteristics for each class were created using the RSM-MCS. Then, the effect of the input variables focusing on irregularity parameters on the vulnerability for each class was examined by performing a comparative study of respective fragility curves. The generation of seismic fragility curves and the completion of cursory sensitivity analyses for each class are detailed as follows.

3.1 Steel building structures

3.1.1 Building characteristics

As shown in Fig. 1a, steel moment-resisting building structures in Shelby County, TN, in the United States, which is located in the New Madrid Seismic Zone (NMSZ) were chosen for this application. Note that the NMSZ is considered the most seismically active area in the Central United States as a series of past high-intensity earthquakes and aftershocks occurred. As a result of infrequent seismic events, the majority of existing structures have been only designed considering the combined gravity and wind loadings without seismic details. Specifically, L-shaped moment-resisting frame structures considered irregular buildings in plan were selected to analyse their seismic vulnerability in this region. As illustrated in Fig. 1b, the most feasible L-shaped plan configurations that were determined from the past work [1] were

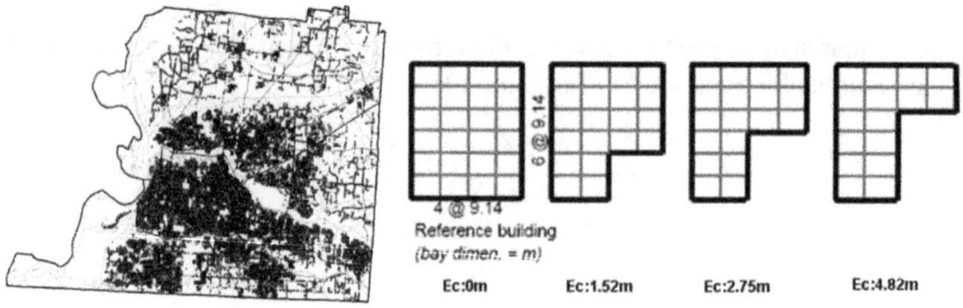

(a) (b)

Figure 1: Region of interest. (a) Building stocks in Shelby County, TN; (b) most feasible L-shaped configurations [1].

utilized in this study. Based on the examination of building inventory in the region of interest (see Table 1), most buildings are made up for industrial occupancy use and consist of less than five stories and different built years. Hence, the story of structures was limited to five-story office buildings with a variety of L-shapes and built years.

After the building shapes in plan were determined, potential variables for the selected structures, which have a significant impact on regional seismic response and fragility, were considered to include their uncertainty. These variables included earthquake direction, steel yield strength, damping ratio, bottom floor height, year built, and eccentricity. Forty-three experimental design combinations (i.e. 43 buildings), which served as the basis for the development of FEM building models, were constructed based upon the CCD with the seven variables. When building the CCD-based combinations, upper, central, and low values corresponding to each variable are required. Detailed information (i.e. values) for each variable was obtained from the past study [1].

3.1.2 Structural design and seismic simulation

Each building out of 43 combinations was designed according to appropriate building design codes. For example, a building with pre-1970 built year that was compatible to the non-seismic design was designed following the 1969 Southern Standard Building Code (SSBC) [13] with consideration of only gravity and wind loads. Another example is that a building with post-1990 year built was designed according to 1991 Southern Building Code Congress International (SBCCI) [14]. For the non-seismic building design, an example of exterior and interior columns and beams selected according to the SSBC building design requirements was shown in Table 2.

Table 1: Building inventory data in Shelby County [12].

Occupancy	Industrial	Residential	Others
	97.3%	2.1%	0.6%
Year Built	Pre-1970	1970–1990	Post-1990
	58.1%	32%	9.9%
Number of stories	Less than 5	6–10	More than 11
	99.9%	0%	0.1%

Table 2: Structural frames for the example LSMF building designed with SSBC 1969.

	Moment resisting frames		
Story	E. Columns	I. Columns	Beams
1	W14X111	W14X111	W24X76
2	W14X111	W14X111	W24X76
3	W14X95	W14X95	W24X76
4	W14X95	W14X95	W24X76
5	W14X68	W14X68	W24X68

An analytical 3D model to each building was generated using frame elements available in Zeus-NL [15] accounting for material and geometric nonlinearities. Geometric and material properties for columns and beams were described using the sections selected from the structural design process. It was assumed that the connections between beam and columns that were idealized as simple joints were fully fixed at all rotational and translational directions. Lumped masses were assigned to each element enabling seismic analyses. Representative analytical models for both rectangular- and L-shaped buildings can be seen in Fig. 2. The rectangular building was used as a benchmarking regular building, so as to compare its fragilities against those resulting from buildings with different L-shapes.

To evaluate seismic vulnerability of the structures subjected to various potential earthquake scenarios, synthetic ground motions that were generated by Rix and Fernandez-Leon [16] for the region were used as input loads in the models. The ground motion pool was created based upon the soil and other site features in the region where there has been a lack of actual ground motion data. In detail, 30 ground motions having different magnitudes and hypocentral distances that were extracted out of the pool were applied to each model. Spectral acceleration (S_a) curves representing the input ground motions can be seen in Fig. 3, and

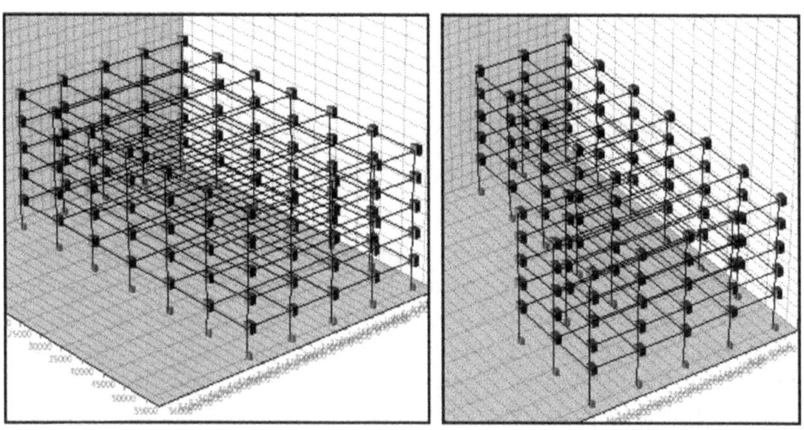

(a) (b)

Figure 2: Sample 3D models. (a) Rectangular-shaped building model; (b) L-shaped building model.

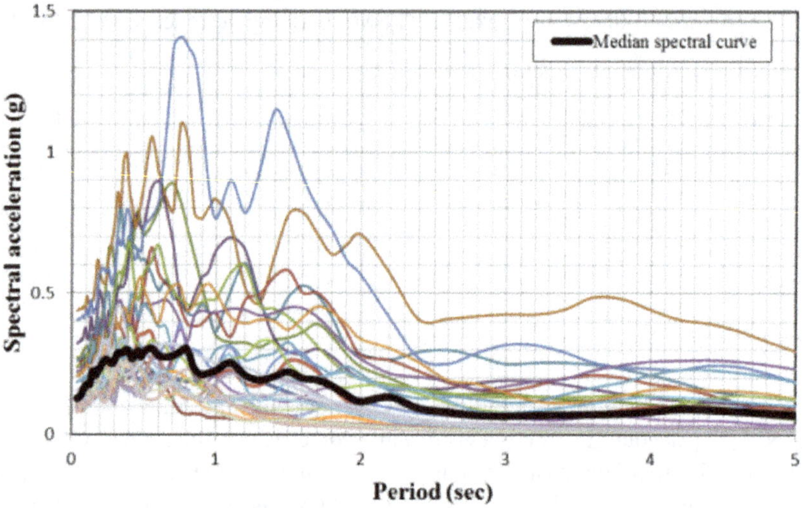

Figure 3: Spectral acceleration [1, 16].

the corresponding median value is also plotted in this figure. It appears that the S_a values vary depending on ground motion characteristics.

Nonlinear time history analysis of each model loaded with one from the ground motion pool was separately performed to estimate its seismic response. Translational displacements resulting from the ground motions were captured during the seismic analyses, and maximum inter-story drift ratios that have been commonly used as a seismic behaviour assessment indicator were calculated using the following equation:

$$Maximum\,Inter-Story\,Drift\,Ratio = \max_{\forall(x,z,t),i}\left|\left(\frac{\delta_{(x,t),i}}{h_i}, \frac{\delta_{(z,t),i}}{h_i}\right)\right| \tag{3}$$

where $\delta_{(x,t),i}$ and $\delta_{(z,t),i}$ are the translational displacement responses in the strong and weak directions at time t at the ith floor when performing the nonlinear time history analysis of the model; h_i is inter-story height of the ith floor [1]. By running the seismic analyses, outputs, which are the max inter-story drift ratios, corresponding to each of the 43 input combinations were obtained.

3.1.3 Building fragilities

A regression analysis for a complete matrix consisting of all inputs and outputs was carried out to construct a RSM to estimate the max inter story drift ratios and generate fragility curves for the structures. The RSM function was able to approximate the drifts for the various structures resulting from a certain level of S_a. MCSs with probability density functions of each input variable that were assumed in the past work [1] were performed on the RSM function. Probability, which exceeds a drift ratio limit at a certain damage state were determined from the responses from the joint MCS-RSM simulations, resulting in a single fragility point given a S_a. In this application, three damage limit states, including IO=immediate occupancy, LS=life safety, CP=collapse prevention, that have been widespread in the fragility assessment of moment-resisting structures were used. By performing the described process multiple times, a number of probability values at different levels of S_a were obtained, and these points were plotted to complete a fragility curve for each damage state. The overall fragility curves of the structures are illustrated in Fig. 4a. Detailed information for the generation of fragility curves can be found elsewhere [1].

To assess the influence of planar building irregularity that can be expressed in terms of eccentricity on seismic fragilities of the structure population, specific fragility curves (see Fig. 4b) accounting for variability in eccentricity (i.e. 0 m, 2.8 m, and 4.82 m) were produced using the MCS-RSMs. A particular eccentricity value used as one governing variable was

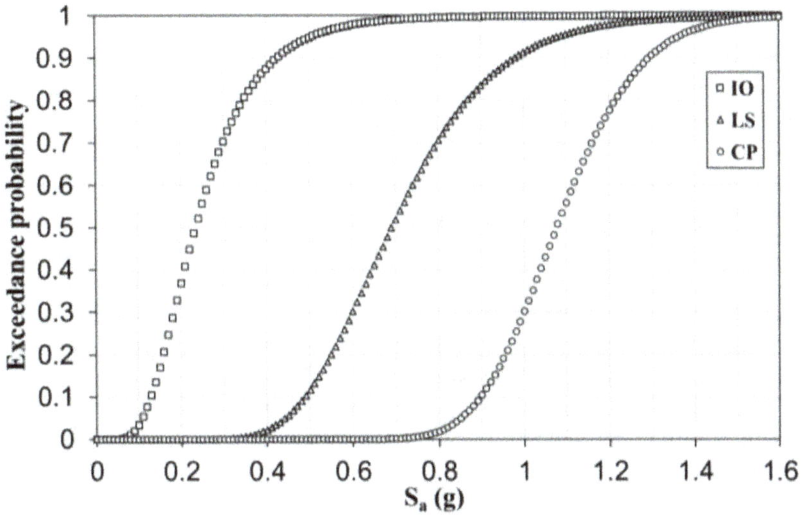

(a)

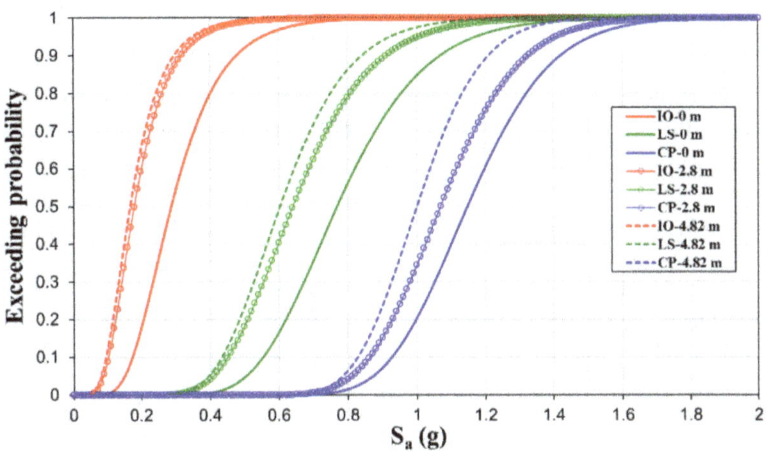

(b)

Figure 4: Sample building fragilities. (a) Overall curves; (b) specific curves with irregularities [1].

treated as the fixed one during the simulations, but the remainder of variables were considered randomly changing variables within the assumed probability range. The resulting curves for three different values of eccentricity were visually compared to explore the effect of building irregularity on seismic vulnerability at each damage state over a range of S_a. It appears that vulnerability for the structures over three damage states becomes higher due to an increase in seismic torisonal responses captured from the seismic analyses of buildings with the increase in planar eccentricity.

3.2 Steel bridge structures

3.2.1 Bridge characteristics

The proposed approach was also applied to horizontally, curved steel I-girder bridges which are classified as irregular bridges, in an attempt to look into their fragilities. For the bridge application, 99 curved steel bridges located across the States of Pennsylvania, Maryland, and New York in the United States were collected and examined. For the inventory analysis of the bridges, the bridge design plans were directly obtained from the respective Departments of Transportation (DOTs). Statistics for significantly structural bridge variables, including number of spans, maximum span length, deck width, maximum column height, radius of curvature, girder spacing, and cross-frame spacing, are listed in Table 3. This table indicates that the majority of the bridges have two spans. The variables with the statistical information were used as input variables for the CCD-based RSM establishment as created in the building fragility study. To account for uncertainties in material and other features for the bridges, steel and concrete strengths and Young's moduli and damping ratio were included in the RSM establishment as well.

3.2.2 Structural modelling and seismic analysis

3D analytical models of the bridges made with different values for each input variable were generated using OpenSees program [17], reflecting their material and geometric nonlinearity. The models were based upon the recommended modelling techniques from the past work [18]. The girders and concrete deck in a superstructure system were idealized using Beam-Column elements available in OpenSees, while pier columns and caps, abutments, and footings were modelled with the OpenSees displacement-based BeamColumn elements. The representative ground motions that were arbitrarily selected, considering a broad range of

Table 3: Statistics for significant bridge variables of the considered curved bridges [4].

Bridge variables	Mean	Standard deviation	Median	Mode
Number of spans	2.53	1.73	1.5	2
Maximum span length, m	43.8	13.7	43.1	Not Unique
Deck width, m	12.9	4.4	12.2	12.8
Maximum column height, m	6.5	4.3	5.3	4.6
Radius of curvature, m	513.5	542.9	304.6	250
Girder spacing, m	2.6	0.4	2.5	Not Unique
Cross-frame spacing, m	5.4	1.2	5.2	7.3

peak ground accelerations (PGAs), from the ground motion pool used in the building fragility analysis was used for the bridge analysis. Nonlinear time history analyses for each model subjected to each ground motions were completed, and seismic responses, including maximum column curvature ductility and bearing and abutment deformations, were extracted during each analysis. These peak responses were considered outputs to each model. The RSM formula was developed within the combined input and output matrix by performing its least-square regression analysis. More details on the modelling approach and the RSM development are able to be found in the past studies [3–4].

3.2.3 Bridge fragilities

This section involves creation of seismic fragility curves for the bridge population using the RSM formula in conjunction with MCSs. The probability of the estimated seismic response using the RSM to surpass a quantitative limit was calculated at each of the damage limit states, including slight, moderate, extensive, and complete that were described and determined by Federal Emergency Management Agency (FEMA). Specifically, the probability was determined using the RSM-MCS results under the probability density distribution for each input variable. This point was deemed to be a probability or fragility value at a given, particular PGA. This procedure was repeated over an entire range of PGAs, and then a whole fragility curve was built.

To examine geometrical irregularity effects of the curved bridge population on seismic vulnerability, a comparison between fragility curves of two bridges per damage state, including the almost straight bridge with radius of curvature 3036 m and tight curved bridge with 871 m curvature, was made as shown in Fig. 5. It shows that the fragilities for the tight curved bridge which has more geometrical irregularity than the almost straight bridge are higher than those for the straight one. It can be interpreted that the curved bridge is more vulnerable to seismic events than straight bridge because the increase in radius of curvature resulted in an increase in global seismic torsional responses, leading to higher seismic fragility.

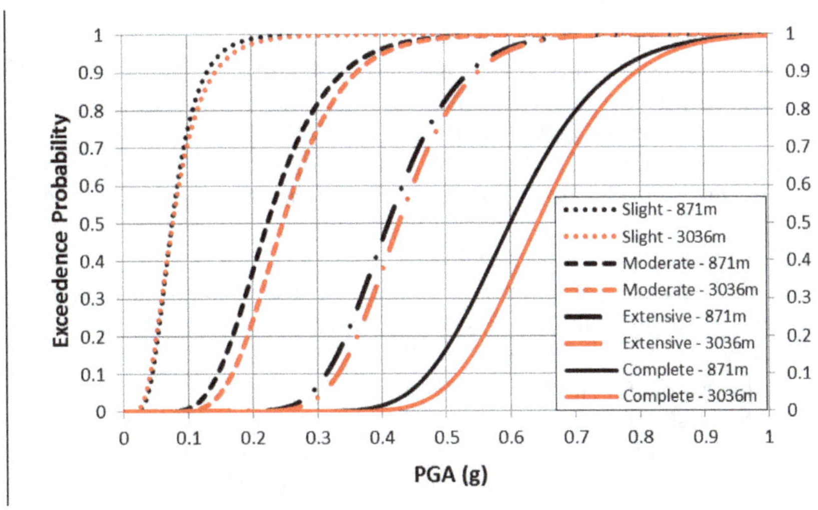

Figure 5: Comparison between fragility curves of two bridges having a radius of curvature of 871 m and 3036 m [4].

4 CONCLUSIONS

Integrated RSMs and MCS methodology was implemented to demonstrate that seismic fragility curves for structural populations with irregularities may be generated in an efficient manner. The methodology was applied to two classes of irregular structures, including steel moment-resisting framing building structures having planar irregularities and horizontally curved steel I-girder bridges. It has been proven that the seismic vulnerabilities of each class were rapidly assessed in the form of fragility curves that were built by performing efficient numbers of simulations through the RSM-MCS process. This methodology helped not only to enhance the computational efficiency to examine both building and bridge vulnerabilities to earthquakes, but also to explore the effects of structural irregularities on each. For further implementation of this methodology, this should be validated with either experimental or empirical fragility data that can be probabilistically produced by performing structural damage data resulting from actual ground motions.

REFERENCES

[1] Seo, J., Dueñas-Osorio, L., Craig, J.I. & Goodno, B.J., Metamodel-based regional vulnerability estimate of irregular steel moment-frame structures subjected to earthquake events. *Engineering Structures*, **45**, pp. 585–597, 2012. https://doi.org/10.1016/j.engstruct.2012.07.003

[2] Seo, J. & Linzell, D.G., Horizontally curved steel bridge seismic vulnerability assessment. *Engineering Structures*, **34**, pp. 21–32, 2012. https://doi.org/10.1016/j.engstruct.2011.09.008

[3] Seo, J. & Linzell, D.G., Nonlinear seismic response and parametric examination of horizontally curved steel bridges using 3D computational models. *ASCE Journal of Bridge Engineering*, **18**(3), pp. 220–231, 2013. https://doi.org/10.1061/(asce)be.1943-5592.0000345

[4] Seo, J. & Linzell, D.G., Use of response surface metamodels to generate system level fragilities for existing curved steel bridges. *Engineering Structures*, **52**, pp. 642–653, 2013. https://doi.org/10.1016/j.engstruct.2013.03.023

[5] Rogers, L.P., & Seo, J., Vulnerability sensitivity of curved precast-concrete I-girder bridges with various configurations subjected to multiple ground motions. *ASCE Journal of Bridge Engineering*, **22**(2), pp. 04016118-1–19, 2017. https://doi.org/10.1061/(asce)be.1943-5592.0000973

[6] Park, J., Towashiraporn, P., Craig, J.I. & Goodno, B.J., Seismic fragility analysis of low-rise unreinforced masonry structures. *Engineering Structures*, **31**(1), pp. 125–137, 2009. https://doi.org/10.1016/j.engstruct.2008.07.021

[7] Jeon, J.S., DesRoches, R., Kim, T. & Choi, E., Geometric parameters affecting seismic fragilities of curved multi-frame concrete box-girder bridges with integral abutments. *Engineering Structures*, **122**, pp. 121–143, 2016. https://doi.org/10.1016/j.engstruct.2016.04.037

[8] Jeon, J.S., DesRoches, R. & Lee, D.H. Post-repair effect of column jackets on aftershock fragilities of damaged RC bridges subjected to successive earthquakes. *Earthquake Engineering and Structural Dynamics*, **45**(7), pp. 1149–1168, 2016. https://doi.org/10.1002/eqe.2700

[9] Ghosh, J., Padgett, J.E. & Dueñas-Osorio, L., Surrogate modeling and failure surface visualization for efficient seismic vulnerability assessment of highway bridges. *Probabilistic Engineering Mechanics*, **34**, pp. 189–199, 2013. https://doi.org/10.1016/j.probengmech.2013.09.003

[10] Seo, J. & Park, H., Probabilistic seismic restoration cost estimation for transportation infrastructure portfolios with an emphasis on curved steel I-girder bridges. *Structural Safety*, **65**, pp. 27–34, 2017. https://doi.org/10.1016/j.strusafe.2016.12.002

[11] Wu, C.F.J. & Hamada, M., *Experiments-Planning, Analysis and Parameter Design Optimization*, John Wiley & Sons, pp. 760, 2000.

[12] Muthukumar, S., *The application of advanced inventory techniques in urban inventory data development of earthquake risk modeling and mitigation in Mid-America*. PhD dissertation, Georgia Institute of Technology, Atlanta, 2008.

[13] Southern Standard Building Code, Southern Building Code Congress, Birmingham, AL, 1969.

[14] Southern Building Code Congress International, Standard Building Code, Birmingham, AL, 1991.

[15] Elnashai, A.S., Papanikolau V. & Lee D.H., *ZEUS Non-linear: A System for Inelastic Analysis of Structures V1.9.0*, UIUC, Urbana-Champaign, IL, 2011.

[16] Rix, G.J. & Fernandez-Leon, J.A., *Synthetic ground motions for Memphis, TN*, Mid-America Earthquake Center, 2008.

[17] Mazzoni, S., McKenna, F., Scott, M.H., Fenves, G.L., *et al. Open system for earthquake engineering simulation (OpenSees)*. Pacific Earthquake Engineering Research Center. Version 1.7.3; 2008.

[18] Nielson, B.G., *Analytical fragility curves for highway bridges in moderate seismic zones*. Ph.D. dissertation, School of Civil and Environmental Engineering, Georgia Institute of Technology, Atlanta, 2005.

Author index

www.ingramcontent.com/pod-product-compliance
Lightning Source LLC
LaVergne TN
LVHW062226200926
848484LV00006B/1484